Global E-waste Management Strategies and Future Implications

Global E-waste Management Strategies and Future Implications

Edited by

SHASHI ARYA

Technology Development Centre, CSIR-National Environmental
Engineering Research Institute (CSIR-NEERI), Nehru Marg, Nagpur,
Maharashtra, India

SUNIL KUMAR

Solid and Hazardous Waste Management Division, CSIR-National
Environmental Engineering Research Institute (CSIR-NEERI), Nehru
Marg, Nagpur, Maharashtra, India

ELSEVIER

Elsevier
Radarweg 29, PO Box 211, 1000 AE Amsterdam, Netherlands
The Boulevard, Langford Lane, Kidlington, Oxford OX5 1GB, United Kingdom
50 Hampshire Street, 5th Floor, Cambridge, MA 02139, United States

Notices

Knowledge and best practice in this field are constantly changing. As new research and experience broaden our understanding, changes in research methods, professional practices, or medical treatment may become necessary.

Practitioners and researchers must always rely on their own experience and knowledge in evaluating and using any information, methods, compounds, or experiments described herein. In using such information or methods they should be mindful of their own safety and the safety of others, including parties for whom they have a professional responsibility.

To the fullest extent of the law, neither the Publisher nor the authors, contributors, or editors, assume any liability for any injury and/or damage to persons or property as a matter of products liability, negligence or otherwise, or from any use or operation of any methods, products, instructions, or ideas contained in the material herein.

ISBN: 978-0-323-99919-9

For Information on all Elsevier publications
visit our website at https://www.elsevier.com/books-and-journals

Publisher: Susan Dennis
Acquisitions Editor: Anita Koch
Editorial Project Manager: Kyle Gravel
Production Project Manager: R. Vijay Bharath
Cover Designer: Matthew Limbert

Typeset by MPS Limited, Chennai, India

Working together to grow libraries in developing countries

www.elsevier.com • www.bookaid.org

Dedication

To my incredible dad, who has consistently been by my side, I am eternally grateful for your guidance, resilience, and boundless affection. I appreciate your unwavering faith in me, and you will forever hold a special place in my heart.

Your daughter

Shashi

Contents

List of contributors

Fuad Ameen
Department of Botany and Microbiology, College of Science, King Saud University, Riyadh, Saudi Arabia

Shashi Arya
Waste Re-processing Division, CSIR–National Environmental Engineering Research Institute (CSIR-NEERI), Nagpur, Maharashtra, India; Academy of Scientific and Innovative Research (AcSIR), Ghaziabad, Uttar Pradesh, India; Centre for Environmental Policy, Faculty of Natural Sciences, Imperial College London, London, United Kingdom

Somvir Arya
Industrial and Production Engineering Department, Dr. B.R. Ambedkar National Institute of Technology, Jalandhar, Punjab, India

Abhishek Awasthi
Tsinghua University, Beijing, P.R. China

Aman Basu
Department of Biology, York University, Toronto, ON, Canada

Arvind Bhardwaj
Industrial and Production Engineering Department, Dr. B.R. Ambedkar National Institute of Technology, Jalandhar, Punjab, India

Sartaj Ahmad Bhat
River Basin Research Center, Gifu University, Gifu, Japan; Waste Re-processing Division, CSIR-National Environmental Engineering Research Institute (CSIR-NEERI), Nagpur, Maharashtra, India

Warren J. Bruckard
CSIRO Mineral Resources, Clayton South, VIC, Australia

Miao Chen
School of Applied Sciences, RMIT University, Melbourne, VIC, Australia; CSIRO Mineral Resources, Clayton South, VIC, Australia

Guangyu Cui
State Key Laboratory of Pollution Control and Resource Reuse, Tongji University, Shanghai, P.R. China

Pranav Prashant Dagwar
CSIR-National Environmental Engineering Research Institute (CSIR-NEERI), Nagpur, Maharashtra, India

Brajesh Kumar Dubey
Environmental Engineering and Management, Department of Civil Engineering, Indian Institute of Technology Kharagpur, Kharagpur, West Bengal, India

Deblina Dutta
Department of Environmental Science, SRM University—AP, Amaravati, Andhra Pradesh, India

Mahesh Game
CSIR-National Environmental Engineering Research Institute (CSIR-NEERI), Nagpur, Maharashtra, India

Sachin Rameshrao Geed
Engineering Science and Technology Division, CSIR-NEIST, Jorhat, Assam, India

Biswajit Gogoi
Engineering Science and Technology Division, CSIR-NEIST, Jorhat, Assam, India

Ajay Gupta
Industrial and Production Engineering Department, Dr. B.R. Ambedkar National Institute of Technology, Jalandhar, Punjab, India

Subrata Hait
Department of Civil and Environmental Engineering, Indian Institute of Technology Patna, Bihta, Bihar, India

Nawshad Haque
CSIRO Energy, Clayton South, VIC, Australia

Rohit Jambhulkar
CSIR-National Environmental Engineering Research Institute (CSIR-NEERI), Nagpur, Maharashtra, India

Arya Anuj Jee
Department of Civil Engineering, National Institute of Technology, Patna, Bihar, India

Kavita Kanaujia
Department of Civil and Environmental Engineering, Indian Institute of Technology Patna, Bihta, Bihar, India

Sandip Karmakar
Department of Civil Engineering, National Institute of Technology, Patna, Bihar, India

Sunil Kumar
Waste Re-processing Division, CSIR-National Environmental Engineering Research Institute (CSIR-NEERI), Nagpur, Maharashtra, India; Academy of Scientific and Innovative Research (AcSIR), Ghaziabad, Uttar Pradesh, India

Dolly Kumari
Department of Chemistry, Dayalbagh Educational Institute, Agra, Uttar Pradesh, India

Debajyoti Kundu
CSIR-National Environmental Engineering Research Institute (CSIR-NEERI), Nagpur, Maharashtra, India

Fusheng Li
River Basin Research Center, Gifu University, Gifu, Japan

J. Senophiyah Mary
Institute for Globally Distributed Open Research and Education, Coimbatore, Tamil Nadu, India; Environmental Consultant, Indian Green Service, Delhi, Delhi, India

Nityanand Singh Maurya
Department of Civil Engineering, National Institute of Technology, Patna, Bihar, India

Unnikrishna Menon
Environmental Engineering and Management, Department of Civil Engineering, Indian Institute of Technology Kharagpur, Kharagpur, West Bengal, India

Ankit Motghare
CSIR-National Environmental Engineering Research Institute (CSIR-NEERI), Nagpur, Maharashtra, India

Srushti Muneshwar
CSIR-National Environmental Engineering Research Institute (CSIR-NEERI), Nagpur, Maharashtra, India

Rumi Narzari
Department of Energy, Tezpur University, Tezpur, Assam, India

Anudeep Nema
Department of Civil Engineering, School of Engineering, Eklavya University, Damoh, Madhya Pradesh, India

Aneri Patel
Waste Re-processing Division, CSIR-National Environmental Engineering Research Institute (CSIR-NEERI), Nagpur, Maharashtra, India

Maneesh Kumar Poddar
Department of Chemical Engineering, National Institute of Technology Karnataka, Surathkal, Karnataka, India

Mark I. Pownceby
CSIRO Mineral Resources, Clayton South, VIC, Australia

Rajnikant Prasad
Civil Engineering Department, G. H. Raisoni Institute of Engineering and Business Management, Jalgaon, Maharashtra, India

Md. Rakibul Qadir
PP&PDC, Bangladesh Council of Scientific and Industrial Research, Dhanmondi, Dhaka, Bangladesh; School of Applied Sciences, RMIT University, Melbourne, VIC, Australia; CSIRO Mineral Resources, Clayton South, VIC, Australia

Loganath Radhakrishnan
Institute for Globally Distributed Open Research and Education, Coimbatore, Tamil Nadu, India; Environmental Consultant, Indian Green Service, Delhi, Delhi, India

Rahul Rautela
CSIR-National Environmental Engineering Research Institute (CSIR-NEERI), Nagpur, Maharashtra, India; Academy of Scientific and Innovative Research (AcSIR), Ghaziabad, Uttar Pradesh, India

Ajishnu Roy
School of Geographical Sciences and Remote Sensing, Guangzhou University, Guangzhou, Guangdong, P.R. China

Simran Sahota
Department of Chemical Engineering, National Institute of Technology Karnataka, Surathkal, Karnataka, India

Dayanand Sharma
Department of Civil Engineering, National Institute of Technology, Patna, Bihar, India; Department of Civil Engineering, Sharda University, Greater Noida, Uttar Pradesh, India

Deval Singh
Waste Re-processing Division, CSIR-National Environmental Engineering Research Institute (CSIR-NEERI), Nagpur, Maharashtra, India

Gunjan Singh
Department of Chemistry, Dayalbagh Educational Institute, Agra, Uttar Pradesh, India

Radhika Singh
Department of Chemistry, Dayalbagh Educational Institute, Agra, Uttar Pradesh, India

Dipeshkumar R. Sonaviya
M.S. Patel Department of Civil Engineering, C. S. Patel Institute of Technology, Charusat, Anand, Gujarat, India

Kumari Sweta
Department of Civil Engineering, National Institute of Technology, Patna, Bihar, India

Mamta Tembhare
Waste Re-processing Division, CSIR-National Environmental Engineering Research Institute (CSIR-NEERI), Nagpur, Maharashtra, India

Anjaly P Thomas
Environmental Engineering and Management, Department of Civil Engineering, Indian Institute of Technology Kharagpur, Kharagpur, West Bengal, India

Anusha Vishwakarma
Department of Civil and Environmental Engineering, Indian Institute of Technology Patna, Bihta, Bihar, India

Shilpa Vishwakarma
Waste Re-processing Division, CSIR-National Environmental Engineering Research Institute (CSIR-NEERI), Nagpur, Maharashtra, India

Bholu Ram Yadav
CSIR-National Environmental Engineering Research Institute (CSIR-NEERI), Nagpur, Maharashtra, India; Academy of Scientific and Innovative Research (AcSIR), Ghaziabad, Uttar Pradesh, India

CHAPTER 1

A global glance on waste electrical and electronic equipments (WEEEs)

Shashi Arya[1,2,3], Dolly Kumari[4], Rumi Narzari[5] and Sunil Kumar[1,2]
[1]Waste Re-processing Division, CSIR-National Environmental Engineering Research Institute (CSIR-NEERI), Nagpur, Maharashtra, India
[2]Academy of Scientific and Innovative Research (AcSIR), Ghaziabad, Uttar Pradesh, India
[3]Centre for Environmental Policy, Faculty of Natural Sciences, Imperial College London, London, United Kingdom
[4]Department of Chemistry, Dayalbagh Educational Institute, Agra, Uttar Pradesh, India
[5]Department of Energy, Tezpur University, Tezpur, Assam, India

1.1 Introduction

According to the "Global E-waste monitor" report published from UNU 2020, approximate 53.6 Mt (million tonnes) of E-waste was generated globally in 2019 as compared to 44.75 Mt in 2016 (Arya & Kumar, 2020a). In E-waste generation, Asia is leading with 24.9 Mt, preceded by 13.1 Mt in America and 2.9 Mt in Africa, 12 Mt in Europe, and 0.7 Mt in Oceania (Forti et al., 2020). E-waste is increasing at a rapid pace and has been the forefront runner to emerge as the significant waste stream in industrialized countries, growing at a steady rate of 3%−5%. The Global E-waste Monitor 2017 report indicated that 40% (18.2 million tons) of the world was generated domestically in Asia. Hence, China has surpassed the United States to become the top producer of E-waste in the world (10.2 million tons). It was also estimated that China will generate 15.5 and 28.4 million tonnes of E-waste in 2020 and 2030, respectively (Zeng & Li, 2016).

E-waste treatment and recovery is severely affected by the presence of hazardous materials such as lead, cadmium, mercury, nickel, and brominated flame retardants (BFRs). They together increase the intricacy of E-waste treatment. Thus the disposal of hazardous waste is a significant challenge in E-waste management (Kaya, 2016). According to the reported data, only 17.4% of E-wastes are formally collected and recycled, and the remaining 82.6% is traded by illegal markets and disposed off without following the waste guidelines (Forti et al., 2020). Many developing countries namely, Benin, Cote d'Ivoire, Liberia, Kenya, South Africa, Uganda, Senegal, Philippines, India, China, Malaysia, Indonesia, Vietnam, Bangladesh, Nigeria, Pakistan, Bhutan, Nepal, and Sri Lanka import E-waste without proper planning and understanding of the consequences associated with the E-waste over a long period. According to recent reports, most of these wastes in general municipal landfills are

Global E-waste Management Strategies and Future Implications
DOI: https://doi.org/10.1016/B978-0-323-99919-9.00018-0

1

stored in warehouses of primitive recycling operations without proper recycling management (Herat & Agamuthu, 2012).

It has been found that the E-waste recovery industry can be a potent source of livelihoods and blue-collar jobs as well as business opportunities (Masud et al., 2019). However, its management can often suffer from a lack of proper infrastructure and effective scientific methodologies (Lakshmi et al., 2017). In third world countries, informal E-waste recycling is more prevalent (e.g., open burning, incineration, acid stripping of metals, acid baths) (Arya & Kumar, 2020b) and generates toxic by-products such as heavy metals, dioxins, and furans (Dai et al., 2020). Many efforts are being made to manage E-waste, ideal roadmaps for sustainable collection, segrega-tion, storage, transportation, and treatment methods (Wath et al., 2011), but still the presence of numerous loopholes is exploited by industrialists; it is expected that proper legislation will be introduced to manage it effectively (Adanu et al., 2020) (Fig. 1.1).

Conventional tools and methods are commonly involved in the open burning of plastic waste, exposure to toxic solders, acid baths to recover saleable materials and components from waste electrical electronic equipment (WEEE) with minimal to no safeguards to human health and the environment, which results in polluting the land, air, and water due to river dumping of acids and widespread general disposal (Manomaivibool & Panate, 2009). Management of E-waste effectively in terms of cost and environmental impact is a complex task. Some countries have organized E-waste management systems; however, several others are still trying to find a solution that ensures minimizing the negative environmental impacts of E-waste treatment and recycling (Srivastava et al., 2020). There are significant considerations that are needed to be fulfilled for developing the E-waste management system. Some of them are as

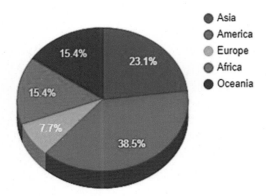

Figure 1.1 Statistics of worldwide generation of E-waste (Forti et al., 2020).

listed: special logistic requirements for collecting the E-waste from its generation sources and transporting it to the disposal site (Bhat & Patil, 2014), minimizing environmental impacts of E-waste disposal, and recovery and reuse of certain metals into the production cycle (Wath et al., 2011). The present review focuses on the ecological implications, technologies, and models related to E-waste management. It also highlights the prospects for sustainable management of E-waste.

1.2 Definition of E-waste?

In the 1970s and 1980s, the term "E-waste" was coined in response to a decline in the environment caused by harmful substances imported into developing countries. Because each country has its own definition of E-waste, there is no good explanation in the world that clearly identifies it. What should be called an E-waste and how do you define it? Electrical or electronic equipment (EEE) that consumes electricity but has extended the end of life (EoL) is signified to as E-waste around the globe. The dumping of electronic or electrical means is stated to as EEE or E-waste. Apart from the fact that E-waste is a complete term that includes a variety of items such as electronic goods (i.e., mobile phones, PCs, processors, printers), electrical domestic devices (i.e., coolers, refrigerators, washing machines, dryers, sound systems), and household appliances (e.g., toys, kettles, toaster) (Jayapradha, 2015; Otieno & Omwenga, 2015). E-waste/WEEE was defined by the European Commission as "the end of a life-enhancing/contrasting product that used to create, balance, and transmit electrical or magnetic current and as handled by providing flow during their administrative existence." A combination of plastic, metal, clay, and nonferrous materials is referred as E-waste, according to the Association of Plastics Manufacturers in Europe (APME) (Srivastava & Pathak, 2020). Because of the development and advanced research in information technologies, it has unconventional considerations everywhere, especially in increasing urbanization and industrialization. In wealthy countries, the annual total amount of E-waste produced ranges from 20 to 50 million tonnes. In the United States of America, total E-waste creation enlarged from 1.90 million tonnes to 3.41 million tonnes in 10 years (from 2000 to 2011). India is the third-largest E-waste generator in the world accounted for 3.2 million metric ton (Mt)/annum and intensely imperiled as a chaotic region for recycling approximately 95% of E-waste through rudimentary techniques that are carried out in small houses or unauthorized areas with low working standards and zero safety measures (Arya, Rautela, et al., 2021). According to 2017 statistics, E-waste output reached about 46 MMT (million metric tonnes) with an annual growth rate of 3%–4%, and 52.2 MMT will be produced in 2021 (Babar et al., 2019). Millions of tons of obsolete computers are being generated all over the world every year, for example, the United States individually is estimated to generate on an average of 130,000 computers/day and is estimated to export approximately 80% of the E-waste to low-income countries (Arya, Patel, et al., 2021). E-waste is not just a waste in real sense but a vast commercial opportunity for those that are functioning in the selling and reutilizing of waste. Subsequently, the massive influx of electronic waste into industrialized countries has been due to weak management practices and unhelpful enforcement of environmental laws regarding the extraction and treatment of essential metals.

The use of EEE has introduced as a key element in everyday life and it is present in a wide range of structures, services, and critical levels. It is constructed of an extensive amalgamation of resources. Notable E-waste sources could be the domestic, that is, large or small household items such as refrigerator and cooler, data scanners used as laptops and PCs, sports related equipment, and consumer equipment wastes such as TVs, cell phones, and clinical equipments (Council of the European Parliament, 2003). EEE components such as lead capacitors, batteries, cathode ray tubes (CRTs), and circuit sheets are also come under the category of E-waste (Choksi, 2001). Conferring to the explanation given in Directive 2012/19/EU of the European Parliament, classifying the EEE into 10 identical classes with the same types of equipment as listed in Table 1.1. According to these classes, 42% of E-waste masses comprise of large household appliances, 34% IT and novelty, 14% hardware purchaser, and six other categories of E-waste make up only 10% of the total e-squander organization (Vadoudi et al., 2015). Table 1.1 presents different categories of E-waste with related examples.

WEEE/E-waste covers all constituents, small portions, and equipment, which is a stage or entire of such objects at the time of dumping. The design of the E-waste varies widely depending on the system and function of the gadget. Typically, it comprises a combination of substances and the components of flexible materials have changed over time and between gadgets, while certain elements are common. The large proportion of these materials is used in the production of E-waste, but the components considered to be the most harmful are consistently reduced. Generally, materials in the E-waste can be grouped into five classes (Fig. 1.2) (Shittu et al., 2020). There are two categories of constituents discovered in E-waste: hazardous and nonhazardous. Heavy metals such as mercury (Hg), chromium (Cr) (VI), cadmium (Cd), lead (Pb), plastics, and circuit boards comprising BFRs are among the constituents found in ward the renowned elements that are producing major worry. In the course of the heating process, BFRs can yield dioxins and furans. The various materials that can be found are nickel (Ni), selenium (Se), and copper (Cu). Lead, phosphorus, antimony, and

Table 1.1 Categories of electrical and electronics waste (Adediran & Abdulkarim, 2012).

S. No.	Category	Typical examples
1.	Large Household Appliances (LHAs)	Refrigerators, washing machines, microwaves, heating appliances, fanning/exhaust, ventilation/conditioning equipment, freezers
2.	Small Household Appliances (SHAs)	Vacuum cleaners, sewing/knitting/weaving textile appliances, toasters, fryers, pressing iron, grinders, knives, hair cutting/drying/shaving devices, clocks, watches
3.	Lighting Equipment	Luminaires for fluorescent lamps, low pressure sodium lamps
4.	Consumer Equipment	Radio and TV sets, video cameras/decoders, Hi-fi recorder, audio amplifiers, musical instruments
5.	IT and Telecommunication Equipment	Mainframes, microcomputers, printers, PC (desktop, notebooks, laptops), photocopiers, typewriters, fax/telex equipment, telephones
6.	Medical Devices	Devices for radiotherapy/cardiology/dialysis, analyzers, freezers, detecting/preventing/monitoring/alleviating illness, injury or disability, ventilators
7.	Electrical and Electronic Tools (excluding large-scale industrial tools)	Welding/soldering tools, spraying/spreading/dispersing tools, turning/milling/sanding/sawing/cutting/shearing/drilling/punch in g/folding/bending equipment, nailing/screwing tools, drills
8.	Toys, Leisure and Sports Equipment	Electric trains, car racing sets, video games, sports equipment, coin slot machines, biking/diving/running/ rowing computers
9.	Monitoring and Control Instruments	Smoke detectors, heating regulators, thermostats, measuring/weighing/adjusting appliances for household or laboratory use, other industrial monitoring and control instruments
10.	Automatic Dispensers	for hot drinks, hot or cold bottles/cans, solid, products, money, and all kinds of products

other chemicals are found in the CRT of a TV or computer screen. Other toxic substances comprised in various electrical materials including Se, antimony trioxide, Cd, Co, Mn, brome, and Ba among many others (Rautela et al., 2021; Vishwakarma et al., 2022). Waste printed circuit boards are typically made up of 40% metal, 30% nonabrasive oxide, and 30% plastic, comprising the majority of the components in the periodic table in addition to the most valuable and

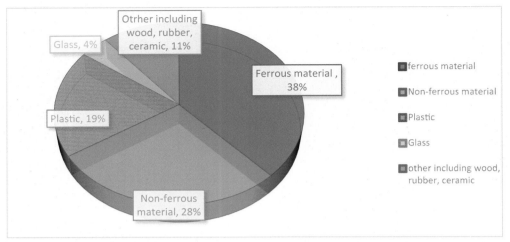

Figure 1.2 The component of E-waste (Rautela et al., 2021; Shittu et al., 2020; Zhou & Xu, 2012).

eco-friendly metals with extreme quality. These nonmetallic and nonferrous metals can be recycled for reuse and recycling, while reinforced resin and epoxy resin (which have some low-cost qualities) can be burned or reused as coating wipes and subsequent powder materials to restore their thermal conductivity. In any event if these small and dangerous substances are not well managed, they can reflect the opposition's environmental and social impacts (Rautela et al., 2021; Robinson, 2009; Zhou & Xu, 2012).

Regional sheets containing large metals such as lead and cadmium, cathode beam tubes containing lead oxide, and barium batteries containing cadmium, mercury at screen level, polychlorinated biphenyl's present in more established capacitors and transformers, brominated fire resistance used printed circuit sheets, connectors and plastic installation, polyvinyl chloride—coating copper connectors, and plastic PC coverings are the materials which emit noxious dioxins and furans when imitated to recover precious metals. According to the basel action network, the world's 500 million PCs contain 286,700 kg of mercury, 716.7 million kg of Pb, and 2.87 billion kg of plastic. Pb can pollute groundwater after it is replenished by sewage systems. When a tube is burned and crushed, it emits harmful fumes into the air (Gupta, 2011). As a result of several difficulties such as specialized technology, strategic, financial, and information failures, E-waste management is a perplexing endeavor. Hence, there is an immense need to accomplish E-waste in eco-friendly, prescribed and straightforward way by removing/reusing valuable metals in streams. Table 1.2 represents common metals present in E-waste with their occurrence and impact on human health.

Table 1.2 Common toxic metals, dioxins and persistent organic pollutants associated with E-waste and their health impacts (Kiddee et al., 2013; Pant et al., 2012; Rautela et al., 2021).

Metal	Occurrence	Impact
Antimony (Sb)	A melting agent in CRT glass, plastic computer housings and a solder alloy in cabling	Antimony has been classified as a carcinogen. It can cause stomach pain, vomiting, diarrhea and stomach ulcers through inhalation of high antimony levels over a long time period
Americium (Am)	Fire detectors, medical equipment, active sensing element in smoke detectors	Cytogenetic damage
Arsenic (As)	Gallium arsenide is used in light emitting diodes	It has chronic effects that cause skin disease and lung cancer and impaired nerve signaling
Barium (Ba)	Sparkplugs, fluorescent lamps and CRT gutters in vacuum tubes	Causes brain swelling, muscle weakness, damage to the heart, liver and spleen though short-term exposure
Beryllium (Be)	Power supply boxes, motherboards, relays and finger clips	Exposure to beryllium can lead to berylliosis, lung cancer and skin disease. Beryllium is a carcinogen
Cadmium (Cd)	Rechargeable Ni–Cd batteries, semiconductor chips, infrared detectors, printer inks and toners	Cadmium compounds pose a risk of irreversible impacts on human health, particularly the kidneys
Chromium (Cr)	Production of metal housings, data tapes, floppy-disks), plastic computer housing, cabling, hard disks and as a colorant in pigments	Defects in neurodevelopment, multiple organ failure, carcinogenic and leads to oxidative-stress, is extremely toxic in the environment, causing DNA damage and permanent eye impairment

(Continued)

Table 1.2 (Continued)

Metal	Occurrence	Impact
Lead (Pb)	Solder, lead–acid batteries, cathode ray tubes, cabling, printed circuit boards and fluorescent tubes	Can damage the brain, nervous system, kidney and reproductive system and cause blood disorders. Low concentrations of lead can damage the brain and nervous system in fetuses and young children.
Lithium (Li)	Li-batteries	Affect gastrointestinal and neurologic system
Mercury (Hg)	Batteries, backlight bulbs or lamps, flat panel displays, switches and thermostats	Mercury can damage the brain, kidneys and fetuses
Nickel (Ni)	Batteries, computer housing, cathode ray tube and printed circuit boards, Electron gun in CRT, Rechargeable Ni–Cd batteries	Can cause allergic reaction, bronchitis and reduced lung function and lung cancers
Selenium (Se)	Older photocopy machines	High concentrations cause selenosis
Zinc (Zn)	Interior or CRT Screens	Cytotoxicity, ischemia and trauma
Chlorofluorocarbons	(CFCs) Cooling units and insulation foam	These substances impact on the ozone layer which can lead to greater incidence of skin cancer
Persistent organic pollutants		
Brominated flame retardants (BFRs)	BFRs are used to reduce flammability in printed circuit boards and plastic housings, keyboards and cable insulation	During combustion printed circuit boards and plastic housings emit toxic vapors known to cause hormonal disorders
Polychlorinated biphenyls (PCBs)	Condensers, transformers and heat transfer fluids.	PCBs cause cancer in animals and can lead to liver damage in humans

(*Continued*)

Table 1.2 (Continued)

Metal	Occurrence	Impact
Polybrominated biphenyls (PBBs) Polybrominateddiphenyl ethers (PBDEs)	Dielectric fluids, lubricants and coolants in generators, capacitors and transformers, fluorescent lighting, ceiling fans. Dishwashers, and electric motors	Ingestion, inhalation or dermal contact and transplacental absorption
Polyvinyl chloride (PVC)	Monitors, keyboards, cabling and plastic computer housing	PVC has the potential for hazardous substances and toxic air contaminants. The incomplete combustion of PVC release huge amounts of hydrogen chloride gas which form hydrochloric acid after combination with moisture which can cause respiratory problems
Dioxins Polychlorinated dibenzodioxins (PCDDs) and polychlorinated dibenzofurans (PCDFs)	Released as combustion by-product	Ingestion, inhalation or dermal contact and transplacental absorption
Dioxins-like polychlorinated biphenyls	Released as combustion by-product but also found in dielectric fluids, lubricants and coolants in generators, capacitors and transformers, ceiling fans. Dishwashers, and electric motors	Ingestion, inhalation or dermal absorption
Polyaromatic hydrocarbons (PAHs)	Released as combustion by-product	Ingestion, inhalation or dermal contact and transplacental absorption

References

Adanu, S. K., Gbedemah, S. F., & Attah, M. K. (2020). Challenges of adopting sustainable technologies in E-waste management at Agbogbloshie, Ghana. *Heliyon*, *6*(8). Available from https://doi.org/10.1016/j.heliyon.2020.e04548.
Adediran, Y. A., & Abdulkarim, A. (2012). Challenges of electronic waste management in Nigeria. *International Journal of Advances in Engineering & Technology*, *4*(1), 640−648.

Arya, S., & Kumar, S. (2020a). E-waste in India at a glance: Current trends, regulations, challenges and management strategies. *Journal of Cleaner Production, 271*, 122707. Available from https://doi.org/10.1016/j.jclepro.2020.122707.

Arya, S., & Kumar, S. (2020b). Bioleaching: Urban mining option to curb the menace of E-waste challenge. *Bioengineered*, 640−660. Available from https://doi.org/10.1080/21655979.2020.1775988.

Arya, S., Patel, A., Kumar, S., Loke, Show, & Pau. (2021). Urban mining of obsolete computers by manual dismantling and waste printed circuit boards by chemical leaching and toxicity assessment of its waste residues. *Environmental Pollution, 283*, 117033. Available from https://doi.org/10.1016/j.envpol.2021.117033.

Arya, S., Rautela, R., Chavan, D., & Kumar, S. (2021). Evaluation of soil contamination due to crude E-waste recycling activities in the capital city of India. *Process Safety and Environmental Protection, 152*, 641−653. Available from https://doi.org/10.1016/j.psep.2021.07.001.

Babar, S., Gavade, N., Shinde, H., Gore, A., Mahajan, P., Lee, K. H., & Garadkar, K. (2019). An innovative transformation of waste toner powder into magnetic g-C_3N_4-Fe_2O_3photocatalyst: sustainable e-waste management. *Journal of Environmental Chemical Engineering, 7*(2), 103041.

Bhat, V., & Patil, Y. (2014). E-waste consciousness and disposal practices among residents of Pune city. *Procedia-Social and Behavioral Sciences, 133*, 491−498.

Choksi, S. (2001). The basel convention on the control of transboundary movements of hazardous wastes and their disposal: 1999 protocol on liability and compensation. *Ecology Law Quarterly, 28*, 509−539.

Dai, Q., Xu, X., Eskenazi, B., Asante, K. A., Chen, A., Fobil, J., Bergman, Å., Brennan, L., Sly, P. D., Nnorom, I. C., Pascale, A., Wang, Q., Zeng, E. Y., Zeng, Z., Landrigan, P. J., Bruné Drisse, M. N., & Huo, X. (2020). Severe dioxin-like compound (DLC) contamination in E-waste recycling areas: An under-recognized threat to local health. *Environment International, 139*, 105731.

Forti, V., Blade, C. P., Kuehr, R., & Bel, G. (2020). The Global E-waste Monitor 2020: Quantities, flows and the circular economy potential. *United Nations University (UNU)/United Nations Institute for Training and Research (UNITAR) − co-hosted SCYCLE Programme, International Telecommunication Union (ITU) & International Solid Waste Association (ISWA)*, ISBN Digital: 978-92-808-9114-0 ISBN Print: 978-92-808-9115-7.

Gupta, S. (2011). E-waste management: Teaching how to reduce, reuse and recycle for sustainable development-need of some educational strategies. *Journal of Education and Practice, 2*(3).

Herat, S., & Agamuthu, P. (2012). E-waste: A problem or an opportunity? Review of issues, challenges and solutions in Asian countries. *Waste Management & Research, 30*(11), 1113−1129. Available from https://doi.org/10.1177/0734242X12453378.

Jayapradha, A. (2015). Scenario of E-waste in India and application of new recycling approaches for E-waste management. *Journal of Chemical and Pharmaceutical Research, 7*(3), 232−238.

Kaya, M. (2016). Recovery of metals and nonmetals from electronic waste by physical and chemical recycling processes. *Waste Management, 57*, 64−90. Available from https://doi.org/10.1016/j.wasman.2016.08.004.

Kiddee, P., Naidu, R., & Wong, M. H. (2013). Electronic waste management approaches: An overview. *Waste Management, 33*(5), 1237−1250.

Lakshmi, S., Raj, A., & Jarin, T. (2017). A review study of E-waste management in India. *Asian Journal of Applied Science and Technology (AJAST), 1*(9), 33−36. Available from https://papers.ssrn.com/sol3/papers.cfm?abstract_id = 3048625.

Manomaivibool, P., & Panate. (2009). Extended producer responsibility in a non-OECD context: The management of waste electrical and electronic equipment in India. *Resources, Conservation and Recycling, 53*(3), 136−144.

Masud, M. H., Akram, W., Ahmed, A., Ananno, A. A., Mourshed, M., Hasan, M., & Joardder, M. U. H. (2019). Towards the effective E-waste management in Bangladesh: A review. *Environmental Science and Pollution Research, 26*(2), 1250−1276.

Otieno, I., & Omwenga, E. (2015). E-waste management in Kenya: Challenges and opportunities. *Journal of Emerging Trends in Computing and Information Sciences, 6*(12), 661−666.

Pant, D., Joshi, D., Upreti, M. K., & Kotnala, R. K. (2012). Chemical and biological extraction of metals present in E waste: A hybrid technology. *Waste Management, 32*(5), 979−990.

Rautela, R., Arya, S., Vishwakarma, S., Lee, J., Ki-Hyun, K., & Kumar, S. (2021). E-waste management and its effects on the environment and human health. *Science of the Total Environment*, *773*, 145623. Available from https://doi.org/10.1016/j.scitotenV.2021.145623.

Robinson, B. H. (2009). E-waste: An assessment of global production and environmental impacts. *Science of the total environment*, *408*(2), 183−191.

Shittu, O. S., Williams, I. D., & Shaw, P. J. (2020). Global E-waste management: Can WEEE make a difference? A review of e-waste trends, legislation, contemporary issues and future challenges. *Waste Management*, *120*, 549−563.

Srivastava, R. R. and Pathak, P. (2020). Policy issues for efficient management of E-waste in developing countries, In Handbook of electronic waste management, International Best Practices and Case Studies 81−99. Butterworth-Heinemann.

Srivastava, R. K., Shetti, N. P., Raghava, K., & Aminabhavi, T. M. (2020). Sustainable energy from waste organic matters via efficient microbial processes. *Science of the Total Environment*, *722*, 137927. Available from https://doi.org/10.1016/j.scitotenv.2020.137927.

Vadoudi, K., Kim, J., Lee, S-J., & Troussier, N. (2015). E-waste management and resources recovery in France. *Waste Management & Research*, *33*(10), 919−929. Available from https://doi.org/10.1177/0734242X15597775.

Vishwakarma, S., Kumar, V., Arya, S., Tembhare, M., Rahul., Dutta, D., & Kumar, S. (2022). E-waste in information and communication technology sector: Existing scenario, management schemes and initiatives. *Environmental Technology & Innovation*, 102797. Available from https://doi.org/10.1016/j.eti.2022.102797.

Wath, S. B., Dutt, P. S., & Chakrabarti, T. (2011). E-waste scenario in India, its management and implications. *Environmental Monitoring and Assessment*, *172*(1), 249−262.

Zeng, X., & Li, J. (2016). Measuring the recyclability of E-waste: An innovative method and its implications. *Journal of Cleaner Production*, *131*, 156−162. Available from https://doi.org/10.1016/j.jclepro.2016.05.055.

Zhou, L., & Xu, Z. M. (2012). Research progress of recycling technology for waste electrical and electronic equipment. *Material Review*, *26*(7), 155−160.

CHAPTER 2

Global scenario of E-waste generation: trends and future predictions

Anusha Vishwakarma, Kavita Kanaujia and Subrata Hait
Department of Civil and Environmental Engineering, Indian Institute of Technology Patna, Bihta, Bihar, India

2.1 Introduction

Over the last two decades, the demands for electrical and electronic equipment (EEE) have continuously risen, expediting the EEE production process. But unfortunately, the practices of production of EEE, along with their consumption and disposal, are highly unsustainable, leading to the accumulation of waste electrical and electronic equipment (WEEE) or electronic waste (E-waste). E-waste is the EEE waste discarded by its user without an intent to reuse (Baldé et al., 2017). Continuous technology upgradation, rise in the purchasing capacity of an individual, lifestyle changes, higher dependency on EEE, and shorter lifespan of EEE are the major factors governing the rise in the number of WEEE and have prompted the growth of E-waste production around the world (Priya & Hait, 2020; Trivedi & Hait, 2020a). Apart from technology upgradation, EEE utilization is closely connected to the overall development of the global economy. Economic growth in developing countries (China, Asia, and Indonesia) has led to the generation of more E-waste than in developed countries, making the situation more alarming. Also, for the past few years, WEEE from Information and Communication Technology (ICT) has seen a yearly growth pace of 4.5% (Qu et al., 2019). As the rise in ICT directly impacts the economy of any country, EEE utilization is of utmost importance for the development of the economy (Vishwakarma et al., 2022). As per the latest report by United Nations University, the global production of E-waste in 2019 was approximately 7.3 kg/capita, that is, 53.6 Mt (Forti et al., 2020). Out of the total amount generated, only 17.4% is formally chronicled and recycled. Global E-waste volume has increased by 9.2 million tones since 2014. Based on the functional similarity, end-of-life (EoL) characteristics, and its use for statistical purposes, EEE is classified into six different categories (as shown in Table 2.1).

Table 2.1 provides details of EEE classification and their share in 2019 E-waste generation (Forti et al., 2020). The total amount of generated E-waste in 2019 has significantly increased as compared to 2014 global E-waste production with a rise of 4%, 5%, 7%, 2%, and 4% in small equipment, large equipment, temperature

Global E-waste Management Strategies and Future Implications
DOI: https://doi.org/10.1016/B978-0-323-99919-9.00013-1

Table 2.1 EEE categories and their share in E-waste generated in 2019.

S. No.	Categories of E-waste based on EU-6 classification	Equipment in each category	E-waste generated in 2019 (Mt)
1.	Small EEE	Vacuum cleaners, microwaves, iron, toaster, hairdryer, water cookers, headphones, radio, video recorder, speakers, cameras, household LED luminaires, drones, music toys, alarm, and thermometer.	17.4
2.	Large EEE	Washing machines, photovoltaic panels, dishwasher, washer dryers, ventilators, copiers, welding machines, electric bike, laboratory equipment, professional medical equipment, money dispensers.	13.1
3.	Temperature exchange EEE	Portable air conditioners, freezers, fridges, cold drink dispensers, dehumidifiers, cooling display.	10.8
4.	Screens and monitors	Laptops, cathode ray tube TVs and monitors, LCD and LED flat display monitors and TVs.	6.7
5.	Small IT and telecommunication EEE	Desktop PCs, mobiles, mice, keyboard, routers, printers, scanners, and game consoles.	4.7
6.	Lamps	LED lamps, CFLs, straight tube fluorescent lamps, special Hg and Na lamps.	0.9

Source: Data sourced from Forti, V., Balde, C. P., Kuehr, R., & Bel, G. (2020). *The global E-waste monitor 2020: Quantities, flows and the circular economy potential.* United Nations University (UNU)/United Nations Institute for Training and Research (UNITAR) — co-hosted SCYCLE Programme, International Telecommunication Union (ITU) & International Solid Waste Association (ISWA).

exchange equipment, small IT and telecom EEE, and lamps, respectively. In contrast, screens and monitors' relative contribution have decreased by 1%. With the development of a country's economy, the purchasing power of its individuals rises simultaneously, leading to an increase in the consumption of all EEE categories and associated E-waste generation (Priya & Hait, 2017a, 2018a). However, the decreased contribution of screens and monitors in 2019 E-waste is due to the reduced sales of heavy cathode ray tube monitors and their replacement by flat panel displays (Forti et al., 2020).

E-waste consists of a heterogeneous mixture of several components, whose diversity varies from product to product concerning the age, type, manufacturer, and category of EEE (Priya & Hait, 2017a, 2017b; Trivedi & Hait, 2020b). The metallic fraction of E-waste consists of base metals, precious metals (Au, Ag, Pt,

Pd), and rare earth elements (Arya et al., 2021). About 50% of E-waste comprises iron and steel, followed by plastics and nonferrous metals at 21% and 13%, respectively (Needhidasan et al., 2014; Pant et al., 2012). The inadequate management and unscientific disposal of E-waste release multiple toxic substances, posing a severe threat to human wellbeing and the surrounding natural environment (Arya & Kumar, 2020a; Priya & Hait, 2018b). The closed-loop approach of the circular economy model emphasizes the restoration and regeneration of raw material from waste. It entails redesigning more inventive electronic devices that use less natural ores and have a longer lifespan. Therefore the effective management of E-waste exacerbation is a major concern and is crucial to achieving the sustainable development goals (SDGs) 2030 Agenda (Baldé et al., 2017).

Although it appears that E-waste is growing by the day, the recent global outburst of the COVID-19 pandemic reduced future E-waste production by a total of 4.9 Mt (Baldé & Kuehr, 2021; Dutta et al., 2022), particularly in the first nine months of 2020, when the majority of countries were on lockdown. A minor level of E-waste reduction has been seen in small IT and telecom equipment, with only 0.06 Mt, followed by screens and monitors with 0.5 Mt. Low- to middle-income nations saw a 30% drop in E-waste weight in the first eight to nine months of 2020. In contrast, high-income countries had a 2%—6% reduction in E-waste weight (Baldé & Kuehr, 2021). Compared to middle- and low-income countries, COVID-19 has a minor influence on high-income countries. Few high-income countries, namely New Zealand and Australia, have shown a 4% increase in EEE consumption and thus future E-waste generation. The lower- and lower-middle-income countries are unable to implement measures to protect their population and economy against COVID-19, showed a significant decrease in E-waste generation (Baldé & Kuehr, 2021). However, this reduction is temporary but provides a breathing space to properly manage the staggering volume of E-waste.

Although the effect of the pandemic is devastating from a social and economic viewpoint, due to the unavailability of respective data, we are not considering the impact of COVID-19 on the generation and future projection part of our chapter. The chapter intends to provide a recent valuation of global E-waste generation, distribution into various major regions and countries worldwide, including the income-wise distribution of E-waste. Multiple factors affecting the E-waste generation rate are also summarized. An estimate in global E-waste generation for the next three decades has been made considering the correlation between generated per capita E-waste data and per capita gross domestic product (GDP) data in each country by using a suitable statistical technique. Lastly, the role of E-waste projection in formulating SDGs that could help achieve the circular economy is explored.

2.2 Factors driving E-waste generation

2.2.1 Urbanization

Urbanization is essentially a progression from a customary rural economy to a state-of-the-art industrial economy. Urban growth is speeding up, and housing is reclassifying as cities offer more employment, better lifestyles, good infrastructure, and safer living conditions, attracting people to migrate from rural areas (Davis, 1965). The rapidly expanding cities and the waste generated in urban agglomerations pose societal and environmental challenges. In 2018, almost 55% of the world's population resided in the urban area, and by 2050, this figure is predicted to rise to 68% (United Nation [UN], 2018). This growth has, in turn, triggered the amount of E-waste in urban areas as only 4% of the rural population has access to computers.

In contrast, it increases to 23% for the urban population (Gohain, 2020). This clearly shows a digital divide between rural and urban communities. In urban areas, the high living standard and availability of disposable income followed by a growth in consumerism are responsible for massive E-waste generation (RS Secretariat, 2011). A remarkable growth in developing countries' economies, including Brazil, China, and India, and rising consumerism in their cities increases product consumption and generation of E-waste (Abalansa et al., 2021).

2.2.2 Shorter obsolescence age of electronic products

Advances in technology, rapid upgradation in features, and increased dependency on electronic products lead to a surge in the sale of EEE (Verma & Hait, 2019). Most telecommunication products, such as cell phones and computers, have a short lifespan of fewer than 2 years (Bhutta et al., 2011). Previous personal computers had an average lifespan of 7 years that reduced to 2−5 years, adding more and more E-waste (RS Secretariat, 2011). A case study by the European Environment Agency (EEA, 2020) shows that cell phones and home appliances such as washing machines and vacuum cleaners have an average lifecycle that is 2.3 years shorter than their intended life. The short lifecycle of electronic devices encourages manufacturers to develop more products.

Additionally, it forces users to feel unsatisfactory and interpret that the product is no longer usable (Ahmed, 2016). It has also been found that it is a marketing strategy that companies are deliberately manufacturing short lifespan products via technical failures (especially in mobile phones) to maintain the pace of consumption of EEE (Makov & Fitzpatrick, 2021; RS Secretariat, 2011). The planned obsolescence policy of electronic devices in the government sector (such as in India) also contributes to the growth of E-waste volume, as they discard equipment after a planned period, even if the device is in working condition (Victor & Kumar, 2012). Electronic products have significant potential to be used for more extended periods. By emphasizing the circular economy concept, we can increase their longevity and delay their obsolescence by subsequent reuse, repair, remanufacturing, or recycling (EEA, 2020).

2.2.3 Industrial development

Electronic devices are the backbone of the industrial revolution with broader application of electronic-based technology in several areas (defense devices, entertainment types of equipment, medical diagnosis, and others). Electronic manufacturing industries are the most emerging and rapidly embryonic industries worldwide. They are constantly engaged in manufacturing innovative electronic products with more user-friendly, more advanced features, increasing their demand globally. The rise in the electronic industry is an opportunistic factor that shows the growth of that particular country in the global market (Vats & Singh, 2014). Correspondingly, the downside of this growth is that it contributes moderately to premature obsolescence. Electronic businesses are attempting to meet current demand by producing more items, resulting in much electrical equipment being discarded as they reach their EoL (RS Secretariat, 2011).

The current pandemic raises global shifts in trade policy, so reliance on any specific country for electronics raw materials can be a risky strategy. Developing countries like India are grabbing this opportunity to enable policies and share to the global market of EEE production. The year 2020 has been a turbulent year for the electronic industry with a whopping rise in online activities; the retail e-commerce sales totaled US$4.28 trillion worldwide (Chevalier, 2021). The evolution in e-retail services leads to increased demand for more innovative products, ultimately generating more E-waste.

2.2.4 Technological advancement

Over the last 20 years, the accelerated growth in technology has transformed our lives. The technological revolution of electronic devices has changed the way we live. It offers a variety of services for entertainment (game apps, social media), security and home security (CCTV cameras), and digitization of medical equipment (portable pulse oximeter, blood pressure meter). In recent years, several advanced electronic tools have come across from kitchen appliances (smart blender, smart Wi-Fi or Alexa-operated fridge, meat thermometer), entertainment products (smartphones, smartwatch, tablets, smart TV, voice assistant devices) to transportation facilities (automotive cars, battery-operated electric vehicles) (Ahmed, 2016). An unprecedented pace has been seen in the growth of the ICT sector, especially in mobile phones; technological advancement has proliferated; how a smartphone replaces many other devices such as cameras, music devices, radio, watches, calculators, and others (Gupta et al., 2021). The technical innovation in mobile phones with high-speed internet connectivity (3G, to 4G and now 5G) has provided numerous services, paving the way for frequently replacing our mobile phones (Chen et al., 2018). By 2020, the 4G network will reach 84.7% of the world's population (International Telecommunication Union [ITU], 2020). Moving to a 5G network, which does not support the older one, will force users to upgrade their older device with a new one. Furthermore, according to a recent report by O'Dea (2021), the current global smartphone subscriber base has surpassed 6 billion, which is predicted to grow by several hundred million in the future year. The figure suggests that there will be more contribution in E-waste from waste mobile phones in the coming year.

The constant innovation in design, upgradation in features, and software update that does not support older models influence consumers to easily upgrade their old gadgets with new ones to maintain parity. This relentless revolution in electronic devices has made our lives faster and easier and changed the lifestyle with more comfort for people, but with a cost of more waste accumulation.

2.2.5 Gross domestic product

In the last decade, various studies have determined how waste generation was affected by a nation's GDP. Several landmark studies witnessed that GDP, purchasing power parity (PPP), and WEEE production all had a significant linear relationship (Baldé et al., 2017; Kumar et al., 2017; Kusch & Hills, 2017; Priya & Hait, 2017a, 2017b). Economic development in terms of GDP is considered a sensitive factor related to E-waste generation, which means that any nation's economic development will increase E-waste. A study by Kumar et al. (2017) indicates that E-waste generated per capita of any country is related to its purchasing power hence individual income. The study also suggested that along with GDP, any country's large population can contribute to total E-waste production (Gundupalli et al., 2018). Still, if they have low purchasing power and GDP, it is needless that they would generate a significant amount of E-waste (Kumar et al., 2017). For example, in 2019, China and the United States had high GDP and PPP values of US$23.444 trillion and US$21.433 trillion, respectively, and they shared more amounts of 10,129 kt and 6918 kt in global E-waste volume (Forti et al., 2020; World Bank, 2020). India (the second highly populated nation worldwide) has comparatively contributed less (3230 kt) in global E-waste amount because of its low GDP and PPP (US$8.907 trillion in 2020) (Forti et al., 2020; World Bank, 2019). A US$1000 increase in GDP in European countries adds 0.5 kg WEEE (Kusch & Hills, 2017). Based on nominal GDP, low GDP countries have high intensity of WEEE; simultaneously, higher GDP countries have low WEEE intensity (Kusch & Hills, 2017). In developing countries (India, China, Brazil, and Indonesia), the rise in purchasing power of inhabitants indicates that very soon, total E-waste volume from developing countries will exceed developed countries (Li et al., 2015).

2.2.6 Legislation and policies

The lack of adequate policies and regulations explicitly regarding WEEE management is one primary reason for unprecedented growth in E-waste (Rautela et al., 2021). Developing countries are mainly dealing with lenient legislation. Over the last two decades, because of the lack of adequate guidelines, Asia and Africa have become the emerging destination or digital dumbs for E-waste importation from developed countries (Davis et al., 2019). To bridge the technological gap, developed nations are shipping their electronic trash to economically poor or developing countries, as they lack strict legislation (Priya & Hait, 2017a, 2017b; Vats & Singh, 2014). The laxity in legislation encourages the unauthorized sector to import E-waste from rich countries illegally, allows informal sectors to treat E-waste unpleasantly, and does not provide proper WEEE collection, recycling, and disposal services (Arya & Kumar, 2020b; Tiseo, 2020). As of 2019, only 78 countries follow the E-waste protocol, comprising 71% of the world's population (Forti et al., 2020). However, over half of the world's nations are not following laws and policies to handle WEEE. Regulation can aid with the successful handling of a large volume of E-waste (Pariatamby & Victor, 2013). To meet this, laws and regulations should be simple, enforceable, and clear to specify the collection cost and take-back of E-waste (ITU, 2021).

2.3 Global E-waste generation scenario

A total of 53.6 Mt of E-waste, representing 7.3 kg per person, was produced in 2019. In Fig. 2.1A and B, a comparison of E-waste generation by region-wise and income level, respectively, is presented. Asia generates 47% of all E-waste, followed by

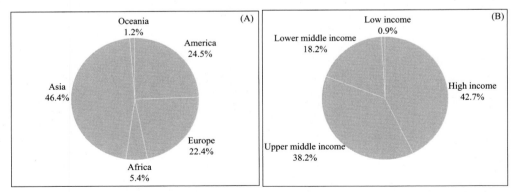

Figure 2.1 Percentage of E-waste generated: (A) region-wise and (B) income group for the year 2019. *Data sourced from Forti, V., Balde, C. P., Kuehr, R., & Bel, G. (2020).* The global E-waste monitor 2020: Quantities, flows and the circular economy potential. *United Nations University (UNU)/United Nations Institute for Training and Research (UNITAR) — co-hosted SCYCLE Programme, International Telecommunication Union (ITU) & International Solid Waste Association (ISWA).*

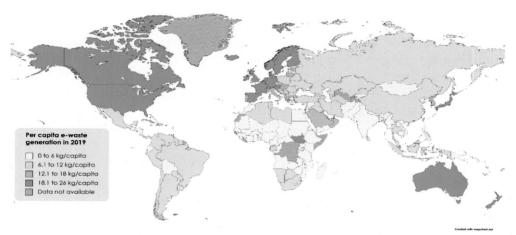

Figure 2.2 Country-wise distribution of E-waste per capita generated in 2019. *Data sourced from Forti, V., Balde, C. P., Kuehr, R., & Bel, G. (2020). The global E-waste monitor 2020: Quantities, flows and the circular economy potential. United Nations University (UNU)/United Nations Institute for Training and Research (UNITAR) — co-hosted SCYCLE Programme, International Telecommunication Union (ITU) & International Solid Waste Association (ISWA).*

America, which produces 25% of all E-waste; Europe, which produces 22% of all E-waste; Africa, which generates 5% of all E-waste; and Oceania, which generates 1% of all E-waste. Based on income, high-income countries account for 43% of E-waste generation, followed by upper-middle-income countries sharing 38% of E-waste generated. The percent share of E-waste in lower-middle-income and low-income countries is 18% and 1%, respectively.

Per capita waste generation has been compared in Fig. 2.2. North America, Australia, with part of Europe generate E-waste ranging from 18 to 26 kg/capita. India and parts of Africa, where most countries are low- and lower-middle-income, produce only 0–6 kg of E-waste per capita. Part of Europe, including Russia and South America, generates 6–12 kg/capita. Few countries of Europe and Saudi Arabia generate 12–18 kg/capita.

2.4 Region-specific E-waste scenario

2.4.1 E-waste status in America

America generated 13.1 Mt of the world's total E-waste in 2019, averaging 13.3 kg/person. The United States produced 6.92 million tons of this, approximately 21 kg/person, and only 15% of the material was recycled. The produced E-waste comprises raw materials worth $7.49 billion. Only 1.2 Mt of the total was collected and properly recycled. Most American nations export their E-waste to Africa, China, and Mexico, where it is processed informally, causing pollution (Patil & Ramakrishna, 2020). One study showed that 80% of US E-waste ends up in Asian countries, trailed by 200 GPS (Lee et al., 2018; Thakur & Kumar, 2021). In 2016, the Americas produced 11.3 Mt of E-waste, with 7 Mt from North America, 3 Mt from South America, and 1.2 Mt from Central America. Only 17%, about 1.9 Mt, was collected and reprocessed. It weighed 11.6 kg of E-waste per person (Kumar & Singh, 2019). North and South America have generated significant amounts of E-waste over the years. America's two highest-income countries, the United States and Canada, produce the most E-waste, followed by Brazil and Mexico (Patil & Ramakrishna, 2020).

Despite such a scenario, the United States has not regulated strict centralized laws governing E-waste management. Instead, some states, like Canadian states and others, have implemented E-waste laws that lack consistency. About 75%–80% of the US population is covered by E-waste management regulations. The legislation, however, is not as stringent and does not prevent landfilling activities. Most Latin American nations lack national E-waste recycling regulations; however, several have attempted to adopt extended producer responsibility (EPR) for some E-waste categories. Likewise, in North America, both the United States and Canada have implemented the EPR program, mandatory in a few states.

Moreover, many states, even covered under the law, do not have convenient collection opportunities due to the difference in the scope of legislation policies. Even though the United Nations sets out regulatory measures to prevent harmful effects resulting from improper disposal and treatment practices. Such disparity in policies across the continent leads to several issues in the disposal of E-waste (Organisation for Economic Co-operation and Development [OECD], 2016; Patil & Ramakrishna, 2020). If the equipment meets specific criteria, it can be managed certified. Hundreds of recycling facilities are in operation, with standards that have been revised and improved regularly since their commencement in 2010. The formal collection is mandatory in a few countries with a specific legal framework. Lack of consistent legislation framework and less contribution from research are the significant challenges associated with the process (Forti et al., 2020).

2.4.2 E-waste status in Europe

In 2019, the production of E-waste was estimated to be around 12.1 Mt, with 16.2 kg of E-waste produced per capita, the highest per capita rate. However, Europe also has the highest formally documented percentage of E-waste collection and recycling (42.5%) (Baldé et al., 2017; Tiseo, 2020). In 2019, Russia, Germany, and the United Kingdom were Europe's top three E-waste producers, each producing about 1.6 Mt of E-waste. That is far less than the amount of E-waste produced by China and the United States (Tiseo, 2020). Out of 44, 37 European countries have formally implemented policies and regulations to manage E-waste. In Europe, the WEEE Directive (2012/19/EU) governs a substantial part of the total volume of E-waste. This directive assures safe and accountable collection, recycling and resource recovery from all the six categories of E-waste. Most European Union (EU) countries have well-developed and effective E-waste management infrastructure. They collect discarded electronic products by private operators from municipalities and stores, retrieve recyclable/reusable components, and dispose of the residuals. Statistics show that Europe ranks highest in E-waste collection, documentation, and recycling, with 59% collection rates in northern Europe and 54% in western Europe.

In 2017, Europe and the United States alone contributed almost half of the total global volume of E-waste generated. In Europe, polybrominated diphenyl ethers (PBDEs) and polybrominated biphenyls (PBBs) have been restricted, as their toxicity and persistence in nature can cause serious health problems. Belarus has more developed E-waste collection and recycling facilities than other European countries. In Belarus, E-waste management complies with the EPR framework of manufacturers and suppliers and also follows Law No. 271-IS, a centralized law, which came into force in 2007. Based on the EU WEEE directive, Ukraine is in the process of establishing an EPR system. In 2017, Russia also launched an EPR program to manage its E-waste (Forti et al., 2020).

Nonetheless, the government must improve waste collection and recycling to meet compliance with authorities' standards. The feasibility of achieving the targets and location was studied in several countries, such as Romania (Magalini et al., 2019) and the Netherlands (Forti et al., 2020), showing a rise in compliance with E-waste collection and reutilization. However, a significant portion of E-waste in the EU is still managed outside compatible recycling zones.

2.4.3 E-waste status in Asia

Asia ranked top in 2019 among E-waste generators with an enormous amount of 24.9 Mt or 2.5 kg per inhabitant; however, only 2.9 Mt of the entire volume was collected and correctly recycled. Among the Asian subregions, East Asia had the highest share of E-waste at 20%. This is followed by Western, Southern, and Central Asia with 6%, 5%, and 0.9% share, respectively. China and Japan have the major share in E-waste production in the Eastern subregion (Tiseo, 2020). It is essential to have an inventory of the country's E-waste flow to design and formulate waste management policies. However, many Asian Pacific nations lack comprehensive records of such E-waste inventories. Many Asia-Pacific nations are experiencing rapid economic growth, resulting in high demand for EEE, thereby increasing the E-waste problem.

In 2019, 17 of the 46 countries come within the framework of E-waste legislation and policies. The lack of appropriate laws, legislation, standards, and strategies in many Asian countries, including Hong Kong, Thailand, China, Pakistan, India, and Cambodia, has led to unethical E-waste recycling (Arya, Rautela, et al., 2021). Countries in South Asia have begun to recognize the necessity for an E-waste management system. India has had policies in place to control E-waste since 2011. Several other countries are exploring similar legislation. In Southeast Asia, some nations are more developed. There is no distinct legal system in the Philippines for E-waste handling, but regulations covering hazardous waste are in force as E-waste is considered hazardous. China has enacted national legislation governing the careful disposal of E-waste. The majority of EEE in Japan is recycled according to a system based on EPR. In central and western Asia, E-waste regulation is absent. However, for the government of countries like Kyrgyz and Kazakhstan, the EPR-based system is under consideration. Furthermore, economically rich countries uncontrollably import WEEE into Asian subregions for economic stability, making the region a worldwide hotspot for E-waste flows. Therefore, to achieve sustainable E-waste management, framing policies are urgently needed to support and execute proper administrative processes (Jang et al., 2021).

2.4.4 E-waste status in Africa

The African continent supplied the second-least quantity of E-waste to the global E-waste in 2019. Compared to other continents, it has generated a minimum amount of about 2.9 Mt, corresponding to 2.5 kg of E-waste generated per capita (Forti et al., 2020). Of this total amount, a significant fraction comes from the transboundary movement of WEEE from economically developed countries (such as the United States, the United Kingdom, Belgium, Germany, and others). Through the two leading ports in Lagos, Nigeria imports approximately 60,000−71,000 tonnes of WEEE yearly (Forti et al., 2020). In 2019, the North African countries contributed the maximum E-waste production of 1.3 Mt, while Central African countries produced a minimum of 0.2 Mt (Forti et al., 2020). Africa ranked last in the context of collection and recycling, with only 0.9% of E-waste formally managed and documented. However, regional WEEE production and disposal figures vary greatly depending on various factors. Of the 49 countries analyzed, only 13 countries (Nigeria, Egypt, Rwanda, South Africa, Madagascar, Côte D'Ivoire, Cameroon, and others) comply with the E-waste legislation (Forti et al., 2020).

Meanwhile, the informal sector dominates most African countries. As a result, establishing a legal framework for collecting and disposing of E-waste is critical. Lack of adequate recycling infrastructure, public ignorance, poor government policy, and uncontrolled dominance of the informal recycling sector in Africa are significant reasons for improper E-waste management.

2.4.5 E-waste status in Oceania

Oceania produced the least amount of E-waste in 2019, with 0.7 Mt, or around 1.31% of the world's E-waste volume. With 16.2 kg/person, Oceania stood in the second position in the per capita production rate. In 2019, 0.06 Mt (8.8% of total) of E-waste was formally collected and treated; the rest, 91.2%, is uncertain. As of 2019, only Australia, out of 12 countries in Oceania, follows the E-waste policy and legislation. Two countries generate more than 92% of Oceania's E-waste (in 2019): Australia (554 kt) and New Zealand (96 kt), while Tonga and Vanuatu each have only 0.3 kt. The raw material value of Oceania's total E-waste (0.7 million tonnes) generated in 2019 is equivalent to US$0.7 billion. The Product Stewardship Regulations 2011 of the Australian government has provided its citizens' access to better collection and recycling services, especially for EoL computers and televisions. This policy also certifies recyclers' safety following AS/NZS 5377:2013 standards. In July 2019, the Victorian government banned E-waste disposal in landfills. It launched a package of $16.5 million to promote the safe management of toxic materials and recover valuable substances (Forti et al., 2020).

2.5 Future projection of E-waste generation

The consumption of EEE is strongly linked with global economic development. GDP per capita is assumed to measure a country's economic growth, which was adjusted for PPP in 2010 to facilitate cross-national comparisons. The technique presented in this chapter is based on a report by Kaza et al. (2018) and was used to calculate future E-waste output. A multilinear regression model is employed to capture the relationship between per capita GDP and E-waste production, represented in Eq. (2.1). The best possible model includes GDP per capita's natural logarithm as an independent variable and per capita kilograms of E-waste as a dependent variable. The p-value is less than 0.05, with R^2 value is 0.947 indicates that the model fits the data well. Choosing 2019 as the base year, the estimated E-waste generation rate is determined using Eq. (2.2).

The model predicted per capita E-waste generation

$$= 128.92 - 47.20(\ln(\text{GDP per capita})) + 5.42(\ln(\text{GDP per capita}))^2$$
$$- 0.18(\ln(\text{GDP per capita}))^3 \tag{2.1}$$

(Projected waste generation rate)$_{\text{target year}}$

$$= \left[\frac{\text{(Model predicted waste generation rate)}_{\text{target year}}}{\text{(Model predicted waste generation rate)}_{\text{base year}}} \right]$$

$$* \text{(Actual waste generation rate)}_{\text{base year}} \qquad (2.2)$$

For calculating model predicted E-waste generation rate of the target year 2030, 2040, and 2050, projected GDP (constant PPP 2010, international dollar) data were taken from OECD, and projected population data were taken from UN for the corresponding target year 2030, 2040, and 2050 (OECD, 2021; UN, 2019). Each country's estimated per capita GDP was determined by dividing the GDP PPP by the projected population for 2030, 2040, and 2050. The same is used as input data for model prediction per capita E-waste generation. Model predicted E-waste generation rate for the base year 2019 and target years 2030, 2040, and 2050 was calculated using Eq. (2.1). The data were obtained from the Global E-Waste Monitor Report 2020 to get the actual waste generation rate for the base year 2019 (Forti et al., 2020). Using the foregoing methods, overall E-waste generation forecasts for 2030, 2040, and 2050 are displayed in Fig. 2.3A, whereas per capita estimates for E-waste are depicted in Fig. 2.3B. The projected E-waste volume in 2030 would be reaching around 82 Mt, and by 2040 would have crossed 90 Mt and is anticipated to rise 100 Mt by the end of 2050. The E-waste generation rate would reach 9.9 kg/person by 2030, rising from earlier reported 7.3 kg/person in 2020. E-waste per capita production would surpass 10 kg/person by the end of 2040 and is expected to exceed 10.5 kg/person by the end of the next decade in the year 2050.

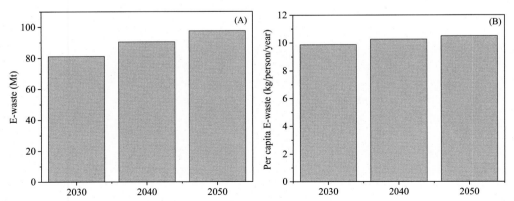

Figure 2.3 Projection of (A) total E-waste generation and (B) per capita E-waste generation for the years 2030, 2040, and 2050.

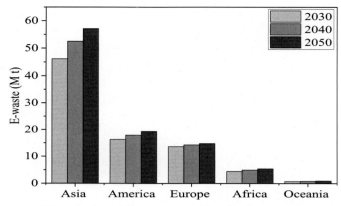

Figure 2.4 Region-specific projection of E-waste generation for the years 2030, 2040, and 2050.

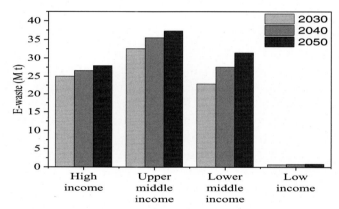

Figure 2.5 Projection of E-waste generation based on income group for the years 2030, 2040, and 2050.

Furthermore, the projected calculated data based on the region and income group are shown in Figs. 2.4 and 2.5, respectively. The key assumptions in the future forecast of E-waste are based on two critical factors: GDP growth and population growth. China and India are the two most populous countries globally with a significant increase in their economies, with more E-waste expected to be generated in the future. On the basis of this, it is estimated that Asian countries will produce double the amount of E-waste in the coming decades (Fig. 2.4). Moreover, African countries are also expected to experience a hike in the production of E-waste. Whereas the Americas, Europe, and Oceania regions are expected to gradually see a rise in their E-waste production.

Fig. 2.5 shows that lower-middle-income countries will experience more growth in E-waste production over the next three decades. This can be due to economic growth, urbanization, trends in more electronic product consumption, and less avail-ability of waste management infrastructure. In contrast, the production rate will increase gradually for the upper-middle-income and high-income countries, as they have already reached a stable position in economic development, at which point mate-rial consumption does not impact GDP growth.

2.6 Role of future projection in formulating sustainable development goals

The vast E-waste creation that exists today has surely grown difficult to control. If cur-rent production rates continue to rise, this will become more complex and pressure existing waste-management amenities. Thus it is necessary to forecast the future amount of E-waste to formulate a well-planned strategy for E-waste management. Future prediction is a long-term perspective that provides future insight and may anticipate the consequence. A reasonably estimated amount will help waste managers, policymakers, and stakeholders devise new approaches and policies that help build an efficient collection, disposal, and management infrastructure. It will also help manufac-turers and retailers redesign their operations and create innovative high-tech products to reduce E-waste. Evaluating E-waste generation and its future prediction are of utmost importance for effective E-waste management. These approaches and strong laws help formulate plans and strategies for recycling activities and implementing urban mining concepts. Ultimately the projected amount will help design sustainability routes to achieve a circular economy model and reduce landfill loads. This will further lead to the achievement of Agenda 2030, the SDGs goals linked with E-waste management.

2.7 Conclusions

The E-waste generation rate and future projections of WEEE are required not only for statistical purposes but also for quantify E-waste generation, assist in the distribution of each category of E-waste in various regions and income groups, assist in the formu-lation of regulations and legal policies for the proper use and disposal practices, and plan collection and recycling activities. E-waste generation is dependent on a country's economic growth in the form of GDP. With the economy's growth, individuals' pur-chasing power also increased, leading to the usage and accumulation of discarded electronic items as E-waste. Various E-waste drivers were discussed. The approach for E-waste projection was developed based on available techniques, and the technology was used to project E-waste for 2030, 2040, and 2050. EEE is a complex system manu-factured using different materials, that is, metal, plastic, glass, and ceramic. One cannot ignore the usefulness of EEE in every sector of human needs. The generated WEEE needs to be disposed of properly, including recycling and recovery activities, which should plan with all the required infrastructure and human resources with regulations and legislation in force to deal with E-waste. The respective agencies formulate goals to achieve sustainable development and plan to implement them. Thus gathering data for generation and carrying out projection seem helpful in solving our problem in the near future.

References

Abalansa, S., El Mahrad, B., Icely, J., & Newton, A. (2021). Electronic waste, an environmental problem exported to developing countries: The GOOD, the BAD and the UGLY. *Sustainability*, *13*(9), 5302.

Ahmed, S. F. (2016). The global cost of electronic waste. *The Atlantic*. Retrieved August 11, 2021, from https://www.theatlantic.com/technology/archive/2016/09/the-global-cost-of-electronic-waste/502019/.

Arya, S., & Kumar, S. (2020a). Bioleaching: Urban mining option to curb the menace of E-waste challenge. *Bioengineered*, *11*(1), 640−660.

Arya, S., & Kumar, S. (2020b). E-waste in India at a glance: Current trends, regulations, challenges and management strategies. *Journal of Cleaner Production*, *271*, 122707.

Arya, S., Patel, A., & Kumar, S. (2021). Urban mining of obsolete computers by manual dismantling and waste printed circuit boards by chemical leaching and toxicity assessment of its waste residues. *Environmental Pollution*, *283*, 117033.

Arya, S., Rautela, R., Chavan, D., & Kumar, S. (2021). Evaluation of soil contamination due to crude E-waste recycling activities in the capital city of India. *Process Safety and Environmental Protection*, *152*, 641−653.

Baldé, C. P., Forti, V., Gray, V., Kuehr, R., & Stegmann, P. (2017). *The global E-waste monitor 2017: Quantities, flows and resources*. United Nations University, International Telecommunication Union, and International Solid Waste Association.

Baldé, C. P., & Kuehr, R. (2021). *Impact of the COVID-19 pandemic on E-waste in the first three quarters of 2020*. United Nations University (UNU)/United Nations Institute for Training and Research (UNITAR)−Co-hosting the SCYCLE Programme.

Bhutta, M. K. S., Omar, A., & Yang, X. (2011). Electronic waste: A growing concern in today's environment. *Economics Research International*, *2011*, Article ID 474230.

Chen, Y., Chen, M., Li, Y., Wang, B., Chen, S., & Xu, Z. (2018). Impact of technological innovation and regulation development on E-waste toxicity: A case study of waste mobile phones. *Scientific Reports*, *8*(1), 1−9.

Chevalier, S. (2021). Global retail e-commerce sales 2014−2024. *Statista*. Retrieved August, 09, 2021, from https://www.statista.com/statistics/379046/worldwide-retail-e-commercesales/#statisticContainer.

Davis, K. (1965). The urbanization of the human population. *Scientific American*, *213*(3), 41−53.

Davis, J. M., Akese, G., & Garb, Y. (2019). Beyond the pollution haven hypothesis: Where and why do E-waste hubs emerge and what does this mean for policies and interventions? *Geoforum; Journal of Physical, Human, and Regional Geosciences*, *98*, 36−45.

Dutta, D., Arya, S., Kumar, S., & Lichtfouse, E. (2022). Electronic waste pollution and the COVID-19 pandemic. *Environmental Chemistry Letters*, *20*, 971−974.

European Environment Agency. (2020). Europe's consumption in a circular economy: the benefits of longer-lasting electronics. Retrieved March 09, 2021, from https://www.eea.europa.eu/publications/europe2019s-consumption-in-acircular/benefits-of-longer-lasting-electronics.

Forti, V., Balde, C. P., Kuehr, R., & Bel, G. (2020). *The global E-waste monitor 2020: Quantities, flows and the circular economy potential*. United Nations University (UNU)/United Nations Institute for Training and Research (UNITAR) − co-hosted SCYCLE Programme, International Telecommunication Union (ITU) & International Solid Waste Association (ISWA).

Gohain, M. P. (2020). 23% of urban population has access to computers, only 4% of rural: survey. *The Times of India*. Retrieved August 09, 2021 from https://timesofindia.indiatimes.com/india/23-of-urban-population-has-access-to-computers-only-4-of-rural-survey/articleshow/77075283.cms.

Gundupalli, S. P., Hait, S., & Thakur, A. (2018). Classification of metallic and non-metallic fractions of E-waste using thermal imaging-based technique. *Process Safety and Environmental Protection, 118*, 32–39.

Gupta, N., Trivedi, A., & Hait, S. (2021). Column leaching of metals from PCB of end-of-life mobile phone using DTPA under oxidising condition. Integrated approaches towards solid waste management, 233–244. Springer.

International Telecommunication Union. (2020). *Measuring digital development facts and figures 2020*. Retrieved September 10, 2021 from https://www.itu.int/en/ITUD/Statistics/Documents/facts/FactsFigures2020.pdf.

International Telecommunication Union. (2021). *Policy practices management for E-waste: Tools for fair and economically viable extended producer responsibility*. Retrieved September 11, 2021 from https://www.itu.int/en/myitu/Publications/2021/04/15/13/01/Policy-practices-for-E-waste-management.

Jang, Y. C., Rhee, S. W., & Kim, J. Y. (2021). *Toward sustainable e-waste management in Asia and the Pacific*. Economic and Social Commission for Asia and the Pacific (ESCAP), United Nations.

Kaza, S., Yao, L., Bhada-Tata, P., & Woerden, F. V. (2018). *What a waste 2.0: A global snapshot of solid waste management to 2050*. World Bank Group.

Kumar, A., Holuszko, M., & Espinosa, D. C. R. (2017). E-waste: An overview on generation, collection, legislation and recycling practices. *Resources, Conservation and Recycling, 122*, 32–42.

Kumar, S., & Singh, V. (2019). E-Waste: Generation, environmental and health impacts, recycling and status of e-waste legislation. *Journal of Emerging Technologies and Innovative Research, 6*(1), 592–600.

Kusch, S., & Hills, C. D. (2017). The link between E-waste and GDP—New insights from data from the Pan-European region. *Resources, 6*(2), 15.

Lee, D., Offenhuber, D., Duarte, F., Biderman, A., & Ratti, C. (2018). Monitour: Tracking global routes of electronic waste. *Waste Management, 72*, 362–370.

Li, J., Zeng, X., Chen, M., Ogunseitan, O. A., & Stevels, A. (2015). "Control-Alt-Delete": Rebooting solutions for the E-waste problem. *Environmental Science and Technology, 49*(12), 7095–7108.

Magalini, F., Thiébaud, E., & Kaddouh, S. (2019). Quantifying WEEE in Romania 2019 vs 2015. *Sofies*. https://www.ecotic.ro/wp-content/uploads/2019/10/Quantifying-WEEE-in-Romania-2019.pdf.

Makov, T., & Fitzpatrick, C. (2021). Is repairability enough? Big data insights into smartphone obsolescence and consumer interest in repair. *Journal of Cleaner Production, 313*, 127561.

Needhidasan, S., Samuel, M., & Chidambaram, R. (2014). Electronic waste—An emerging threat to the environment of urban India. *Journal of Environmental Health Science and Engineering, 12*(1), 1–9.

O'Dea, S. (2021). Smartphone subscriptions worldwide 2016–2026. *Statista: Technology and Telecommunication*. Retrieved August 09, 2021 from https://www.statista.com/statistics/330695/number-of-smartphone-users-worldwide/.

Organisation for Economic Co-operation and Development. (2016). *Extended producer responsibility: Updated guidance for efficient waste management*. OECD Publishing. Available from https://doi.org/10.1787/9789264256385-en.

Organisation for Economic Co-operation and Development. (2021). *Real GDP long-term forecast (indicator)*. Organisation for Economic Co-operation and Development. Retrieved September 04, 2021 from https://data.oecd.org/gdp/real-gdp-long-term-forecast.htm.

Pant, D., Joshi, D., Upreti, M. K., & Kotnala, R. K. (2012). Chemical and biological extraction of metals present in E waste: A hybrid technology. *Waste Management, 32*(5), 979–990.

Pariatamby, A., & Victor, D. (2013). Policy trends of E-waste management in Asia. *Journal of Material Cycles and Waste Management, 15*(4), 411–419.

Patil, R. A., & Ramakrishna, S. (2020). A comprehensive analysis of e-waste legislation worldwide. *Environmental Science and Pollution Research, 27*(13), 14412–14431.

Priya, A., & Hait, S. (2017a). Comparative assessment of metallurgical recovery of metals from electronic waste with special emphasis on bioleaching. *Environmental Science and Pollution Research, 24*(8), 6989–7008.

Priya, A., & Hait, S. (2017b). Qualitative and quantitative metals liberation assessment for characterization of various waste printed circuit boards for recycling. *Environmental Science and Pollution Research, 24* (35), 27445–27456.

Priya, A., & Hait, S. (2018a). Comprehensive characterization of printed circuit boards of various end-of-life electrical and electronic equipment for beneficiation investigation. *Waste Management, 75,* 103–123.

Priya, A., & Hait, S. (2018b). Toxicity characterization of metals from various waste printed circuit boards. *Process Safety and Environmental Protection, 116,* 74–81.

Priya, A., & Hait, S. (2020). Biometallurgical recovery of metals from waste printed circuit boards using pure and mixed strains of *Acidithiobacillus ferrooxidans* and *Acidiphilium acidophilum*. *Process Safety and Environmental Protection, 143,* 262–272.

Qu, Y., Wang, W., Liu, Y., & Zhu, Q. (2019). Understanding residents' preferences for E-waste collection in China—A case study of waste mobile phones. *Journal of Cleaner Production, 228,* 52–56.

Rautela, R., Arya, S., Vishwakarma, S., Lee, J., Kim, K. H., & Kumar, S. (2021). E-waste management and its effects on the environment and human health. *Science of the Total Environment, 773,* 145623.

RS Secretariat. (2011). *E-waste in India.* India Research Unit (Larrdis), Rajya Sabha Secretariat, New Delhi. https://greene.gov.in/wp-content/uploads/2018/01/e-waste_in_india-Document.pdf.

Thakur, P., & Kumar, S. (2021). Evaluation of e-waste status, management strategies, and legislations. *International Journal of Environmental Science and Technology, 19,* 6957–6966.

Tiseo, I. (2020). Countries covered by electronic waste legislation, policy, and regulation from 2014 to 2019. Statista. Retrieved August 08, 2021 from https://www.statista.com/statistics/1154905/projection-ewaste-generation-worldwide/.

Trivedi, A., & Hait, S. (2020a). Efficacy of metal extraction from discarded printed circuit board using *Aspergillus tubingensis*. In S. Ghosh, R. Sen, H. Chanakya, & A. Pariatamby (Eds.), *Bioresource utilization and bioprocess* (pp. 167–175). Springer.

Trivedi, A., & Hait, S. (2020b). Bioleaching of selected metals from E-waste using pure and mixed cultures of Aspergillus species. *Measurement, analysis and remediation of environmental pollutants* (pp. 271–280). Springer.

United Nation. (2018). *Department of economic and social affairs.* Retrieved September 10, 2021, from https://www.un.org/development/desa/en/news/population/2018-revision-of-world-urbanization-prospects.html.

United Nation. (2019). *World population prospects 2019: Total population (both sexes combined) by region, subregion and country, annually for 1950–2100.* United Nation, Department of Economic and Social Affairs. Retrieved August 17, 2021, from https://population.un.org/wpp/Download/Standard/Population/.

Vats, M. C., & Singh, S. K. (2014). Status of E-waste in India—A review. *Transportation, 3*(10).

Verma, A., & Hait, S. (2019). Chelating extraction of metals from E-waste using diethylene triamine pentaacetic acid. *Process Safety and Environmental Protection, 121,* 1–11.

Victor, S. P., & Kumar, S. S. (2012). Planned obsolescence: roadway to increasing E-waste in Indian government sector. *International Journal of Soft Computing and Engineering, 2*(3), 554–559.

Vishwakarma, S., Kumar, V., Arya, S., Tembhare, M., Dutta, D., & Kumar, S. (2022). E-waste in information and communication technology sector: Existing scenario, management schemes and initiatives. *Environmental Technology & Innovation, 27,* 102797.

World Bank. (2019). *GDP per capita, PPP (current international $).* Retrieved August 11, 2021, from https://data.worldbank.org/indicator/NY.GDP.PCAP.PP.CD.

World Bank. (2020). *GDP PPP (current international $).* Retrieved September 12, 2021, from https://data.worldbank.org/indicator/NY.GDP.MKTP.PP.CD?locations = CN-US.

CHAPTER 3

Challenges and extended business opportunity associated with E-waste management options

Rahul Rautela[1,2], **Deblina Dutta**[3], **Pranav Prashant Dagwar**[1], **Mahesh Game**[1], **Ankit Motghare**[1], **Srushti Muneshwar**[1], **Rohit Jambhulkar**[1] and **Debajyoti Kundu**[1]

[1]CSIR-National Environmental Engineering Research Institute (CSIR-NEERI), Nagpur, Maharashtra, India
[2]Academy of Scientific and Innovative Research (AcSIR), Ghaziabad, Uttar Pradesh, India
[3]Department of Environmental Science, SRM University—AP, Amaravati, Andhra Pradesh, India

3.1 Introduction

E-waste incorporates a varied range of electronic devices, like broadcastings and information technology equipment, lighting, medical devices, monitoring and control devices, electronic and electrical tools, electronic toys, mobiles, and computers (Shahabuddin et al., 2022). A huge amount of E-waste is sent to countries like China, India, and Kenya where the working conditions are better and the environmental standards are lower, making it more profitable to process E-waste there. However, take-back laws of E-waste have been passed, such as Basel Convention, to stop the transboundary movements of E-waste (Selvaraj, 2021). It is estimated that globally generated E-waste have a monetary value of about $57.0 billion where only $10.0 billion means of E-waste is recycled and recovered sustainably, which is accounted around 15.0 million tons (Mt) of negative CO_2 footprint (Shahabuddin et al., 2022).

According to report of Tata Strategic Analysis, it was predicted that India will be among the largest markets worldwide for consumer electronics. Electronic equipment such as mobile phones, personal computers, and televisions are on the rise, which makes this assumption possible. For every 2 years, the number of mobile subscribers increases by 30%, and television production increases by 0.5% (Gomathi & Rupesh, 2018). There are several challenges and opportunities in E-waste management. The foremost challenge is lack of E-waste legislation; 61 countries only had E-waste legislation until 2014 and in 2019 the number has increased to 78 (Forti et al., 2020). Though populated countries like India and China, have e-waste legislation at present, in Pakistan and Bangladesh it has not yet established. Legislations are

Global E-waste Management Strategies and Future Implications
DOI: https://doi.org/10.1016/B978-0-323-99919-9.00005-2

also not effectively implemented in many regions of south-east Asian countries, and some parts of Northern Africa have inadequate or do not have any e-waste legislation (Balde et al., 2017).

However, in spite of the legislation on E-waste collection and recycling, handling and processing of E-waste are still challenging owing to its complex nature, that is, it includes 40% metal, 30% plastic, and 30% oxides of different materials (Kaya, 2016). Hence, along with managing hazards in E-waste management, recovery of precious, rare earth, and useable materials is really very challenging. Major challenges like collection, maintenance of homogeneity, low energy recovery and emission are faced during treatment and handling of E-waste in order to recover the valuable and precious metals. Other bottlenecks in E-waste management are deficiency of infrastructures amenities, cost-effective recovery option, policy, and variation in legislation between countries (Shahabuddin et al., 2022). Conventional treatment process of E-waste like open burning, acid pretreatment, and incineration resulted in release of heavy metals, furans, and dioxins (Dai et al., 2020; Rautela et al., 2021), whereas urban mining of precious metals can facilitate recovery option, job creation, and commercial opportunities. For instance, from one metric ton of circuit board, 1.5 and 210 kg of gold and copper, respectively, can be recovered (Bazargan et al., 2012). It is estimated that one ton of circuit board contains 300 and 40 times higher gold and copper than in their ores, respectively. Hence, urban mining can be a promising approach in resource recovery from metal (Arya & Kumar, 2020). In contrast, global E-waste production is three times higher economical than global silver mining and many country's GDP combined (World Economic Forum, 2019).

Hence, based on the valuable importance of E-waste in commercial sector, this book chapter discusses about the challenges, barriers, business opportunities, and benefits. Thus, to build an effective E-waste management facility, country-specific standards and legislation are required. Along with this a massive public awareness, practical implementation of rules, government enticements, and development of cost-effective technologies are a serious concern.

3.2 Challenges and barriers in E-waste management

The advancement of people's lifestyle, technological development, and economic progress worldwide have led to the ever-increasing production and generation of E-waste throughout the world, which interims causes significant rise in environmental and health concerns and its mismanagement extends the risk (Tue et al., 2016; Wang et al., 2017; Wen et al., 2006; Yang et al., 2008; Bhuiya et al., 2020; Matsakas et al., 2017). The need of a befitting waste management technique and safe disposal of E-waste has turned to be a global priority due to the constant discarding of it to the landfills, which is a common approach worldwide to dispose of E-waste, which certainly causes degradation in environment (Song & Li, 2015). Safe and proper management of E-waste is a great challenge which is further intricated by ambiguous legislations, prevalent informal sector, poor infrastructure, lack of tech-based operations and skilled human resource that is shown in Fig. 3.1 (Banzal, 2022).

The challenges which the E-waste management sector is facing are enormous, which are facing up it in the end of life management of this waste. Due to all of these challenges, it is a growing tough task to manage E-waste in various nations around the world, and as there is no reliability on data on amount of E-waste generated in different parts of world due to which the end-of-life of many electronic products cannot be calculated with accuracy, this interim leads to the improper accumulation of waste product mostly at landfills (Pariatamby & Victor, 2013).

Legislations are the most important foundation for smooth functioning of any management system; the ambiguity in legislation, one of the key challenges which affects the E-waste management system in the developing nations, is a major concern for the enforcing authority. Apart from the developed nations there were still no rules and legislations for managing the E-waste in many parts of the world, until 2014 it was seen that 44% of the countries had legislations for managing the waste which included around 61 countries and till 2019 around 78 countries had their legislations for managing E-waste (Balde et al., 2017). In India, the rules and regulations are being formulated by Ministry of Environment and Forests, which provides us with legislation regarding E-waste management and environmental protection. So being the authority it approves the guidelines for the source identification of E-waste in India and also supports appropriate handling procedures in environment-friendly manner (Andeobu et al., 2021; MoEF, 2008). Some of the rules which have been implemented in India have been provided in Table 3.1.

Despite the formation of extended producer responsibility (EPR) an approach toward reducing the E-waste impact, its implementation was not effectively done. So

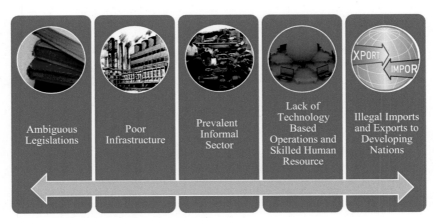

Figure 3.1 Challenges in E-waste management.

Table 3.1 Implementation of waste management rules in India (Agoramoorthy & Chakraborty, 2012; Andeobu et al., 2021; Song & Li, 2015).

Rules	Year of implementation
Municipal Solid Waste Waste Management Rules	2004
Hazardous and Waste Management rules	2008
E-waste Management and Handling Rules	2010
The Hazardous and Waste Management rules	2011
E-waste Management and Handling Rules	2016

this attributes to the peculiarities in India's E-waste management system. For instance people still have unwillingness to hand out their old and obsolete electronic gadgets or even pay for its recycling, which thus cause the piling up of unnecessary E-waste (Aras et al., 2015; Zeng & Li, 2016).

The need for a proper infrastructure plays a very major role after the formation and implementation of legislation for E-waste management. Due to the absence or limited availability of these infrastructures for recycling and managing the E-waste, it becomes more prevalent and starts to release its aftermath effects. In India, the lack of infrastructure for the E-waste management and inadequate information channel about the quantity of E-waste has possessed risk infused environment for humans (Andeobu et al., 2021; Islam et al., 2016; Kumar & Dixit, 2018). As there is a shortage of formally approved recycle centers due to the improperly managed supply chain which again is tough spot to connect to people and spread awareness for the same (Hindrise, 2022).

Alongside the prevalence of informal sectors rising in this field it becomes difficult to manage the E-waste safely; the informal operations are carried out in many countries like China, Indonesia, India, and a few other Asian and African countries (Andeobu et al., 2021). In India the E-waste recycling is a market-motivated industry, and is predominantly dominated by the various numbers of informal players. It is about 90% of E-waste which gets recycled illegally in this chain of informal activities mostly done by women and children, who lack awareness and education about the adverse impact that E-waste causes severely to human's health; the recycling is usually done for the availability of precious elements like silver, gold, palladium, and many other elements from the waste (Dwivedy & Mittal, 2012; Niza et al., 2014; Pandey & Govind, 2014). The informal sector provides livelihoods to millions of untrained workers who are settled in the urban slums and carry out more than 95% of recycling works by risky procedures (Patibanda et al., 2020).

Technology-based operation is completely neglected in the informal sector so as to maximize profits by not opting the tech-based extraction of metals/materials from the waste; this causes a lot of adverse impacts both on environment and human health. To increase profits, the precious elements can be extracted from the E-waste via urban mining; this process requires modern machines and technology to successfully but safely recover the precious, rare earth and useable materials and manage the hazardous materials alongside (Shahabuddin et al., 2022). For these operations the need of skilled and reliable human resource has increased, because the people who work currently in the informal sector for removal of such elements are exposed to a lot of health hazards, and due to their unskilled practices it can also cause problems to environment, so to avoid such issues the current need of tech-based operations along with skillful human resource has risen in demand in recent years. In India, under the scheme of National Skill Development Mission the Indian government has started an innovative program to provide short courses and training to the E-waste collectors, handlers, and dismantlers, designed by Electronic Sector Skill Council of India, in association with Skill Council for Green Jobs and regulatory agencies like Central Pollution Control Board (CPCB) and State Pollution Control Boards (SPCBs) (Hindrise, 2022).

The imports and exports of E-waste to developing nations illegally have been seen in rise in recent years. Many countries have banned the import of E-waste as a donation to their respective nations as they sense to cause environmental and health hazards; for example Pakistan has recently banned the imports of E-waste from other countries (Imran et al., 2017). Due to the increasing hazardous impacts on human health and environment many developing countries have drafted their rules for restricting the hazardous substances in Electronic and Electrical Equipment (EEE) products, For instance, the Bangladesh's Department of Environment has drafted rules in 2019 for such avoidance of hazardous materials (WTO, 2020).

The E-waste management process considers these as challenges and barriers to its proper and safe disposal as well as management. Today many countries including both developed and developing are changing the rules and regulations and also working towards changing the ground realities because of the adverse environmental and health impacts; therefore certain steps are ensured to protect the population and biodiversity in the respective regions. E-waste management is a growing sector with a lot of opportunities which are supposed to be explored and the present conditions to be improved; this will help to achieve sustainable development goals leading towards circular economy (CE).

3.3 Business opportunity within E-waste management

With increasing electronics market and rising obsolescence rates of electronic equipment, E-waste is growing faster than any other waste stream. E-waste contributes almost 5% of the total generated Municipal Solid Waste worldwide (Meenakshi & Harini, 2012). According to the World Economic Forum estimates, the volume of

E-waste generated globally has been steadily rising; nearly 50 Mt of E-waste were generated in 2018, which is expected to reach 120 Mt by 2050 (World Economic Forum, PACE, 2019). According to the estimate, India generates almost 800,000 tons of E-waste annually and is expected to grow at an average rate of 10%—15% (Meenakshi & Harini, 2012). The material composition of E-waste contains plastic (polypropylene, acrylonitrile, butadiene, styrene, and polystyrene), additives (flame retardants and adhesives), and metals (Babu et al., 2007; Mao et al., 2020). Considering the recyclability of E-waste components, it becomes a source of income for the industries and opens the door for new jobs.

The metals like gold, silver, copper, palladium, platinum, aluminum, and rare earth metals present in the E-waste are quite enough to reuse and provide extended business opportunities. The metals in E-waste are also resources, and if these metals can be recycled with scientific methods, it can alleviate the shortage of metal resources around the world (Zeng et al., 2017). Some materials of E-waste have a capacity for reuse like the different capacitors and circuit boards. The plastic content of electronic appliances' bodies serves as a good raw material for pyrolysis, as it provides efficient energy and material recycling (Debnath et al., 2015). Pyrolysis of plastics can be used to dispose of the waste and to recover synthetic fuel. The oil recovered from this process can be used as diesel generator fuel for burners (Kantarelis et al., 2011). The plastic from E-waste serves as a raw material for hydrogen production through a two-stage reaction system of pyrolysis-gasification (Acomb et al., 2013). Some studies illustrated that a Liquid Crystal Coated Polaroid Glass Electrode material collected from disposed liquid crystal display computer monitors can be used as electrodes in a microbial fuel cell for electricity production (Gangadharan et al., 2015). Advanced research on efficient technology and reduction in toxic chemicals and treatment of these toxic chemicals is necessary so that a cost-effective and eco-friendly process can be established. This will attract innovation and business as well as eliminate the incentive to dump waste. The E-waste recycling business is a green business opportunity that supports entrepreneurs to sustainably conserve available resources that would otherwise be spent on the manufacture of recovered materials.

3.3.1 Circular economy approach

A linear economic system was based on exploiting natural resources while converting them into products that were eventually disposed untreated (Haibin & Zhenling, 2010). However, the CE approach focuses on the recycling and reusing of discarded waste putting into circularity, which ultimately addresses the global crises of climate change, pollution, biodiversity loss, and the management of waste, especially E-waste. There has been a growing attention given to the concept of the CE by policymakers around the world, as it appears to be capable of reducing the overconsumption of natural resources and at the same time enhancing economic growth. The aim of CE in E-waste is to restore the value of E-waste through maximum reuse, repair, recovery, and re-manufacturing at the end-of-life cycle (Bridgens et al., 2019). A large amount of metals are required for the production of EEE, including iron, copper, silver, gold, aluminum, manganese, chromium, and zinc along with various rare earth elements. The rate at which these abiotic resources are being extracted for EEE manufacturing is significantly higher than the rate at which they are being formed in nature (Lèbre et al., 2017). Thus the approach of CE will be essential to accomplish the resource demand for the nation.

In developing countries particularly in India, due to lack of understanding of the CE, the majority of recycling activity is carried out by unskilled rag-pickers. Involvement of governments and policymakers in the concept of CE can eventually enhance the sustainable E-waste management. Thus, it can help to reduce the burden on the environment and also to boost the economy, and it will result in the achievement of Sustainable Development Goals (SDGs) (Nandy et al., 2022).

3.3.2 Circular business models

The E-waste business model evolves from the recycling sector, which operates according to the norms and regulations of the nation in which it is created. But, consumer habits and routines created by linear business models operate as a behavioral barrier to shifting purchasing behaviors (Parajuly et al., 2020). To shift consumer behaviors, a move to the CE would necessitate harnessing the force of both rational (e.g., economic) and nonrational (e.g., moral) incentives (Planing, 2015). This probably applies to mainstreaming alternative business models, such as leasing for consumer e-devices and creating demand for repaired and re-manufactured items (de Vicente Bittar, 2018; Van Weelden et al., 2016). Regardless of the potential economic gains of re-manufacturing and/or product-service systems, their adoption in industries is limited (Linder & Williander, 2017). Re-manufacturing is especially uncommon for consumer items, a problem that cannot be remedied by the typical "green marketing" strategy without addressing customer behavior (Vogtlander et al., 2017). Furthermore, the design of strategies and businesses must go beyond physical attributes to include the human components of consumption (Wastling et al., 2018).

In India, E-waste policies and regulations are now shifting from environmental protection to pollution prevention, bolstering the popularity of the E-waste business model. A circular business model (CBM) can be established with such a toolkit, as this model encourage the innovative E-waste management approach by putting E-waste in circularity with a reverse system. However, organizations that specialize in E-waste processing can provide a value proposition. These organizations are most likely to participate in the collecting of their end-of-life electronic equipment because they see re-manufacturing and refurbishing techniques as a viable business model for profiting on leftover resources. The CBM should be designed to enable a long-term value chain by using information technology and digitalization concepts. This model aids in e-scrap asset assessment and anticipating future E-waste statistics. An integrated and long-term business approach facilitates appropriate E-waste management. Recently, CBMs have also targeted residential E-waste management. A commercial model connecting governing authorities and nongovernmental organizations is now showing interest in E-waste management. The EPR is key to the business concept as it includes a defined technique for developing collecting systems. It focuses on the implementation of new advancements in order to reap economic benefits. This model is extremely intended in terms of the entity's ability to create income.

3.4 Sustainability benefits in business

An appropriate definition of sustainability is doing business without negatively affecting the environment, community, or society at large. In business, sustainability generally refers to two main areas: impact of business on the environment and on society. Sustainability is about achieving a positive impact on at least one of these aspects. A lack of responsibility by businesses may lead to a variety of issues such as climate change, social inequality, and discrimination as a result of the failure to assume responsibility. A sustainable business makes business decisions based on a wide variety of factors, including the environment, economics, and society. To ensure financial sustainability, these organizations watch the impact of the operations they carry out in order to ensure that short-term profits do not turn into long-term liabilities. An environmental friendly business will reduce the impact on the environment and preserve natural resources (D'Adamo et al., 2019) (Fig. 3.2).

The following things can be carried out to enhance the sustainability in business (Spiliakos, 2018):

1. consider investing in products that reduce dependence on natural resources,

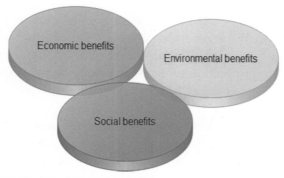

Figure 3.2 Sustainability benefits of E-waste business.

2. consider buying recycled products to reduce the carbon footprint, and
3. consider all the business activities to see if there is anything that can be changed to improve those activities.

A green business not only benefits the environment, but can also save money in the long run by reducing waste and improving efficiency. A reduction in the impact of business on the environment will result in a more sustainable business that is more profitable. In order to have a long-term success in the business, one must be less dependent on natural resources than the competitors and have plans to deal with rising costs due to global warming in order to remain competitive in the future.

1. *Waste management financial sustainability*

The management of noxious wastes in the environment has gained a lot of attention around the world as a result of the unbridled disposal into the ecosystem in a reckless way, leading to the possibility of serious adverse environmental consequences. The global amount of waste generated yearly is estimated to reach approximately 1.3 billion tons, which is expected to increase to almost 4.3 billion tons by the year 2025, a significant global growth rate. It is estimated that the rate of generation of electronic waste (E-waste) around the world is three times higher than any other type of waste, and it is steadily growing around the world. This is one of the most significant factors contributing to pollution of the environment, and it is therefore imperative that effective recycling techniques be developed so it can be mitigated in the future. According to the forecast, it is projected that by the year 2020, the number of E-waste generated from old computers will rise by 500%, while the amount of E-waste generated from discarded mobile phones will be almost 18 times higher than in 2007 (Abdelbasir et al., 2018).

According to the latest data, India is currently one of the four countries that produce more E-waste than any other and it is the second most populated country in the world. India's macroeconomic fundamentals have led it to one of the most promising emerging markets in the world. Due to the acceleration of urbanization, building of international business trades, and rapid advances in the field of EEE, the population is beginning to have higher aspirations for a luxurious lifestyle. There was a Digital India Flagship program launched by the Prime Minister of India in 2015. There are a number of new and advanced appliances that have been created as a result of this initiative that have improved gross value added by approximately 388 billion dollars (Arya & Kumar, 2020).

2. *Developing sustainable manufacturing practices for the management of E-waste*

Presently, most of the sustainability issues related to product design or production are focused primarily on ensuring that products are environmentally sustainable and therefore fall short of building sustainability across all aspects of society, economy, and ethics (Ravindra & Mor, 2019).

3. *Green manufacturing*

During the manufacture of computers and other subsystems, it is important to minimize waste generated during the manufacturing process in order to minimize any adverse environmental impact. An objective of green manufacturing is to make electronic devices more energy-efficient and to improve the way that they are used in daily life. There are several ways in which E-waste can be managed, including the development of environmental sustainability methods for production, the development of energy-efficient electronics, and the improvement of disposal and recycling procedures (Fig. 3.3).

The following approaches have been employed in order to ensure that green manufacturing concepts are promoted at all levels:

4. *Green use*

Using computers, smartphones, peripheral devices, and other computers in a manner that minimizes the electricity consumption of the devices and makes use of them in an eco-friendly manner.

5. *Green disposal*

In the case of electronic equipment that is no longer needed, it can either be repurposed or properly disposed of, or recycled.

6. *Green design*

Assuring that computers, servers, printers, projectors, and other digital devices are designed in an energy-efficient manner. As part of the government's commitment to

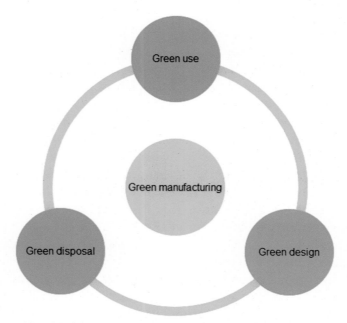

Figure 3.3 Sustainable manufacturing practices for E-waste management.

promoting green manufacturing concepts, various voluntary programs and regulations have also been introduced by regulatory authorities to ensure their implementation (Ahemad & Shrivastava, 2013).

3.4.1 Economic benefits

By definition, economic benefits are tangible benefits which can be measured in terms of revenue generated or money saved by implementing policies which will result in tangible economic benefits. Consider the concept and definition of economic benefits, surplus, and how net income can be utilized to determine new policies and procedures in the business world. Additionally, the recycling of E-waste not only recycles it into useful resources for manufactures to use, but it also reduces the burden on the economy of the nation by making the scarce materials available to manufacturers. In addition, this will also save the cost of extraction when it comes to the mining and processing of the minerals and raw materials that are extracted.

The numerous advantages that can be gained from recycling E-waste make it one of the most important sectors of the recycling industry. However, the benefits of recycling E-waste go far beyond the environmental advantages. Creating and generating revenue from E-waste businesses such as Protec provides employment and stimulates the economy in a positive way. Fortunately, there is no sign that the E-waste recycling industry will disappear anytime soon. There is a projection that the region will generate approximately 49.4 billion dollars in revenue by the year 2020, which is a significant number. Additionally, manufacturers have offered buyback incentives as a result of E-waste recycling. As a result of participating in buyback programs, customers receive lower costs, while manufacturers save money on raw materials when they take part in such programs. Recycling E-waste not only offers a cost-effective method of disposing of waste, but it also eliminates the need to maintain waste sites and transport E-waste to landfills. A large portion of E-waste is sent overseas to landfill sites, so the taxpayers are forced to pay hefty sums for this process.

3.4.2 Environmental benefits

3.4.2.1 Ensures the preservation of natural resources

In addition to reducing our carbon footprint and increasing energy efficiency, recycling E-waste is a great way to preserve our finite natural resources on Earth, including the ability to recover valuable materials and reuse them in manufacturing new products.

3.4.2.2 Prevents toxic chemicals from polluting the ecosystem

A number of toxic chemical substances are contained in electronic components, including nickel, cadmium, lithium, mercury, and lead. Whether buried in landfills or disposed in water bodies, these toxic chemicals flow into our soil, waterways, and ecosystem, contaminating crops, livestock, and sea life, ultimately affecting human health and ecosystems. These toxic chemicals contaminate our food and pose long-term health and environmental hazards.

3.4.2.3 Promotes mindful consumerism

By making E-waste recycling a regular part of our consumer behavior and decisions, we are giving our environment a much-needed reminder of the impact we can have on our environment. The time has come to embrace mindful consumerism as opposed to contributing to the throw-away culture that contributes to waste and irresponsibility, for us to think before we buy and repair/recycle before we throw anything away.

3.4.3 Social benefits

3.4.3.1 Conserves natural resources

In order to make new products from old electronics, it is necessary to recycle valuable materials from the old electronics. In turn, this will result in energy savings, pollution reductions, reduced greenhouse gas emissions, and a reduction in natural resources being extracted from the earth because fewer raw materials have to be extracted.

3.4.3.2 Protects environment

Recycling of E-waste provides a safe and efficient way to deal with toxic chemicals found in E-waste streams, such as mercury, lead, and cadmium, which need to be managed properly.

3.4.3.3 Creates jobs

The E-waste recycling process has the capability of protecting the ecosystem from harmful toxins, but it is also an important part of supporting the economic development of a country by generating employment that can be relied upon on for a long time to come. To be able to perform the recycling processes, a large amount of manpower is required, which provides many skilled people with good-paying jobs and sustains the economy as a whole.

3.4.3.4 Saves landfills

As a result of E-waste recycling, unnecessary landfills and dumps can be avoided. Sustainability is being prioritized as a result of incorporating a number of factors into all development efforts, including social, financial, economic, technical, cultural, and gender considerations. E-waste is likely to have negative effects on the environment as well as various health consequences as a result of its development. Therefore the creation of a multilateral agreement for the disposal of E-waste, whether through landfilling or incineration, is an urgent need. International negotiation and cooperation is the only viable solution to this problem because of the global character of the problem and the challenges of creating a sustainable and environmentally sound E-waste processing system in low-income nations, which means global negotiations and cooperation are the only viable solutions. Moreover, comprehensive international E-waste management and legislation might be a significant step toward mitigating the risks associated with E-waste, which is the best method for achieving sustainable development.

3.5 Industrial challenges and opportunities for the implementation of circular business models

Because of the take-make-dispose business model, the worldwide electronics and electrical equipment sector is set up in a linear way, with externalities associated to products not internalized. The net zero transformation of the industry necessitates the design of products with low carbon footprints that have longer lifespans, higher levels of repairability, fewer toxicants, higher material efficiency, and better recovery. Also, the government policies and corporate actions should support the CE by utilizing the potential of product and value chain digitalization through digital product passports, eco-design, longer product lifespans, addressing planned obsolescence, higher repairability, recycling and achieving higher recycling rates, as well as better resource recovery and value. A paradigm shift is required to focus on change as a policy package for the electronics and electrical industry in order to enable CE in E-waste. The most relevant challenges for the implementation of sustainable CBMs in the E-waste management are discussed in the below section.

3.5.1 Accurate balance of costs and benefits

Some CE strategies can be expensive to develop in the near run. For instance, implementing CBMs for cleaner production may require significant financial investment in new industrial technologies. Similarly, furnaces or smelters demand high processing energy, collection and transportation, and emission control systems, all of which are expensive to construct (Mendoza et al., 2022). The development of more environmentally friendly electronics products by manufacturers is one tactic to stop the depletion of scarce resources. The E-waste management authorities have chosen to achieve this by holding producers financially liable for the waste that is produced by their products. EPR is the practice of making producers responsible for the expenses of any recovery or disposal processes necessary for products they sell after their useful lives have finished. Reusing or recycling what is produced is in the producer's best interest (Clemente et al., 2012).

3.5.2 Technical constraints for the implementation of circular economy strategies

While some CE techniques (such as reuse, refurbishing, and remanufacturing) can help extend the lifespan of EEE and their components, they can also have an impact on how well the technology performs in subsequent use cycles, which can result in less resource, environmental, and financial savings than anticipated. Also, to track the progress of CE implementation in the nation, evaluation indicator systems, such as comprehensive indicators, work indicators, and reference indicators for CE development, should be created. The Comprehensive Indicators illustrate the fundamental idea that a CE places an emphasis on resource efficiency as well as resource recycling and reuse.

3.5.3 Complexity of the forward and reverse logistics management

Extremely large, heavy, or bulky products, as well as those that may contain hazardous substances, may be simple for manufacturers to legally recover, but they might be challenging to move and expensive to recondition. For example, moving a washing machine versus an ink cartridge in a reverse supply chain, will undoubtedly be far more difficult and costly (Isernia et al., 2019). It is particularly challenging to recover value from products with complex construction. Modern smartphones and laptops, for example, are more difficult to reconstitute than machines with coarser-grained modularity, such desktop PCs. Finally, the viability of value recovery will depend on the availability of items that can be reformulated at a reasonable price. When manual effort that takes a lot of time is necessary, worn goods must still be valuable enough to be worth the expense (Atasu et al., 2021).

3.5.4 Policy development and incentives

Effective CE policies can encourage the adoption and spread of suitable business model, technical, and social innovation by offering the essential impulses to unleash their potential for the transition. Policies can promote technological innovation, such as establishing reverse logistics, social innovation, such as launching new collaborations and social initiatives to improve cooperation in the value chain, and technical innovation, such as influencing design process or production process standards (Gillabel et al., 2021). A financial incentive structure might be created to encourage and reward those producers who believe in producing their goods in a more sustainable manner. Standards and labeling should be developed for wider market acceptance. In this regard, a collection of industry-recognized best practices, international norms, and guidelines for the extending of life span can be created. Standards to prevent forced obsolescence in electronics will also be necessary to verify compliance and promote faith in manufacturers' circular claims. For example, the Indian government may implement a financial incentive program to encourage the establishment of recycling facilities at the state level, similar to Eco-parks. The eco-park is a collection of small and medium-sized businesses where it is intended for the informal and formal sectors to collaborate and complete end-to-end processing with zero landfill. Activities in the eco-park include manual disassembly for material separation, including structural metal parts, heat sinks, ferrous metal, ferrite, and ceramic components, nonferrous metal scrap, primarily Cu and Al, glass components, etc. for processing at nearby smelters.

3.5.5 Design for recycling

Promoting eco-design is more successful with a worldwide framework than with individual domestic laws. To maintain the materials and value in use for as long as possible, the EEE must be developed in a way that initially provides a longer life of the products. Second, the products should be designed such that they can be recovered, refurbished, remanufactured, and recycled. Because electrical and electronic equipment is becoming more complex and new items are entering the market, recycling and recovery infrastructure is currently playing catch-up despite advances in mechanical and chemical recycling technology. Notably, 80% of a product's environmental impact is already known. Therefore CE regulations should be implemented to make sure that manufacturers adhere to eco-design principles in order to improve first the products' durability, reparability, and upgradeability, and also to design products that are easily separated into different parts and materials that can be recycled, repaired, or upgraded. The design should include sustainability standards and other life-cycle stages both at the level of the overall product and the materials. By improving product design and facilitating product and material recycling, regions like the EU have created rules that are intended to increase resource efficiency and the CE. Promoting the design for recyclability, extending product life, standardizing interfaces for accessories like chargers and headphones, creating products with modular structures that allow for easy component and part replacement, allowing for ICT/CE device upgrading, and evaluating the potential for an eco-design program are all possible components of an electronics manufacturing promotion scheme (MeitY, 2021).

3.6 Conclusion and recommendations

E-waste has become a global problem across the world. Mining of primary materials for electronics has become unsustainable for businesses as the world shifts from a linear to a CE. By recycling E-waste, billions of dollars can be saved in materials costs and carbon footprints can be reduced significantly. E-waste resource recovery can be an advantage in a CE if it is harnessed as a strategic business advantage.

To deal with the present challenges related to E-waste and to extend business opportunities some recommendations are provided:

1. The adverse effects of E-waste are still unknown to a major part of the population. Awareness programs are to be created to spread the knowledge of toxic effect of E-waste and the management strategies. The spread of awareness among people can protect human health and the environment. Nongovernmental organizations

with the help of government should reach every nook and corner of the country for the betterment of the society.

2. People should understand the consequences of using electronic products for a short period of time and dump them into the environment. To save the environment and the society, people should select those products which constitute less toxic substances, designed with upgradation facility, and are in take-back options.

3. The formal E-waste recycling facilities should have proper collection facilities so that the major part of E-waste can be recycled properly. For this, building of bridge between the formal and informal E-waste recycling facilities would be of great help.

4. The government should help the recycling facilities to develop their processes so that they can opt for a circular economic approach, thus aiding to economic growth of the country.

References

Abdelbasir, S. M., El-Sheltawy, C. T., & Abdo, D. M. (2018). Green processes for electronic waste recycling: A review. *Journal of Sustainable Metallurgy*, *4*, 295–311. Available from https://doi.org/10.1007/s40831-018-0175-3.

Acomb, J. C., Nahil, M. A., & Williams, P. T. (2013). Thermal processing of plastics from waste electrical and electronic equipment for hydrogen production. *Journal of Analytical and Applied Pyrolysis*, *103*, 320–327.

Agoramoorthy, G., & Chakraborty, C. (2012). Control electronic waste in India. *Nature*, *485*, 309. Available from https://doi.org/10.1038/485309b.

Ahemad, M., & Shrivastava, R. (2013). Green manufacturing (GM): Past, present and future (a state of art review). *World Review of Science*, *10*, 17–55. Available from https://doi.org/10.1504/WRSTSD.2013.050784.

Andeobu, L., Wibowo, S., & Grandhi, S. (2021). A systematic review of E-waste generation and environmental management of Asia Pacific countries. *International Journal of Environmental Research and Public Health*, *18*, 9051. Available from https://doi.org/10.3390/ijerph18179051.

Aras, N., Korugan, A., Büyüközkan, G., Şerifoğlu, F. S., Erol, I., & Velioğlu, M. N. (2015). Locating recycling facilities for IT-based electronic waste in Turkey. *Journal of Cleaner Production*, *105*, 324–336.

Arya, S., & Kumar, S. (2020). E-waste in India at a glance: Current trends, regulations, challenges and management strategies. *Journal of Cleaner Production*, *271*, 122707.

Atasu, A., Dumas, C., & Wassenhove, L. N. V. (2021). The circular business models. Harvard Business Review. Available from https://hbr.org/2021/07/the-circular-business-model.

Babu, B. R., Parande, A. K., & Basha, C. A. (2007). Electrical and electronic waste: A global environmental problem. *Waste Management & Research*, *25*(4), 307–318.

Balde, C. P., Forti V., Gray, V., Kuehr, R., & Stegmann, P. (2017). *The global E-waste monitor 2017* (pp. 1–116). Bonn/Geneva/Vienna: United Nations University (UNU), International Telecommunication Union (ITU) & International Solid Waste Association (ISWA).

Banzal, P. (2022). *E-waste management in India: Challenges and opportunities*. <https://www.communicationstoday.co.in/article-on-E-waste-management-in-india-challenges-and-opportunities/> Accessed 24.07.22.

Bazargan, A., Lam, K. F., & McKay, G. (2012). *Challenges and opportunities in E-waste management* (pp. 39–66). Nova Science Publishers.

Bhuiya, M. M. K., Rasul, M., Khan, M., Ashwath, N., & Mofijur, M. (2020). Comparison of oil extraction between screw press and solvent (n-hexane) extraction technique from beauty leaf (*Calophyllum inophyllum* L.) feedstock. *Industrial Crops and Products*, *144*, 112024. Available from https://doi.org/10.1016/j.indcrop.2019.112024, ISSN 0926-6690.

Bridgens, B., Hobson, K., Lilley, D., Lee, J., Scott, J. L., & Wilson, G. T. (2019). Closing the loop on E-waste: A multidisciplinary perspective. *Journal of Industrial Ecology*, *23*(1), 169—181.

Clemente, A., Franzluebbers, B., LaRochelle, B. (2012). *Cost calculating model for electronic waste management*. Worcester, MA: Worcester Polytechnic Institute. <https://web.wpi.edu/Pubs/E-project/Available/E-project-050612-085500/unrestricted/RenoSam_Final_Report.pdf>.

D'Adamo, I., Ferella, F., & Rosa, P. (2019). Wasted liquid crystal displays as a source of value for E-waste treatment centers: A techno-economic analysis. *Current Opinion in Green and Sustainable Chemistry*, *19*, 37—44. Available from https://doi.org/10.1016/j.cogsc.2019.05.002.

Dai, Q., Xu, X., Eskenazi, B., Asante, K. A., Chen, A., Fobil, J., & Huo, X. (2020). Severe dioxin-like compound (DLC) contamination in E-waste recycling areas: An under-recognized threat to local health. *Environment International*, *139*, 105731.

de Vicente Bittar, A. (2018). Selling re-manufactured products: Does consumer environmental consciousness matter? *Journal of Cleaner Production*, *181*, 527—536.

Debnath, B., Baidya, R., Biswas, N. T., Kundu, R., & Ghosh, S. K. (2015). E-waste recycling as criteria for green computing approach: analysis by QFD tool. *Computational advancement in communication circuits and systems* (pp. 139—144). Berlin: Springer.

Dwivedy, M., & Mittal, R. K. (2012). An investigation into e-waste flows in India. *Journal of Cleaner Production*, *37*, 229—242.

Forti, V., Baldé, C. P., Kuehr, R., Bel, G. (2020). *The global E-waste monitor 2020: Quantities, flows and the circular economy potential*. Bonn/Geneva/Rotterdam: United Nations University (UNU)/United Nations Institute for Training and Research (UNITAR).

Gangadharan, P., Nambi, I. M., & Senthilnathan, J. (2015). Liquid crystal polaroid glass electrode from E-waste for synchronized removal/recovery of Cr + 6 from wastewater by microbial fuel cell. *Bioresource Technology*, *195*, 96—101.

Gillabel, J., Manshoven, S., & Grossi, F. (2021). *Business models in a circular economy*. European Topic Centre on Waste and Materials in a Green Economy. <file:///C:/Users/HP/Downloads/2.1.2.4.%20etc%20Eionet%20Report%20Circular%20Business%20Models_final_edited%20for%20website.pdf>.

Gomathi, N., & Rupesh, P. L. (2018). Study on business opportunities extracted from E-waste A Review. *International Journal of Engineering & Technology*, 7, 1106—1109.

Haibin, L., & Zhenling, L. (2010). Recycling utilization patterns of coal mining waste in China. *Resources, Conservation and Recycling*, *54*(12), 1331—1340.

Hindrise (2022). Hindrise Social Welfare Foundation. Available from https://hindrise.org/resources/E-waste-management-in-india/. Accessed 24.07. 22.

Imran, M., Haydar, S., Kim, J., Rizwan Awan, M., & Ali Bhatti, A. (2017). E-waste flows, resource recovery and improvement of legal framework in Pakistan. *Resources, Conservation and Recycling*, *125*, 131—138. Available from https://doi.org/10.1016/j.resconrec.2017.06.015, ISSN 0921-3449.

Isernia, R., Passaro, R., Quinto, I., & Thomas, A. (2019). The reverse supply chain of the E-waste management processes in a circular economy framework: Evidence from Italy. *Sustainability*, *11*(8), 2430.

Islam, M. T., Abdullah, A. B., Shahir, S. A., Kalam, M. A., Masjuki, H. H., Shumon, R., & Rashid, M. H. (2016). A public survey on knowledge, awareness, attitude and willingness to pay for WEEE management: Case study in Bangladesh. *Journal of Cleaner Production*, *37*, 728—740. Available from https://doi.org/10.1016/j.jclepro.2016.07.111, ISSN 0959-6526.

Kantarelis, E., Yang, W., Blasiak, W., Forsgren, C., & Zabaniotou, A. (2011). Thermochemical treatment of E-waste from small household appliances using highly pre-heated nitrogen-thermogravimetric investigation and pyrolysis kinetics. *Applied Energy*, *88*(3), 922—929.

Kaya, M. (2016). Recovery of metals and non-metals from electronic waste by physical and chemical recycling processes. *Waste Management*, *57*, 64—90.

Kumar, A., & Dixit, G. (2018). An analysis of barriers affecting the implementation of E-waste management practices in India: A novel ISM-DEMATEL approach. *Sustainable Production and Consumption*, *14*, 36—52. Available from https://doi.org/10.1016/j.spc.2018.01.002, ISSN 2352-5509.

Lèbre, É., Corder, G., & Golev, A. (2017). The role of the mining industry in a circular economy: A framework for resource management at the mine site level. *Journal of Industrial Ecology, 21*(3), 662—672.

Linder, M., & Williander, M. (2017). Circular business model innovation: inherent uncertainties. *Business Strategy and the Environment, 26*(2), 182—196.

Mao, S., Gu, W., Bai, J., Dong, B., Huang, Q., Zhao, J., & Wang, J. (2020). Migration characteristics of heavy metals during simulated use of secondary products made from recycled E-waste plastic. *Journal of Environmental Management, 266*, 110577.

Matsakas, L., Qiuju, G., Stina, J., Ulrika, R., & Paul, C. (2017). Green conversion of municipal solid wastes into fuels and chemicals. *Electronic Journal of Biotechnology, 26*, 69—83. Available from https://doi.org/10.1016/j.ejbt.2017.01.004, ISSN 0717-3458.

Meenakshi, D. T., & Harini, V. (2012). Entrepreneurship opportunities in managing E-waste. *Asia Pacific Journal of Management & Entrepreneurship Research, 1*(3), 47.

MeitY (2021). *Circular economy in electronics and electrical sector.* New Delhi: Ministry of Electronics and Information Technology, Government of India. <https://www.meity.gov.in/writereaddata/files/Circular_Economy_EEE-MeitY-May2021-ver7.pdf>.

Mendoza, J. M. F., Gallego-Schmid, A., Velenturf, A. P., Jensen, P. D., & Ibarra, D. (2022). Circular economy business models and technology management strategies in the wind industry: Sustainability potential, industrial challenges and opportunities. *Renewable and Sustainable Energy Reviews, 163*, 112523. Available from https://doi.org/10.1016/j.rser.2022.112523.

MoEF (2008). *Guidelines for environmentally sound management of E-waste* (as approved Vide Ministry of Environment and Forests (MoEF)) (No. 23-23/2007-HSMD). <https://www.Yumpu.com/En/Document/View/6274477/GuidelinesFor-Environmentally-Sound-Management-Of-E-waste> Accessed 24.02.22.

Nandy, S., Fortunato, E., & Martins, R. (2022). Green economy and waste management: An inevitable plan for materials science. *Progress in Natural Science: Materials International, 32*, 1—9.

Niza, S., Santos, E., Costa, I., Ribeiro, P., & Ferrão, P. (2014). Extended producer responsibility policy in Portugal: A strategy towards improving waste management performance. *Journal of Cleaner Production, 64*, 277—287. Available from https://doi.org/10.1016/j.jclepro.2013.07.037, ISSN 0959-6526.

Pandey, P., & Govind, M. (2014). Social repercussions of E-waste management in India: A study of three informal re-cycling sites in Delhi. *The International Journal of Environmental Studies, 71*, 241—260.

Parajuly, K., Fitzpatrick, C., Muldoon, O., & Kuehr, R. (2020). Behavioral change for the circular economy: A review with focus on electronic waste management in the EU. *Resources, Conservation & Recycling: X, 6*, 100035.

Pariatamby, A., & Victor, D. (2013). Policy trends of E-waste management in Asia. *Journal of Material Cycles and Waste Management, 15*, 411—419. Available from https://doi.org/10.1007/s10163-013-0136-7.

Patibanda, S., Bichinepally, S., Yadav, B. P., Bahukandi, K. D., Sharma, M. (2020). E-waste management and its current practices in India. Advances in industrial safety; Springer: Singapore; pp. 191—202.

Planing, P. (2015). Business model innovation in a circular economy reasons for non-acceptance of circular business models. *Open Journal of Business Model Innovation, 1*(11), 1—11.

Rautela, R., Arya, S., Vishwakarma, S., Lee, J., Kim, K. H., & Kumar, S. (2021). E-waste management and its effects on the environment and human health. *Science of The Total Environment, 773*, 145623.

Ravindra, K., & Mor, S. (2019). E-waste generation and management practices in Chandigarh, India and economic evaluation for sustainable recycling. *Journal of Cleaner Production, 221*, 286—294. Available from https://doi.org/10.1016/j.jclepro.2019.02.158.

Selvaraj, N. (2021). E-waste management and its business opportunities at Tamilnadu. *Journal of Entrepreneurship & Organization Management, 10*, 72021.

Shahabuddin, M., Uddin, M. N., Chowdhury, J. I., Ahmed, S. F., Uddin, M. N., Mofijur, M., & Uddin, M. A. (2022). A review of the recent development, challenges, and opportunities of electronic waste (E-waste). *International Journal of Environmental Science and Technology*, 1—8.

Song, Q., & Li, J. (2015). A review on human health consequences of metals exposure to E-waste in China. *Environmental Pollution, 196*, 450–461.

Spiliakos (2018). *What does "sustainability" mean in business?* Harvard Business School. <https://online.hbs.edu/blog/post/what-is-sustainability-in-business>.

Tue, N. M., Goto, A., Takahashi, S., Itai, T., Asante, K. A., Kunisue, T., & Tanabe, S. (2016). Release of chlorinated, brominated and mixed halogenated dioxin-related compounds to soils from open burning of E-waste in Agbogbloshie (Accra, Ghana). *Journal of Hazardous Materials, 302*, 151–157.

Van Weelden, E., Mugge, R., & Bakker, C. (2016). Paving the way towards circular consumption: exploring consumer acceptance of refurbished mobile phones in the Dutch market. *Journal of Cleaner Production, 113*, 743–754.

Vogtlander, J. G., Scheepens, A. E., Bocken, N. M., & Peck, D. (2017). Combined analyses of costs, market value and eco-costs in circular business models: Eco-efficient value creation in re-manufacturing. *Journal of Remanufacturing, 7*(1), 1–17.

Wang, W., Tian, Y., Zhu, Q., & Zhong, Y. (2017). Barriers for household E-waste collection in China: Perspectives from formal collecting enterprises in Liaoning province. *Journal of Cleaner Production*. Available from https://doi.org/10.1016/j.jclepro.2017.03.202.

Wastling, T., Charnley, F., & Moreno, M. (2018). Design for circular behaviour: Considering users in a circular economy. *Sustainability, 10*(6), 1743.

Wen, X. F., Li, J. H., Liu, H., Yin, F. F., Hu, L. X., Liu, H. P., & Liu, Z. Y. (2006). An agenda to move forward E-waste recycling and challenges in China. *Proceedings of the 2006 IEEE international symposium on electronics and the environment, 2006* (pp. 315–320). IEEE.

World Economic Forum. (2019) *We generate 125,000 jumbo jets worth of E-waste every year. Here's how we can tackle the problem.*

World Economic Forum, PACE. (2019). *A new circular vision for electronics: Time for a global reboot.* World Economic Forum. Available from: <http://www3.weforum.org/docs/WEF_A_New_Circular_Vision_for_Electronics.pdf>.

WTO. (2020). *Committee on technical barriers to trade—notification—Bangladesh—electrical and electronic products.* World Trade Organisation. Accessed 24.07.22.

Yang, J., Lu, B., & Xu, C. (2008). WEEE flow and mitigating measures in China. *Waste Management, 28*(9), 1589–1597.

Zeng, X., & Li, J. (2016). Measuring the recyclability of E-waste: An innovative method and its implications. *Journal of Cleaner Production, 131*, 156–162. Available from https://doi.org/10.1016/j.jclepro.2016.05.055, ISSN 0959-6526.

Zeng, X., Yang, C., Chiang, J. F., & Li, J. (2017). Innovating e-waste management: From macroscopic to microscopic scales. *Science of The Total Environment, 575*, 1–5.

CHAPTER 4

Understanding the existing trends in the E-waste management (1993—2021) research domain and its future with a focus on India

Ajishnu Roy[1] and Aman Basu[2]
[1]School of Geographical Sciences and Remote Sensing, Guangzhou University, Guangzhou, Guangdong, P.R. China
[2]Department of Biology, York University, Toronto, ON, Canada

4.1 Introduction

The fastest-growing waste stream is electronic waste (E-waste). Over the last two decades, the production of E-waste has increased rapidly in tandem with economic growth. Electrical and electronic equipment (EEEs) have become a necessity in modern society. Because of its widespread availability and use, it has enabled a large portion of the world's population to enjoy higher living standards. EEEs include household appliances, IT and telecommunications, consumer and lighting equipment, electrical and electronic tools, toys, leisure and sports equipment, and medical equipment that rely on electric currents or electromagnetic fields to function properly (European Union, 2003).

A total of 53.6 million tons (Mt) (or 7.3 kg per capita) of E-waste were generated worldwide in 2019, which is a 21% rise from 2014 when it was 44.4 Mt (or 6.4 kg per capita) (Forti et al., 2020). Also, the global E-waste will reach 74.7 Mt (or 9 kg per capita) by 2030, that is, almost double in the coming 9 years. This makes E-waste the fastest-growing residential waste stream in the world. Its key drivers are greater EEE consumption rates, rapid product end of life (EoL), and a lack of available repair solutions. However, only 17.4% of the E-waste generated in 2019 was collected and recycled, meaning that almost US$57 billion worth of Au, Ag, Cu, Pt, and other valuable, recoverable elements were lost in 2019 alone (most of which were dumped or burnt rather than being collected for treatment and reuse). Though this recycling amount grew by 1.8 Mt from 2014, the E-waste generation (total) increased by 9.2 Mt. This indicates an ever-widening gap between necessary and actual recycling activities of E-waste. As per the latest global data (2019), in 2019, Asia produced the

Global E-waste Management Strategies and Future Implications
DOI: https://doi.org/10.1016/B978-0-323-99919-9.00012-X

most E-waste (24.9 Mt or 5.6 kg per capita), followed by the Americas (13.1 Mt or 13.3 kg per capita), Europe (12 Mt or 16.2 kg per capita), Africa (2.9 Mt or 2.5 kg per capita), and Oceania (0.7 Mt or 16.1 kg per capita). However, as per collected and properly recycled amount of E-waste, Europe tops the list with 5.1 Mt (42.5%), followed by Asia (2.9 Mt, 11.7%), Americas (1.2 Mt, 9.4%), Oceania (0.06 Mt, 8.8%), and Africa (0.03 Mt, 0.9%) (Forti et al., 2020). In 2019 small equipment made up the majority of the E-waste (17.4 Mt), followed by big equipment (13.1 Mt), temperature exchange equipment (10.8 Mt), screens and monitors (6.7 Mt), lighting (0.9 Mt), and large equipment (13.1 Mt) (4.7 Mt).The fastest-growing E-waste categories (based on total weight) in 2014 were temperature exchange equipment (up to 7%); big equipment (up to 5%); both lighting and small equipment (up to 4%); and small IT and telecommuting equipment (+ 2%). Although the number of nations with national E-waste policies, laws, or regulations has climbed from 78 (or 71% of the world's population) in 2019 to 61 (or 44%), many others (115 countries) are yet to do so.

In 2002 E-waste emerged as an environmental issue in India. In the following years, guidelines for E-waste management came (MoEF, 2008), and then the waste law came into effect ["E-waste (Management and Handling Rules), 2011"] (MoEF, 2011), which was revised in 2016 ["E-waste (Management) Rules, 2016] (MoEF, 2016). For India, E-waste generated has increased from 1.5 to 2.4 kg per capita or 1973 to 3230 kt in 5 years 2014—19. The rate of electronic trash collection has remained constant at 1% of the total quantity formally collected (30 kt). The top nine states in India that contribute the most E-waste are Maharashtra (19.8%), Tamil Nadu (13%), Andhra Pradesh (12.5%), Uttar Pradesh (10.1%), West Bengal (9.8%), Delhi (9.5%), Karnataka (8.9%), Gujarat (8.8%), and Madhya Pradesh (7.6%) (calculated in % of a ton per annum) (ASSOCHAM-NEC, 2018). Though there are a significant number of authorized recyclers of E-waste in India (312 in number), with a cumulative capacity for treating and recycling nearly 800 kt E-waste annually, about 95% of Indian E-waste recycling is performed by the informal sector (Arya & Kumar, 2020a).

There have been a handful of articles, specifically focused on waste, either on a global or national scale (see Table 4.1). A work gave a comprehensive account of Waste Electrical and Electronic Equipment (WEEE) research to define and analyze the main areas of research (Pérez-Belis et al., 2014). Another study has opined on the apparent disinterest of researchers to address E-waste-related issues in India (Borthakur & Govind, 2017). In a work (Li et al., 2018), the authors argued that developing countries entered a rapid development period in the field of solid waste reuse and recycling. There is a significant gap between developed and developing countries in construction and demolition waste and the organic fraction of municipal solid waste. A group of authors (Andrade et al., 2019) have discussed that although there is rapid development concerning E-waste has occurred primarily during the past decade (i.e.,

Table 4.1 Comparative recent literature about bibliometric analyses of the E-waste-related research domain.

S. No.	Study	Overview period and region	Central focus	Database	Documents
1.	Pérez-Belis et al. (2014)	1992–2014	Electronic waste	Various	307
2.	Borthakur and Govind (2017)	1986–2016 (global)	Electronic waste	Scopus	4201
3.	Li et al. (2018)	1992–2016 (global)	Solid waste, reuse or recycling or recycle	WoS	6289
4.	Andrade et al. (2019)	1998–2018 (global)	Electronic waste	WoS	3311
5.	Gao et al. (2019)	1981–2018 (global)	Electronic waste	WoS	2800
6.	Zhang et al. (2019)	1986–2018 (global)	Electronic waste	WoS	2847
7.	Borthakur (2020)	2004–20 (global)	Electronic waste, policy, governance	WoS, Scopus	6226 (Scopus) and 4361 (WoS)
8.	Borthakur and Singh (2020)	2000–18 (global)	Electronic waste	Scopus	5524
9.	Ismail and Hanafiah (2020)	2005–19 (global)	Electronic waste	Various databases	130
10.	Andeobu et al. (2021a)	2005–20 (selected countries from the Asia Pacific region)	Electronic waste	ProQuest, Emerald, ScienceDirect, WoS	185
11.	Andeobu et al. (2021b)	2005–20 (Canada, France, the United States, the United Kingdom, Nigeria, South Africa)	Electronic waste	ScienceDirect, WoS	205
12.	Chen et al. (2021)	2008–17 (global)	Electronic waste	WoS	100
13.	de Albuquerque et al. (2021)	2000–20 (global)	Electronic waste	WoS, Scopus	44
14.	Ranjbari et al. (2021)	2001–20 (global)	Waste management, circular economy	WoS	962

(Continued)

Table 4.1 (Continued)

S. No.	Study	Overview period and region	Central focus	Database	Documents
15.	Singh et al. (2021)	2008−20 (global)	Electronic waste, circular economy	Scopus	326
16.	Velasco-Muñoz, Aznar-Sánchez, Pozas-Ramos, et al. (2021)	2000−19 (global)	Sustainable management, WEEE	Scopus	553
17.	Velasco-Muñoz, Aznar-Sánchez, Manzano-Archilla, et al. (2021b)	2000−19 (global)	WEEE, environment	Scopus	2585
18.	Roy and Basu (this study)	1993−2021 (national—India)	Electronic waste	WoS	606

WoS, Web of Science.

>100 articles annually), due to a large amount of E-waste generated each year and the complexity of E-waste, this research domain still has a long way to go. Another group (Gao et al., 2019) has concluded that management and recycling of E-waste in developing countries, health risk assessment after exposure to organic pollutants, degradation and recovery of waste metal materials, and the impact of heavy metals on children's health were the main hot topics in the E-waste field, while degradation was the frontier topic. Some authors (Zhang et al., 2019) in their E-waste study have established that electronic equipment, extended producer responsibility (EPR), sediment, environment and design, and risk assessment have all been replaced by life cycle assessment, mobile phones, and behavior. Researchers have shown that the historic WEEE Directive and RoHS Directive of the EU has a significant impact on the policies of the majority of emerging economies (Borthakur, 2020). As a few envois from the well-known emerging economies are underrepresented in the research on E-waste, the author advises that it is crucial to have a locally coordinated, systematic, organized, step-by-step approach needed to formulate and implement the E-waste policies in the corresponding countries. In another work (Borthakur & Singh, 2020), the authors have done an overview of research activities on E-waste across the years. A study (Ismail & Hanafiah, 2020) has concluded that the number of publications based on origin countries was still low; the lack of in-depth, current evaluations in this E-waste study field; etc. Another study (Chen et al., 2021) analyzed the top 100 cited studies on E-waste and found that the keywords "E-waste" and "recycling" held the highest occurrences, while environment and health, management and economics, technique and processing, and characteristic and property are the most popular research topics. A different study (Singh et al., 2021) has observed that various issues such as - E-waste, reuse, recycling, and sustainability surround the topic of the circular economy. A study (Dutta et al., 2021) has discussed the issues of E-waste pollution during the Covid-19 pandemic. Some studies (Arya & Kumar, 2020b) have explained the postprocessing of E-waste from urban mining of obsolete computers. Another study (Arya, Rautela, et al., 2021) has estimated the soil degradation from crude E-waste pollution in Delhi. Another work has described the inappropriate recycling protocols of E-waste and its adjoined toxic effects on the environment and human health. Another work (Vishwakarma et al., 2022) has discussed E-waste in Information and Communication Technology (ICT) sector. Thus it can be assumed that there is a need for a study that focuses on a national scale, covers the whole period of E-waste research, dissects out gray on nonpeer-reviewed literature, etc. As China is already the top leading research country on E-waste, India was chosen for this case study due to two focus points—a rampant increase of E-waste and associated adverse effects and similarity with a significant number of emerging economies from three continents (viz., Africa, Asia, and Latin America) (Arya, Patel, et al., 2021). The comparative

analysis of 17 bibliometric analyses of E-waste and related domains from previous literature has been tabulated here.

4.2 Materials and methods

A set of processes are followed in systematic bibliometric literature reviews. To begin, the database must be searched using a topically relevant keyword(s). The most common databases are Web of Science (WoS) and Scopus. Then, to recognize significant elements of the chosen publications, tools for literary analysis are used. The application of network analysis comes next. It elucidates complicated linkages between publications, origin countries, and collaborations, among other things. This study uses bibliometric analysis to conduct a comprehensive overview of studies on E-waste. Iterative processes of defining relevant search keywords, searching the literature, and completing the analysis are used to accomplish systematic literature reviews. Systematic reviews differ from typical narrative reviews in that they use a repeatable, scientific, and transparent procedure to reduce selection bias by searching the literature extensively (Vrabel, 2015).

The analysis and classification of scientific work and publications that relate to E-waste and India from 1993 (i.e., from the initial publications on E-waste) to 2021 are presented in this paper. In the current study, the Web of Science core collection (WoSCC), (by Clarivate Analytics) was used to recollect publications. The "bibliometrix" package in R (v.4.0.5) was used (Aria & Cuccurullo, 2017) for this bibliometric analysis. VOSviewer (v. 1.6.16) and Microsoft Office 2019 have also been used for this study.

The search topic was chosen based on keywords in the title, abstract, and/or keywords linked to the E-waste study. To get a comprehensive search result, it is critical to use a few umbrella terms. To ensure that all of the keywords were included, a "OR" was used to divide them. The time was selected from 1993 to 2021 (20th June—date of access of the WoSCC). The keywords chosen and used in this study are

TS = ("electrical and electronic equipment" OR "electrical and electronic waste" OR "electrical garbage" OR "electrical rubbish" OR "electrical scrap" OR "electrical waste" OR "electrical wastes" OR "electronic garbage" OR "electronic rubbish" OR "electronic scrap" OR "electronic waste" OR "electronic wastes" OR "end-of-life electrical equipment" OR "end-of-life electronic equipment" OR "E-waste" OR "E-wastes" OR "waste electrical and electronic equipment" OR "waste electrical" OR "waste electronic" OR "wastes electrical" OR "wastes electronic" OR "WEEE")

Region: India (in WoSCC)

This yields a total of 606 documents. All categories of published documents have been chosen in this study.

4.3 Results

4.3.1 Characteristics of publication outputs

The number of E-waste-related publications start from 1 in 1993 to 72 in 2021 (till the date of WoSCC access) (Fig. 4.1(1)). The top 5 years of scientific production of E-waste were 2018 (101), 2020 (99), 2019 (81), 2021 (72), and 2017 (49). A total of 606 publications were collected from the WoSCC relating to E-waste research, which consist of 397 articles (65.51%; of this, articles—369 or 60.89%, book chapter articles—19 or 3.13%, early access articles—5 or 0.82% and proceedings paper articles—4 or 0.66%), 120 proceeding papers (19.8%), 74 reviews (12.21%; of this, reviews—69 or 11.38%, book chapter reviews—3 or 0.49% and early access reviews—0.33%), and 15 editorial materials, letters, and meeting abstracts (2.47%; of this, editorial materials—5 or 0.82%, book chapter editorial materials—2 or 0.33%, letters—6 or 0.99%, and meeting abstracts—2 or 0.33%). The 606 publications originate from 310 sources (books, journals, etc.). The average of years from publication is 4.09, the average number of citations per document is 14.67, and the average number of citations year^{-1} documents^{-1} is 2.71. There are 1426 authors and 2177 author appearances for these publications. There are 0.425 documents author^{-1}, 2.35 authors document^{-1}, 3.59 coauthors document^{-1}, and the collaboration index is 2.48. There are 32 authors of single-author documents and 1394 authors of multiauthor documents. There are 1714 author keywords and 1335 Keywords Plus. Average article citations were highest in 2005 (7.5), 2.15 (4.9), and 2.12 (4.8) (Fig. 4.1(2)).

4.3.2 Subject categories

Common themes of current research in the field can be understood through the analysis of subject categories. WoS categories have been used for this purpose. The E-waste-related publications belong to 39 categories, of which the top 10 broad categories with ≥ 9 publications contain 532 (87.78%) publications on E-waste. These are *Engineering* (187 or 30.85%), *Environmental Sciences* (81 or 13.36%), *Green & Sustainable Science & Technology* (59 or 9.73%), *Materials Science* (56 or 9.24%), *Chemistry* (42 or 6.93%), *Computer Science* (36 or 5.94%), *Multidisciplinary Sciences* (21 or 3.46%), *Biotechnology & Applied Microbiology* (20 or 3.3%), *Metallurgy & Metallurgical Engineering* (12 or 1.98%), *Construction & Building Technology* (9 or 1.48%), and *Polymer Science* (9 or 1.48%) (Fig. 4.1(3)).

4.3.3 Top output analysis

4.3.3.1 Journals' analysis

Of the selected 606 publications under consideration, the most relevant top 10 sources compose 166 publications (27.39% of the total share). They are *Materials Today* (5.94%), *Waste Management* (4.78%), *Journal of Cleaner Production* (4.29%), *Waste*

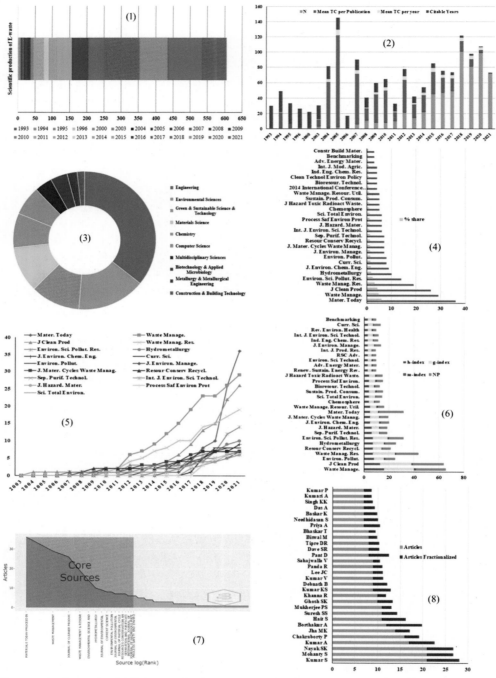

Figure 4.1 Characteristic features of publications, Web of Science subject categories, and performance of journals on E-waste research related to India. (1) Annual scientific production trends in E-waste research related to India (1993—2021). (2) Annual citation characteristics. (3) WoS subject categories in E-waste research. (4) Most relevant sources in E-waste publications. (5) Source dynamics of E-waste research (cumulative). (6) Source Local Impact. (7) Source clustering through Bradford's Law. (8) Most relevant authors in E-waste research.

Management & Research (3.13%), *Environmental Science and Pollution Research* (2.31%), *Hydrometallurgy* (1.65%), *Journal of Environmental Chemical Engineering* (1.48%), *Current Science* (1.32%), *Environmental Pollution* (1.32%), and *Journal of Environmental Management* (1.15%) (Fig. 4.1(4)). Among the sources, the top five cumulative publications (2003–21) are *Materials Today, Waste Management, Journal of Cleaner Production, Waste Management & Research*, and *Environmental Science and Pollution Research*. *Waste Management* used to maintain top production from 2012 to 2019 which was then surpassed by *Materials Today* in 2020 (Fig. 4.1(5)). The top five sources with the most local impact (via h-index) are *Waste Management* (16), *Journal of Cleaner Production* (13), *Environmental Pollution* (8), *Environmental Science and Pollution Research* (7), and *Hydrometallurgy* (7) (Fig. 4.1(6)). However, the top eight, with >200 total citations (TC) are *Waste Management* (1149), *Journal of Cleaner Production* (748), *Waste Management & Research* (397), *Resources, Conservation & Recycling* (369), *Separation and Purification Technology* (278), *Environmental Pollution* (242), *Environmental Impact Assessment Review* (240), and *Chemosphere* (206). Bradford's law, also known as Bradford's scattering law or Bradford distribution, is a pattern that calculates the exponentially decreasing returns of searching for references in scientific journals. In this study, 16 journals occupy zone 1 or the core area, that is, journals that are with the highest frequent citation in the literature of this subject area. Among them, the top five ranked in frequency are *Materials Today* (36), *Waste Management* (29), *Journal of Cleaner Production* (26), *Waste Management & Research* (19), and *Environmental Science and Pollution Research* (14). Also, zone 2 contains 95 journals and zone 3 contains 199 journals (Fig. 4.1(7)).

4.3.3.2 Authors' analysis

The most relevant authors (top 10), as per fractionalized publications, are Borthakur A (7.83), Kumar S (7.08), Mohanty S (5.73), Nayak SK (5.73), Kumar A (5.59), Hait S (5.17), Pant D (4.42), Kumar KS (3.87), Jain A (3.67), and Priya A (3.5). However, the top five authors of most publications (>15) are Kumar S (21), Mohanty S (21), Nayak SK (21), Kumar A (17), and Chakraborty P (16) (Fig. 4.1(8)). Among the authors, the top five authors with local impact (via h-index) are Kumar A (10), Kumar S (9), Chakraborty P (8), Mukherjee PS (8), and Sahajwalla V (8) (Fig. 4.2(1)). The top five authors with the most consistent productivity with time are Kumar S, Mohanty S, Nayak SK, Kumar A, and Chakraborty P (Fig. 4.2(2)). Lotka's Law describes the frequency with which authors' patterns in a given field and over a given period are published. Here, 78.8% and 11.5% of authors produce single and double-authored documents, respectively. Authors with a higher number of publications have lower contributions, such as—3 (4.2%), 4 (1.8%), and 5 (1.1%) (Fig. 4.2(3)).

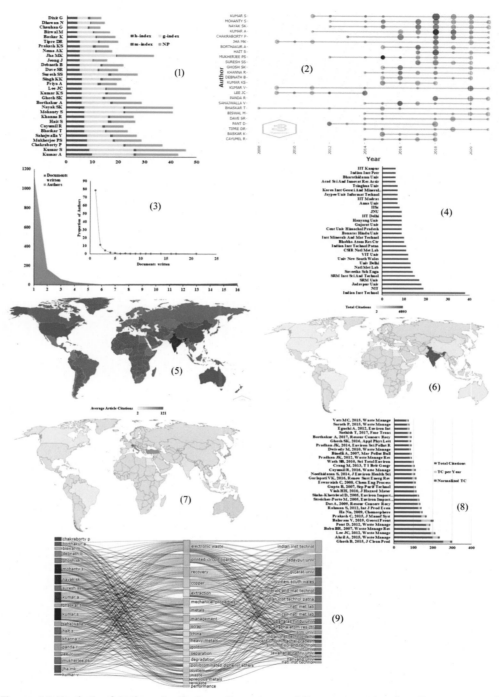

Figure 4.2 Analysis of Authors, Institutions, Countries, and Documents related to E-waste research in India. (1) Authors' Local Impact. (2) Top Authors' Production over time. (3) Author Productivity through Lotka's Law. (4) Most relevant affiliations. (5) Spatial distribution of total scientific production. (6) Spatial distribution of total citations. (7) Spatial distribution of average article citations. (8) Most global cited documents. (9) Three fields plot of E-waste publications of India.

4.3.3.3 Institutions' analysis

The most relevant 10 affiliations related to publications on E-waste from or collaborated with India are the Indian Institutes of Technology (38), National Institute of Technology (19), Jadavpur University (18), SRM University (17), SRM Institute of Science and Technology (15), Saveetha School of Engineering (14), National Metallurgical Laboratory (13), University of Delhi (13), University of New South Wales (12), and Vellore Institute of Technology (12) (Fig. 4.2(4)).

4.3.3.4 Country analysis

Among all the countries that have published E-waste-related documents from or related to India, the top 10 are India (961), South Korea (55), China (41), the United States (32), Australia (27), Japan (21), the United Kingdom (19), Vietnam (14), Saudi Arabia (9), and Italy (8) (Fig. 4.2(5)). This means among the top 10, 40% are from Asia. The top five countries to host corresponding authors who have published E-waste-related documents from or related to India are India (526), South Korea (15), Australia (14), China (13), and Japan (5). Among the publications from India, single-country publications are 470 (89.35%) and multicountry publications are 56 (10.64%). Among the countries that published E-waste-related documents, the top 10 most cited ($\geq$50 citations) countries other than India (TC 6893), as per TC count, are—Australia (383), South Korea (332), China (249), Japan (229), Turkey (226), the United Kingdom (141), Switzerland (121), Italy (81), and the United States (65) (Fig. 4.2(6)). However, as per average article citations, the top five are Switzerland (121), Italy (81), Turkey (75.3), Japan (45.8), the United Kingdom (28.2), Australia (27.4), South Korea (22.1), China (19.2), Denmark (18), and the United States (16.2) (Fig. 4.2(7)).

4.3.3.5 Document analysis

When the most globally cited documents are considered, the top 10 are Ghosh B, 2015, *J Clean Prod* (255, TC per year, TCpy—36.42, normalized TC, nTC—8.75); Akcil A, 2015, *Waste Manage* (200, TCpy—28.57, nTC—6.87); Lee JC, 2012, *Waste Manage* (190, TCpy—19, nTC—4.43); Babu BR, 2007, *Waste Manage Res* (168, TCpy—11.2, nTC—2.56); Pant D, 2012, *Waste Manage* (166, TCpy—16.6, nTC—3.87); Balaram V, 2019, *Geosci Front* (142, TCpy—47.33, nTC—14.78); Prakash C, 2015, *J Manuf Syst* (142, TCpy—20.28, nTC—4.87); Ha NN, 2009, *Chemosphere* (133, TCpy—10.23, nTC—3.63); Rahman S, 2012, *Int J Prod Econ* (125, TCpy—12.5, nTC—2.91); and Das A, 2009, *Resour Conserv Recy* (122, TCpy—9.38, nTC—3.33) (Fig. 4.2(8)). The interrelated nature of authors, Keywords Plus, and affiliations regarding E-waste research can be seen by a three-field plot (Fig. 4.2(9)).

4.3.4 Academic cooperation

In the collaboration network of countries, 17 countries are in cluster 1 with India (Fig. 4.3(1)). Among them, the top five are South Korea, Australia, France, Singapore, and Switzerland. Similarly, three countries are in cluster 2 (Vietnam, Saudi Arabia, and Denmark), and 10 countries are in cluster 3. The top three of this third cluster are the United States, China, and Japan. Most of the countries with higher levels of intercountry collaboration are found in these three clusters. In the collaboration network of authors, there are 13 clusters. The number of authors in the respective cluster is 4 (C1), 7 (C2), 2 (C3), 2 (C4), 2 (C5), 2 (C6), 3 (C7), 3 (C8), 2 (C9), 10 (C10), 2 (C11), 2 (C12), and 2 (C13). The top authors in clusters 1 and 2 are Mohanty S, Nayak SK, Suresh SS (C1), and Kumar S, Kumar A, Mukherjee PS (C2). In the collaboration network of institutions (i.e., affiliations), there are seven clusters. Among them except for clusters 1 and 2, every cluster contains two institutes. Three institutes in clusters 1 and 2 are NIT, BARC, University of Southern Denmark (C1), and University of New South Wales, Institute of Minerals and Materials Technology, CSIR IMMT (C2). Hence, these two clusters (1 and 2) contain most of the institutions with intercountry collaboration.

Bibliographic coupling is a similarity measure that establishes a similarity link between documents using citation analysis. If two works in their bibliographies refer to the same third work, it is known as bibliographic coupling. It is a sign that the two works are likely to be about the same thing. When two documents share one or more citations, they are said to be bibliographically connected. The more citations to other texts that two documents share, the stronger their coupling strength. Similarly, two authors are bibliographically coupled if their cumulative reference lists both contain a reference to a common document, and their coupling strength grows with the number of citations to other documents they share. The bibliographic coupling of E-waste for different aspects—authors, countries, documents, organizations, and sources—(Fig. 4.3(2)−(3)) has been composed.

4.3.5 Keyword analysis and research hotspots

Through the analysis of Keywords Plus, the top five most frequently used words as per their occurrences (>50) are "electronic waste" (73), "recovery" (69), "printed-circuit boards" (59), "management" (55), and "copper" (54) (Fig. 4.3(4)). From the authors' keywords analysis, the top five are "E-waste" (170), "recycling" (59), "electronic waste" (48), "India" (24), and "weee" (24). From the title analysis, the top five are "waste" (202), "E-waste" (178), "electronic" (111), "India" (78), and "management" (78). Lastly, by the analysis of the content of abstracts, the top five are "E-waste" (1179), "waste" (894), "electronic" (563), "recycling" (479), and "metals" (400).

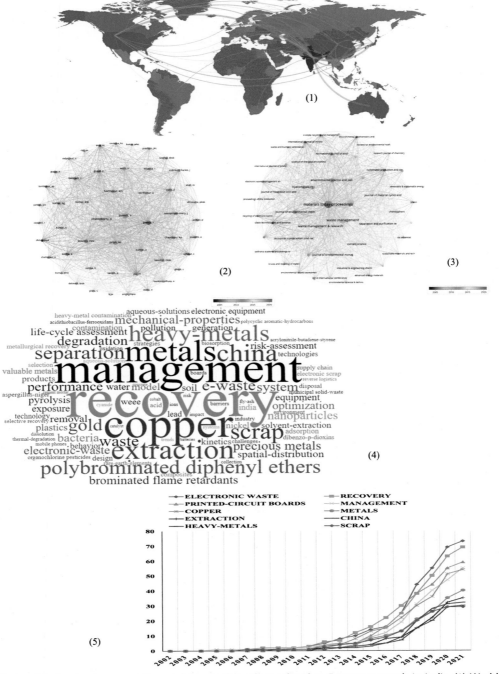

Figure 4.3 Academic cooperation and keywords analysis related to E-waste research in India. (1) World map of collaboration of publications on E-waste in India. (2) Bibliographic coupling network of authors. (3) Bibliographic coupling network of sources. (4) Word Cloud of most frequent keywords by Keywords Plus. (5) Word dynamics (cumulative) by Keywords Plus.

When delving deeper into the temporal trends through cumulative word dynamics, it can be seen that for Keywords Plus, "degradation" was the highest used from 2003 to 2011, but it was exceeded by "recovery" from 2013 to 2017, and then again lost its position to "electronic waste" from 2018 onward. At present (2021), the consecutive four most used keywords are "recovery," "printed circuit boards," "management," and "copper" (Fig. 4.3(5)). Now, for the keywords used by authors, "E-waste" has always maintained the most used position from 2007 to 2021. Likewise, "recycling," "electronic waste," and "weee" has maintained the following positions from the same time. However, from 2018, the rapid growth of "heavy metals" can be seen. From the analysis of titles, "waste" was most used from 2002 to 2010, but was briefly surpassed by "E-waste" (2011–14) and then again assumed the top-used position from 2015 onward. Since 2017, "recovery" has been increasing continuously. From the analysis of the content of abstracts, "waste" has always maintained the most used position from 1993 to 2008, then exceeded by "E-waste" from 2009 onward in the top used position. Likewise, "recycling" was the third most used from 2011 to 2016 but was replaced by "electronic" from 2017 onward.

The collective interconnection of terms based on their matched presence within a specific unit of text is known as a cooccurrence network. Networks are created by connecting pairs of phrases based on a set of cooccurrence criteria. Cooccurrence networks are commonly used to visualize probable linkages between individuals, organizations, concepts, and other entities in a text. With the advent of electronically stored text that is text mining compliant, the development and display of cooccurrence networks have become possible. From the analysis of the cooccurrence network of Keywords Plus, as per betweenness, it can be seen that there are four clusters with the highest cowords of each cluster—"electronic waste" (230.28, C1), "printed circuit boards" (166.36, C2), "polybrominated diphenyl ethers" (32.77, C4), and "nanoparticles" (2067, C3). From the content of abstracts, they are "waste" (1.78, C2), "rights" (0.59, C3), and "metals" (0.18, C1). For authors' keywords, they are "E-waste" (227.23, C2), "weee" (93.44, C4), "electronic waste" (56.15, C3), and "coarse aggregate" (11.24, C1). Through the analysis of titles, they are "waste" (323.32, C3), "E-waste" (210.1, C1), and "recovery" (24.82, C2).

4.3.6 Future research focus analysis

In recent years, as per Keywords Plus, the top trending topics of each year with the highest frequency are "plastics" (12), "dibenzo-p-dioxins" (8), "electronic scrap" (8), "acrylonitrile-butadiene-styrene" and "fly-ash" (6) in 2016; then "degradation" (17), "model" (13), "exposure" (12), "kinetics," and "products" (11, both). In 2018 they were "electronic waste" (73), "recovery" (69), "printed-circuit boards" (59), "management" (55), and "copper" (54). In 2019 these were "extraction" (35), "performance" (17), "nanoparticles" (15), "removal" (13), and "aqueous-solutions" (10). Lastly, in

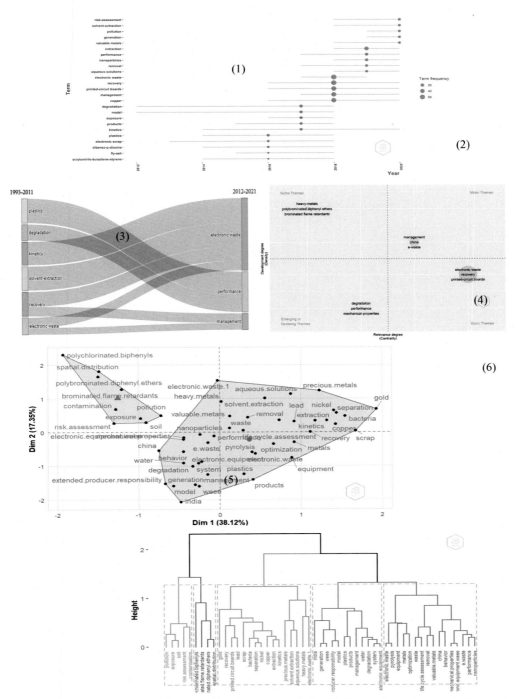

Figure 4.4 Cooccurrence and future research focus analysis. (1) Thematic evolution of Keywords Plus (time slice 2011). (2) Trending topics of Keywords Plus. (3) Thematic map of Keywords Plus. (4) Factorial map of Keywords Plus via Multiple correspondence analysis (MCA). (5) Topic dendrogram of Keywords Plus (five clusters).

2020, they were "risk-assessment" (12), "generation," "pollution" and "solvent-extraction" (11), and "valuable metals" (10) (Fig. 4.4(1)). Through the analysis of the content of the abstract, the trending topics are "green" (113, 2016), "paper" (2013, 2017), "E-waste" (1179, 2018), "concrete" (196, 2019), and "aggregate" (73, 2020). From the analysis of the authors' keywords, they are "reverse logistics" (f = 9, 2016), "weee" (24, 2017), "E-waste" (170, 2018), "leaching" (19, 2019), "compressive strength" (12, 2020), and "circular economy" (11, 2021). Lastly, as per the analysis of titles, they were "metal" (33, 2016), "recycling" (69, 2017), "waste" (202, 2018), "sustainable" (32, 2019), "concrete" (29, 2020), and "circular" (12, 2021).

From the thematic network analysis, the themes that arise from Keywords Plus with the number of keywords associated are "electronic waste" (74), "heavy-metals" (46), "management" (65), and "degradation" (51). Likewise, for authors' keywords, there are nine keywords "weee" (24), "compressive strength" (13), "E-waste" (29), "metal extraction" (9), "environment" (7), "India" (13), "electronic waste" (30), "composite" (3), and "green computing" (5). For titles, there are six keywords "E-waste" (65), "recovery" (60), "review" (19), "mobile" (28), "study" (53), and "waste" (25). Lastly, from the analysis of the content of abstracts, there are three keyword clusters "waste" (119), "rights" (95), and "study" (36).

Thematic evolution refers to a collection of themes that have evolved over time and across subphases. The temporal trend (1993–2021) has been divided into two time-slices: (1) 1993–2011 (18 years) and (2) 2012–21 (9 years). Since dense research packing started effectively in 2004, the first phase becomes 7 years duration, which is almost equivalent to the following phase. It can be seen the themes that used to dominate the E-waste research field (1993–2011) were "degradation," "electronic waste," "kinetics," "plastics," "recovery," "solvent," "extraction" (from Keywords Plus) (Fig. 4.4(2)); "copper," "developing countries," "E-waste," "edta," "electronic waste," "lead," "recycling" (from authors' keywords); "analysis," "electronic," "found," "reserved," "waste" (from abstracts); "application," "based," "E-waste," "effect," "green," "metal," "recovery," "solution," "waste" (from titles). However, the currently trending themes (2012–21) that lead this domain are "performance," "electronic waste," "management" (from Keywords Plus); "E-waste," "E-waste management," "environment," "weee," "electronic waste" (from authors' keywords); "metals," "waste," "rights" (from abstracts); "review," "study," "waste," "circuit," "E-waste" (from titles).

From the thematic map, the basic themes that arise are E-waste, recovery, printed circuit boards (from Keywords Plus); weee, E-waste management, recycle, E-waste, bioleaching, leaching, India, sustainability, and circular economy (from authors' keywords). Niche themes that arise are heavy metals, polybrominated diphenyl ethers, brominated flame retardants (from Keywords Plus); composite, gra, topsis, green computing, cloud computing, carbon footprint, metal extraction, waste printed circuit board, biological leaching (from authors' keywords). Motor themes are management,

China, E-waste (from Keywords Plus); environment, energy (from authors' keywords); study, concrete, based, E-waste, India, management, recovery, circuit, printed (from titles); waste, E-waste, electronic (from abstracts). Emerging or declining themes are degradation, performance, mechanical properties (from Keywords Plus); compressive strength, flexural strength, pollution (from authors' keywords); mobile, model, critical, waste, electronic, development, review, green, energy (from titles) (Fig. 4.4(3)).

Multiple correspondence analysis (MCA) is a data analysis technique used to perceive and epitomize underlying structures in nominal categorical data. From the conceptual structure map of Keywords Plus generated using MCA, two clusters emerge, in two dimensions (38.12% and 17.35%) (Fig. 4.4(4)). Likewise, two clusters in two dimensions (34.43% and 12.51%) for authors' keywords; two clusters in two dimensions (51.89% and 16.84%) for titles; two clusters in two dimensions (38.96% and 27.65%) for the content of abstracts. These can also be visualized by topic dendrograms (Fig. 4.4(5)).

4.4 E-waste and sustainable development goals

There is still the issue of integrating the E-waste research domain into specific sustainable development goals (SDGs) by the UN. E-waste and its management are directly connected to a minimum of six of the SDGs, viz. SDG 3 (Good health and Well-being), 6 (Clean Water and Sanitation), 8 (Decent Work and Economic Growth), 11 (Sustainable Cities and Communities), 12 (Responsible Consumption and Production), and 14 (Life Below Water).

Target 3.9 aims to substantially reduce the number of deaths and illnesses from hazardous chemicals and air, water, and soil pollution and contamination. E-waste comprises several dangerous components that, if improperly deconstructed and processed, can endanger human health by contaminating water, land, and air. Environmentally sound actions should be considered when removing hazardous compounds during the design and production of EEE, as well as when dismantling and processing E-waste. Many demolition activities are currently carried out in some locations using basic and crude methods that endanger human health. Furans and dioxins are released during the process of open cable burning to gain access to copper, for example.

Target 6.1 aims to achieve universal and equitable access to safe and affordable drinking water for all. Target 6.3 aims to improve water quality by reducing pollution, eliminating dumping, and minimizing the release of hazardous chemicals and materials. Heavy metals from E-waste, such as mercury, lithium, lead, and barium, leach even further into the earth after soil contamination, eventually reaching groundwater (Rautela et al., 2021). These heavy metals eventually find their way into ponds, streams, rivers, and lakes after reaching groundwater. Even if they are miles distant from a recycling site, these channels cause acidification and toxification in the water, which is harmful to animals, plants, and communities. It becomes difficult to find safe drinking water.

Target 8.3 aims to promote development-oriented policies that support productive activities, decent job creation, entrepreneurship, creativity, and innovation, and encourage the formalization and growth of micro-, small-, and medium-sized enterprises. The informal sector, both unorganized and organized in different nations, collects and processes a considerable amount of E-waste in developing economies. These occupations are not decent, and formalization of the sector is essential to give these people rights as well as to ensure ecologically sound E-waste management. Target 8.8 aims to the protection of labor rights and promotes safe and secure working environments for all workers, including migrant workers, particularly women migrants, and those in precarious employment. First and foremost, formalization will necessitate state recognition and the incorporation of these workers into a waste management system. Labor rights are more likely to be maintained if this is accomplished. E-waste worker groups have been formed in some circumstances as a result of worker organization, collectively, and social solidarity economics. In certain circumstances, this has lessened the precariousness of these people's jobs.

By focusing particularly on air quality and municipal and other waste management, Target 11.6 seeks to lessen the negative per capita environmental impact of cities. Cities are home to more than half of the world's population, and they utilize 75% of the world's natural resources. Rapid urbanization is causing environmental and human health problems to become more concentrated around the world. Unsound E-waste management in cities is a problem that needs to be addressed, as there are now pressing issues such as poor collection rates, E-waste disposal in normal household bins with limited mandatory separate collections, and open garbage burning and dumping. A shift to smart infrastructure and the use of ICTs to connect communities and improve waste collection efficiency is happening in several cities.

Target 12.4 aims to achieve sustainable management of all waste and chemicals throughout their life cycles following accepted international frameworks and significantly reduce their release into the air, water, and soil to reduce their negative effects on human health and the environment. Open dumping or the use of additional chemical processes such as acid baths and amalgamation to separate valuable components in E-waste is now the most popular E-waste management strategy in developing economies. There is minimal attention paid to eco-design throughout the production of EEE, implying a lack of lifecycle thinking. As a result, many EEE still contains toxic substances like mercury or lead, which reduce product durability. There are nonhazardous substitutes (alternatives) for some of these compounds. However, this does not apply to all compounds at present time. Target 12.5 aims to reduce waste generation through prevention, reduction, repair, recycling, and reuse. It is feasible to avoid trash formation at the EoL by designing EEE with parts that are easily separable, made of recycled metals, and are not dangerous. Manufacturers must move away from deliberate and perceived obsolescence design, and customers must demand more durable products. Manufacturers should also be pushed to create products that are easily repairable and allow for the replacement of broken components. Furthermore, if manufacturers were required to satisfy EPR goals, recycling and reuse would be easier to achieve. EEE is currently not created with circularity in mind, but rather with linearity, which fails to enable prevention, reduction, repair, recycling, and reuse in favor of a "throw-away civilization."

Target 14.1 aims to prevent and significantly reduce marine pollution of all kinds, in particular from land-based activities. Target 14.2 aims to sustainably manage and protect marine and coastal ecosystems to avoid significant adverse impacts. Acidification has the potential to kill marine and freshwater creatures, as well as disrupt biodiversity and destroy ecosystems. If there is acidity in water, it can harm ecosystems to the point where recovery is difficult, if not impossible. When it comes to fauna, mercury destroys plant structure, reduces chlorophyll, and can even cause mortality in certain plants. Mercury has a destructive effect on marine life, reducing fish reproductive capacity, impairing their growth and development, causing behavioral abnormalities, altering their blood and oxygen exchange, damaging sensory processes, and even killing them. Contaminated plankton leads to contaminated fish, contaminated birds, and even, in the future, contaminated humans.

4.5 Conclusion

This work has shown how the E-waste research field has evolved from its emergence in 28 years (1993–2021) in India. Much has been developed and learned about the management of E-waste over the last three decades. A general review of the body of literature on E-waste reveals that rapid progress in this area has primarily occurred since 2004.

Simultaneously, it must be acknowledged that, given a large amount of E-waste generated each year and the complexity of E-waste, this research topic still has a long way to go before it can be completed. With the vast amount of scientific literature available, the use of bibliometric mapping to investigate the relationships between the occurrences of the most prominent terms in the E-waste research domain proved to be a useful tool. The developing world is starting to catch up on concerns. Once mentioned by the world's most powerful countries, they are no longer seeing electronic garbage in the realm of E-waste trash but as a source of resources. Understanding the roots of the E-waste problem necessitates a comprehensive effort, hence ensuring that interdisciplinary research techniques on the matter is equally critical. The emerging economies' economic trajectory is expected to continue soon. As a result, it is critical to address these countries' environmental problems (along with substantial initiatives toward responsible and sustainable E-waste disposal) at the same time as they prosper socio-economically. Lastly, it can be seen that E-waste and its management have been connected to all the three pillars of sustainable development (environment, economy, and society). Hence, to realize sustainable development in this domain, future research needs to come up with intersecting assessment methods, preferably by the ensemble of all three aspects.

CRediT authorship contribution statement

Conceptualization: A.R.; data curation: A.R.; methodology: A.R.; investigation: A.R.; formal analysis: A.R.; software: A.R. and A.B.; interpretation: A.R.; resources: A.R.; project administration: A.R.; writing—original draft preparation: A.R.; writing—review and editing: A.R.; visualization: A.R.; supervision: A.R.

Data availability statement

The authors confirm that the data, used in this study, is available in a public database (Web of Science).

Declaration of competing interests

None.

Funding

This research did not receive any specific grant from funding agencies in the public, commercial, or not-for-profit sectors.

References

Andeobu, L., Wibowo, S., & Grandhi, S. (2021a). A systematic review of E-waste generation and environmental management of Asia Pacific countries. *International Journal of Environmental Research and Public Health, 18*(17), 9051. Available from https://doi.org/10.3390/ijerph18179051.

Andeobu, L., Wibowo, S., & Grandhi, S. (2021b). An assessment of E-waste generation and environmental management of selected countries in Africa, Europe and North America: A systematic review. *Science of the Total Environment, 792,* 148078. Available from https://doi.org/10.1016/j.scitotenv.2021.148078.

Andrade, D. F., Romanelli, J. P., & Pereira-Filho, E. R. (2019). Past and emerging topics related to electronic waste management: Top countries, trends, and perspectives. *Environmental Science and Pollution Research, 26,* 17135—17151. Available from https://doi.org/10.1007/s11356-019-05089-y.

Aria, M., & Cuccurullo, C. (2017). Bibliometrix: An R-tool for comprehensive science mapping analysis. *Journal of Informetrics, 11*(4), 959—975. Available from https://doi.org/10.1016/j.joi.2017.08.007.

Arya, S., & Kumar, S. (2020a). E-waste in India at a glance: Current trends, regulations, challenges and management strategies. *Journal of Cleaner Production*, *271*, 122707. Available from https://doi.org/10.1016/j.jclepro.2020.122707.

Arya, S., & Kumar, S. (2020b). Bioleaching: urban mining option to curb the menace of E-waste challenge. *Bioengineered*, *11*(1), 640−660. Available from https://doi.org/10.1080/21655979.2020.1775988.

Arya, S., Patel, A., & Kumar, S. (2021). Urban mining of obsolete computers by manual dismantling and waste printed circuit boards by chemical leaching and toxicity assessment of its waste residues. *Environmental Pollution*, *283*, 117033. Available from https://doi.org/10.1016/j.envpol.2021.117033.

Arya, S., Rautela, R., Chavan, D., & Kumar, S. (2021). Evaluation of soil contamination due to crude E-waste recycling activities in the capital city of India. *Process Safety and Environmental Protection*, *152*, 641−653. Available from https://doi.org/10.1016/j.psep.2021.07.001.

ASSOCHAM-NEC (2018). *Electricals and electronics manufacturing in India 2018*. Available from: <https://in.nec.com/en_IN/pdf/ElectricalsandElectronicsManufacturinginIndia2018.pdf> Accessed 15.07.22.

Borthakur, A. (2020). Policy approaches on E-waste in the emerging economies: A review of the existing governance with special reference to India and South Africa. *Journal of Cleaner Production*, *252*, 119885. Available from https://doi.org/10.1016/j.jclepro.2019.119885.

Borthakur, A., & Govind, M. (2017). How well are we managing E-waste in India: Evidences from the city of Bangalore. *Energy, Ecology and Environment*, *2*, 225−235. Available from https://doi.org/10.1007/s40974-017-0060-0.

Borthakur, A., & Singh, P. (2020). Mapping the emergence of research activities on E-waste: A scientometric analysis and an in-depth review. In M. N. V. Prasad, M. Vithanage, & A. Borthakur (Eds.), *Handbook of electronic waste management* (pp. 191−206). Elsevier, https://doi.org/10.1016/B978-0-12-817030-4.00017-6 (ISBN: 978-0-12-817030-4).

Chen, X., Liu, M., Xie, B., Chen, L., & Wei, J. (2021). Characterization of top 100 researches on E-waste based on bibliometric analysis. *Environmental Science and Pollution Research*, *28*(43), 61568−61580. Available from https://doi.org/10.1007/s11356-021-15147-z.

de Albuquerque, C. A., Mello, C. H. P., de Freitas Gomes, J. H., & Dos Santos, V. C. (2021). Bibliometric analysis of studies involving E-waste: A critical review. *Environmental Science and Pollution Research*, *28*(35), 47773−47784. Available from https://doi.org/10.1007/s11356-021-15420-1.

Dutta, D., Arya, S., Kumar, S., & Lichtfouse, E. (2021). Electronic waste pollution and the COVID-19 pandemic. *Environmental Chemistry Letters*, 1−4. Available from https://doi.org/10.1007/s10311-021-01286-9.

European Union. (2003). *Waste Electrical and Electronic Equipment Directive (2002/96/EC)*. Available from: <https://www.ciwm.co.uk/ciwm/knowledge/waste-electrical-and-electronic-equipment-directive-2002-96-ec.aspx> Accessed 15.07.22.

Forti, V., Baldé, C.P., Kuehr, R., & Bel, G. *The global E-waste monitor 2020: Quantities, flows and the circular economy potential*. Bonn/Geneva/Rotterdam: UNU/UNITAR − cohosted SCYCLE Programme, ITU & ISWA. Available from: <https://collections.unu.edu/view/UNU:7737> Accessed 15.07.22.

Gao, Y., Ge, L., Shi, S., Sun, Y., Liu, M., Wang, B., Shang, Y., Wu, J., & Tian, J. (2019). Global trends and future prospects of E-waste research: A bibliometric analysis. *Environmental Science and Pollution Research*, *26*, 17809−17820. Available from https://doi.org/10.1007/s11356-019-05071-8.

Ismail, H., & Hanafiah, M. M. (2020). A review of sustainable E-waste generation and management: Present and future perspectives. *Journal of Environmental Management*, *264*, 110495. Available from https://doi.org/10.1016/j.jenvman.2020.110495.

Li, N., Han, R., & Lu, X. (2018). Bibliometric analysis of research trends on solid waste reuse and recycling during 1992−2016. *Resources, Conservation & Recycling*, *130*, 109−117. Available from https://doi.org/10.1016/j.resconrec.2017.11.008.

MoEF (2008). *Guidelines for environmentally sound management of E-waste* (as approved Vide Ministry of Environment and Forests (MoEF)) (Letter No. 23-23/2007-HSMD) Dated March 12, 2008. Available from: <https://kspcb.karnataka.gov.in/sites/default/files/inline-files/NewItem_109_Latest_19_E_Waste_GuideLines.pdf> Accessed 15.07.22.

MoEF (2011). *E-waste (management and handling) rules*. Dated 12th May 2011. Available from: <https://www.meity.gov.in/writereaddata/files/1035e_eng.pdf> Accessed 15.07.22.

MoEF (2016). *E-waste (management) rules*. Dated 23rd March 2016. Available from: <https://greene.gov.in/wp-content/uploads/2018/01/EWM-Rules-2016-english-23.03.2016.pdf> Accessed 15.07.22.

Pérez-Belis, V., Bovea, M. D., & Ibáñez-Forés, V. (2014). An in-depth literature review of the waste electrical and electronic equipment context: Trends and evolution. *Waste Management & Research*, 1—27. Available from https://doi.org/10.1177/0734242X14557382.

Ranjbari, M., Saidani, M., Esfandabadi, Z. S., Peng, W., Lam, S. S., Aghbashlo, M., Quatraro, F., & Tabatabaei, M. (2021). Two decades of research on waste management in the circular economy: Insights from bibliometric, text mining, and content analyses. *Journal of Cleaner Production*, *314*, 128009. Available from https://doi.org/10.1016/j.jclepro.2021.128009.

Rautela, R., Arya, S., Vishwakarma, S., Lee, J., Kim, K.-H., & Kumar, S. (2021). E-waste management and its effects on the environment and human health. *Science of the Total Environment*, *773*, 145623. Available from https://doi.org/10.1016/j.scitotenv.2021.145623.

Singh, S., Trivedi, B., Dasgupta, M. S., & Routroy, S. (2021). A bibliometric analysis of circular economy concept in E-waste research during the period 2008—2020. *Materials Today*. Available from https://doi.org/10.1016/j.matpr.2021.03.525.

Velasco-Muñoz, J. F., Aznar-Sánchez, J. A., Pozas-Ramos, D., & López-Felices, B. (2021). Advances in global research on the sustainable management of waste electrical and electronic equipment. In C. M. Hussain (Ed.), *Environmental management of waste electrical and electronic equipment* (pp. 241—267). Elsevier, https://doi.org/10.1016/B978-0-12-822474-8.00013-1 (ISBN: 978-0-12-822474-8).

Velasco-Muñoz, J. F., Aznar-Sánchez, J. A., Manzano-Archilla, M. J., & López-Felices, B. (2021). Waste electrical and electronic equipment and environment: Context, implications, and trends. In C. M. Hussain (Ed.), *Environmental management of waste electrical and electronic equipment* (pp. 23—48). Elsevier, https://doi.org/10.1016/B978-0-12-822474-8.00002-7 (ISBN: 978-0-12-822474-8).

Vishwakarma, S., Kumar, V., Arya, S., Tembhare, M., Rahul., Dutta, D., & Kumar, S. (2022). E-waste in Information and Communication Technology Sector: Existing scenario, management schemes and initiatives. *Environmental Technology and Innovation*, 102797. Available from https://doi.org/10.1016/j.eti.2022.102797.

Vrabel, M. (2015). Preferred reporting items for systematic reviews and meta-analyses. *Oncology Nursing Forum*, *42*(5), 552—554. Available from https://doi.org/10.1188/15.ONF.552-554.

Zhang, L., Geng, Y., Zhong, Y., Dong, H., & Liu, Z. (2019). A bibliometric analysis on waste electrical and electronic equipment research. *Environmental Science and Pollution Research*, *26*, 21098—21108. Available from https://doi.org/10.1007/s11356-019-05409-2.

CHAPTER 5

Formal and informal E-waste recycling methods for lithium-ion batteries: advantages and disadvantages

Md. Rakibul Qadir[1,2,3], Nawshad Haque[4], Miao Chen[2,3], Warren J. Bruckard[3] and Mark I. Pownceby[3]
[1]PP&PDC, Bangladesh Council of Scientific and Industrial Research, Dhanmondi, Dhaka, Bangladesh
[2]School of Applied Sciences, RMIT University, Melbourne, VIC, Australia
[3]CSIRO Mineral Resources, Clayton South, VIC, Australia
[4]CSIRO Energy, Clayton South, VIC, Australia

5.1 Introduction

Recycling, or particularly e-waste recycling, serves multiple interests to the economy and the ecology. Electronic appliances as well as batteries consume a considerable volume of valuable materials such as transition metals, precious metals, and rare earths in their manufacture as well as significant amounts of steel, aluminum, plastics, and glass. Besides the plastics component, the metals and glass–making components are originally claimed from minerals that have been rigorously treated via ore processing. As a result, e-wastes are often referred to as "urban mines," as they often contain the element of interest in much greater concentrations than in the original mined ore itself. Mineral processing industries thus understand the financial attractiveness of recycling materials for their contained value quite well. Besides recovering embedded value, the other end of the spectrum is the potential environmental advantages offered by recycling of e-wastes. In the case of batteries, either normal alkaline or more advanced Li-ion types, recycling reduces potential electrical and explosion hazards, and primary and secondary pollution of air–water–soil ecosystems. In addition, recycling reduces the need to mine and process virgin raw material and greatly reduces the significant energy requirements related to the manufacture of new batteries. Global efforts toward developing recycling technologies for e-waste have already demonstrated the considerable commercial benefits and environmental gains.

Lithium-ion batteries (LIBs) are secondary or rechargeable batteries that are created with chemical compounds that can be charged, discharged, and recharged repeatedly over many cycles. LIBs were introduced commercially by Sony in 1991 (Qiao & Wei, 2012) and with the fast-growing sector of portable electronic devices, the growth in

Global E-waste Management Strategies and Future Implications
DOI: https://doi.org/10.1016/B978-0-323-99919-9.00017-9

LIBs has increased significantly. LIB production is expected to grow from 95.3 GWh in 2020 to 410.5 GWh in 2024 (Globaldata, 2020). This growth is largely the result of "giga–factories" that have been established to assist the growing electric vehicle (EV) market by technological giants including Tesla, Panasonic, and BYD Auto which are anticipated to accelerate LIB production as well as the demand for constituent metals. End–of–life (EoL) LIB recycling and waste management hence require intensive attention to reduce primary metal mining demand and to reduce potentially hazardous materials being diverted to landfill.

LIBs contain a significant number of metallic elements in various forms (see Fig. 5.1). Steel (often coated with Ni) is commonly used as the outer casing for most button, pouch, and cylindrical types of LIBs. The connectors, wires, and solder may include Al, Zn, Pb, Sn, and Cu. Larger batteries may have associated charging control printed circuit boards (PCB), thermocouples, and wirings introducing a multitude of other elements. The core construction of the cell itself uses metallic Cu and Al of electronic grade purity, while the solvent contains Li. The charge storage material, also known as the active cathode material, usually contains Li and Co in the form of lithium cobalt oxide, LCO ($LiCoO_2$), while other varieties of the active cathode material include, but are not limited to, LMO ($LiMnO_4$), LNO ($LiNiO_2$), NMC

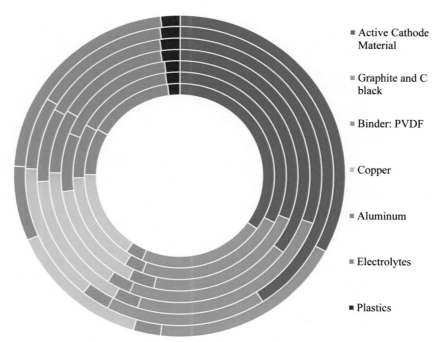

Figure 5.1 Constituents of different lithium-ion battery types (Dai et al., 2019): from center to outward: NMC (111), NMC (622), NMC (811), LCO, NCA, LMO, and LFP.

$(LiMn_xNi_yCo_zO_2)$, NCA $(LiNi_xCo_yAl_zO_2)$, LFP $(LiFePO_4)$, and LTO (Li_2TiO_3). As a result, an industrial raw material of waste LIBs for recycling may contain a mixture of multiple metallic elements of commercial value.

Recycling practices of LIBs greatly depend on the socio-economic conditions, government legislation, waste management supply chains, mass awareness, capital, and the collection infrastructure in a country. While a handful of developed countries have been commercially recycling LIBs for quite some time, many others have redirected their waste streams to new destinations or simply disposed of them in landfill. Nations such as the United Kingdom, Germany, Switzerland, Belgium, France, Netherlands, and Finland in Europe; Japan, China, Korea, and Singapore in Asia; plus, the United States and Australia, already have commercial e-waste recycling industries, many of which have LIB recycling included in the process. Such formal initiatives largely depend on the constant supply of a sorted e-waste feed for high productivity, with recyclers often working in tandem with electronic and automobile companies or original equipment manufacturers (OEM). In contrast, the populous regions of South America, Africa, and much of Asia are less structured in their e-waste recycling activities of unsorted and various size and shaped LIB units, with activities primarily depending on reuse, repurpose, and dumping with municipal waste.

The purpose of this chapter is to provide an overview of LIB e-waste recycling in both the "formal" and "informal" sectors, with an emphasis on recycling activities in developing countries such as India and Bangladesh. In particular, the barriers and challenges existing for these largely informal sectors are described and a few recommendations are made.

5.2 Formal recycling sectors and methods

The initial challenge for waste LIB recycling comes in the form of waste collection practices and infrastructure given the wide variety in sizes and shapes. For example, button cells (watches, headphones, and wearable devices), small pouch, and cylindrical cells (phones, camera, power banks, and peripherals) are particularly difficult to segregate from municipal waste. In comparison, larger LIB units like laptop batteries and EV power packs are mostly recollected after use through recognized channels and also by the OEM. Most commercial recycling facilities have therefore grown in countries where existing systematic electronic waste collection trends are already in practice.

The globally practiced philosophy of the 4Rs—reduce, reuse, recycle, and repurpose (sometimes described as "recover") may also be applied to e-wastes, or particularly to LIBs. Most applications of power storage devices require a rated minimum power capacity, below which the battery unit is labeled "non-functional." Alternatively, other applications could be suited with lesser power requirements giving it a second life or repurpose. EoL LIBs are diagnosed after formal collection for their potential second life depending on the type of battery, condition, and intended future applications. Repurposing processes therefore may include a certain amount of professional disassembly, repair, and reassembly of individual cells. On the contrary, remanufacturing involves revitalizing the consumed component of the battery unit to again suit the original application. Recycling starts with manual or automated dismantling and additional discharging which is often met by submerging the LIBs into a brine solution.

Mechanical comminution may be achieved by a number of methods. These are often associated with operations conducted under an inert atmosphere, submergence in solutions and/or cryogenic operations that minimize effluent emissions, charge aggravation, and explosion hazards. Different modes of comminution are practiced, and options include knife milling, shearing, and shredding, depending on the required level of fineness and liberation. The complex intercalation in the LIB construction between Cu, Al, polymers, and actives usually dictates a high level of fineness (commonly less than 2 mm) required for suitable liberation of the components.

Direct recycling on the other hand utilizes high-pressure supercritical fluids such as liquid CO_2 to crack open the cells for separation of the phases. Separation can be difficult as the active cathode mixture ($LiCoO_2/LiCo_xNi_yMn_zO_2$ and graphite) is typically bonded to the Al foil and polymeric separators with industrial binders like polyvinylidene difluoride (PVDF), which hinders the complete liberation of the constituent phases. Approaches have been made with N-methyl-2-pyrrolidone (NMP) and other solvents to remove (dissolve) the binder and liberate the active mixture, commonly known as "black mass." A roasting or pyrolyzing treatment ($<700°C$) may also be used to remove organics, volatiles, and polymeric portions. In certain cases, temperatures can be as high as $1000°C$ so that graphite is also removed. The supercritical fluids treatment has the ability to achieve a degree of separation of the active components so that the material can then be reduced in size and liberation improved.

Sorting as well as separation (and collection) of low value products like steel, plastic, Cu, and Al foils comes next, by means of a range of techniques which include magnetic separation (for Fe removal), eddy current or electrostatic separation (Cu and Al removal), gravity separation (shaker table), sieve shaking, and froth floatation. The remaining black mass contains mainly graphite, Li metal salts, and low levels of impurities such as Cu, Al, and Fe.

The next step in the formal recycling process is the "deep recovery" of the LIBs components/elements and this process may have several substeps and importantly must be robust and able to accommodate differences in feedstock make, origin, type, etc. Hydrometallurgy and pyrometallurgy are the two most popular processing options for deep recovery, each with certain advantages and disadvantages as noted in Fig. 5.2. Hydrometallurgical options involve the use of reagents (lixiviants) to either selectively extract elements or collectively separate them from unwanted nonmetallics. Selective

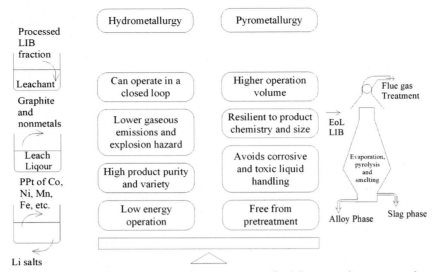

Figure 5.2 Advantages of the two deep recovery processes for lithium-ion battery recycling.

dissolution is not always readily achieved and so pretreatments are often needed to enhance selectivity and to reduce the number of subsequent steps in the process. These steps need to operate under moderate temperatures and produce high purity outputs. Some low value by products may also be recovered in the pretreatment phase. Inevitably, hydrometallurgical processing requires the efficient handling of large volumes of acids, bases, other reagents, and additives under different operating conditions and is required to deal with recycling loops, product refinement, and any aqueous or gaseous emissions. Pyrometallurgical treatment options on the other hand, are often free from pretreatment phases, and can accommodate more robust feedstocks. The raw materials are smelted under controlled conditions such that the metallic elements (Ni, Co, Fe, Cu, etc.) form an alloy and the nonmetallics deport to a slag phase. The resulting alloys and slags may be sold as is, or further refined by hydrometallurgical means such as electrowinning. The advantage of pyrometallurgical treatment is that it is largely free from size limitations and sensitivity toward raw material composition, but on the flip side it requires high energy consumptions and can produce pollution through gaseous emissions.

Fig. 5.3 summarizes the formal processes and subprocesses commercially applied for LIB recycling, and in the following sections we briefly summarize the hydrometallurgical and pyrometallurgical processes. Note that recently, much focus has also been given to greener technologies like bioleaching, the regeneration of actives, and closed-loop operations/facilities. Though bioleaching may have the lowest environmental

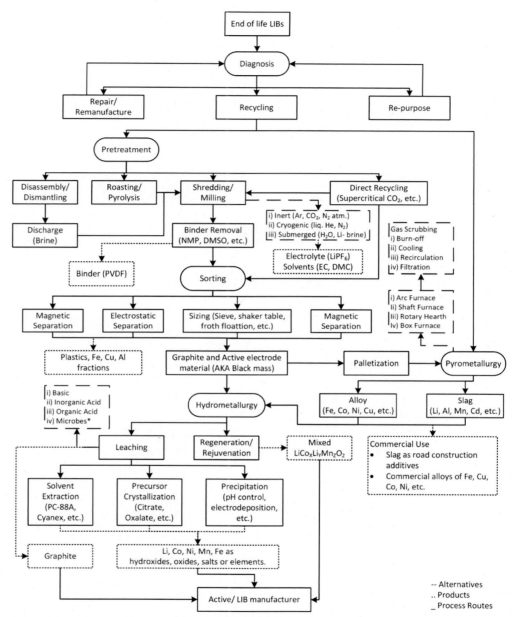

Figure 5.3 Summary of formal recycling trends of lithium-ion batteries.

footprint, it typically has lower yields and finds limited commercialization due to longer residence times, higher maintenance costs, and the selection of appropriate biological organisms.

5.2.1 Hydrometallurgical processes

Fundamentally, hydrometallurgical processes involve the selective leaching of a particular component into the liquid phase using an appropriate lixiviant followed by a solid—liquid separation step. Such chemical processes typically depend on a number of interdependent leaching parameters such as the liquid-to-solid (L/S) ratio, particle size surface area and morphology, concentration of reagents and additives, reaction time and temperature, and degree of agitation. In the case of LIBs several decades of scientific effort has developed promising leaching routes using either organic and inorganic acids, or various bases and solvents to selectively leach and precipitate elements of interest with certain boundary conditions. These systems have been well studied but are always evolving in line with changes in the type of LIB e-waste being treated. Table 5.1 summarizes some of the key hydrometallurgical methods described in the literature, the reagents and conditions used, and the typical elemental recoveries obtained.

1. Acid leaching

Acids, with their high reactivity and oxidizing ability, can react with most metals and many oxides. The phases of interest in the case of LIBs—metallic Al, Cu, steel, and lithium metal salts ($LiCo_xNi_yMn_zO_2$ or $LiFePO_4$) are readily leachable by most organic and inorganic acids under certain conditions. Rigorous pretreatments can separate out most of the Cu, Al, and Fe metals before leaching, but combined processes where these are leached with the metal salts have also been explored. Aluminum metal in particular has the ability to react with bases like NaOH, NH_4OH, and $NaHCO_3$ to form soluble phases. The selection of the acid type depends on the precipitation method, the desired end product, and the emission control of volatiles (namely Cl_2, SO_x, NO_x, etc.). Hydrochloric acid (HCl) has been reported to be the most efficient acid for processing LIBs. Additives like peroxide (H_2O_2) have been noted to increase the leaching efficiency of both organic and inorganic acids and, as a general rule, higher temperatures and more intensive agitation will also improve the leaching kinetics.

After leaching, the graphite and remaining impurities such as plastic fines are separated out by filtration. For higher purity leach feeds (i.e., $LiCoO_2/LiNiO_2/LiMnO_2$ only) the leach solutions can be further subjected to crystallizers to produce oxalates, citrates, and so on. Nickel, Fe and Co all have a close range of pH values for precipitation which can be handled by sorting the battery types (as Co rich or Ni rich), by solvent extraction (SX) or by series of preferential precipitation reactions. Common bases such as NaOH, NH_4OH, and $NaHCO_3$ may be used to control the pH of the leach liquor to precipitate and reclaim Co and Ni as hydroxides which can later be heated to produce pure metal oxides. Electrodeposition has also been applied for higher purity metallic outputs. The Li which remains after the Co/Ni separation can be collected as carbonate or hydroxide, though its recovery is sluggish and may be hindered by any remnant Na content.

Table 5.1 Different hydrometallurgical routes previously examined for lithium-ion battery recycling.

Method	Leaching conditions	Additive	Recovery
OA (Li et al., 2015)	1.5 M succinic acid, 70°C, 2/3 h	1 vol.% H_2O_2 as reducing agent	96% Li, 100% Co
OA (Li et al., 2012)	1.25 M ascorbic acid, 70°C, 1/3 h	H_2O_2 as reducing agent	98.5% Li, 94.8% Co
OA (Li et al., 2012)	0.5–2.0 M citric/malic/aspartic acid, 90°C, 2 h	1–6 vol.% H_2O_2 as reducing agent	100% Li, 90% Co
OA (He et al., 2016)	L–Tartaric acid	4 vol.% H_2O_2 as reducing agent	99% Li, 99% Co
OA (Golmohammadzadeh et al., 2017)	DL–Malic/citric acid	1.25%–2% H_2O_2 as reducing agent	91% Li, 84% Co
OA (Gao et al., 2017)	Formic acid	H_2O_2 as reducing agent	99% Li
OA (Zeng et al., 2015)	1 M oxalate, 95°C, 2.5 h	-	98% Li, 97% Co
OA (Sun & Qiu, 2012)	1 M oxalate, 80°C, 2 h	H_2O_2 as reducing agent	Li/Co >98%
OA, OP (Nayaka, Pai, Santhosh, et al., 2016)	0.5 M glycine + 0.02 M ascorbic acid, 80°C, 6 h	Oxalic acid precursor	95% Co
OA, OP (Nayaka, Pai, Manjanna, et al., 2016)	1 M iminodiacetic + 0.02 M maleic acid + ascorbic acid, 80°C, 6 h	Oxalic acid precursor	99% Li, 91% Co
IP (Guo et al., 2017)	Li effluent	Na_2CO_3, Na_3PO_4 precipitant	74.7% as Li_2CO_3, 92.21% as Li_3PO_4 / 97.2% Li
IA (Zheng et al., 2016)	2.5 M H_2SO_4, L/S = 10/1, 60°C for 4 h	-	
IA (Wang et al., 2009)	4 M HCl ($LiCoO_2$, $LiMn_2O_4$, and $LiCo_{1/3}Ni_{1/3}Mn_{1/3}O_2$), L/S = 20/1, at 80°C, 1 h	-	Li, Co, Mn, Ni >99%
IA (Shin et al., 2005)	H_2SO_4, 75°C, 10 min	15 v % H_2O_2 as reducing agent	-
IA (Joulié et al., 2014)	1–4 M H_2SO_4/HNO_3/HCl, L/S = 20/1 at 20°C–90°C, 2–18 h	NaClO and NaOH as precipitant	>80% Li, 100% Co, 99.99% Ni
IA (He et al., 2017)	1 M H_2SO_4, 40°C, 1 h ($LiNi_{1/3}Co_{1/3}Mn_{1/3}O_2$ type)	1 vol.% H_2O_2 as reducing agent	Co, Li, Mn, and Ni >99.7%
IA (Ferreira et al., 2009)	10 wt.% NaOH (Al), 6 vol.% H_2SO_4 ($LiCoO_2$) at 60°C	1 vol.% H_2O_2 as reducing agent	90% Li, 90% Co
IA (Contestabile et al., 2001)	4 M HCl, 80°C, 1 h	NaOH	-

Process (Reference)	Leaching conditions	Reagents	Recovery
IA (Castillo et al., 2002)	2 M HNO$_3$, 80°C, 2 h	NaOH	-
IA (Bok et al., 2004)	35%HNO$_3$ + 18%H$_2$SO$_4$ + 5%HCl, 24 h, room temperature	-	-
IA (Yang et al., 2017)	H$_3$PO$_4$ (LiFePO$_4$ batteries)	H$_3$PO$_4$	93% Fe, 82.55% Li
IA, SX (Pagnanelli et al., 2016)	H$_2$SO$_4$	H$_2$O$_2$ as reducing agent, DE$_2$HPA, Cyanex 272 with 10% TBP at pH 3.8	-
IA, SX (Virolainen et al., 2017)	H$_2$SO$_4$	Cyanex 272 and PC–88A with TOA/TBP	99.6% Co and Ni
IA, SX (Wang et al., 2016)	3 M H$_2$SO$_4$, at 70°C, L/S = 7:1, for 2.5 h	1.6 M H$_2$O$_2$ as reducing agent, D$_2$EHPA, PC–88A, oxalic acid as precursor	-
IA, SX (Nan et al., 2005)	3 M H$_2$SO$_4$, 70°C, 4 h	Acorga M5640, Cyanex 272	-
IA, SX (Zhang, Yokoyama, Itabashi, Suzuki, et al., 1998)	4 M HCl, 80°C, 1h	0.9 M PC–88A at pH 6.7	-
IA, SX (Swain et al., 2007)	2 M H$_2$SO$_4$, L/S = 10:1 at 75°C, 30 min	5 vol.% H$_2$O$_2$ as reducing agent, 1.5 M Cyanex 272	94% Li, 93% Co
IA, SX (Kang et al., 2010)	2 M H$_2$SO$_4$ at 60°C	6 vol.% H$_2$O$_2$ as reducing agent, 50% saponified 0.4 M Cyanex 272	99% Li, 99% Co
IA, SX, E (Lupi & Pasquali, 2003)	H$_2$SO$_4$	H$_2$O$_2$	-
IA, OP (Meshram et al., 2015)	1 M H$_2$SO$_4$, L/S = 50/1 at 95°C, 4 h	0.075 M NaHSO$_3$ as reducing agent, oxalic acid precursor	96.7% Li, 91.6% Co, 96.4% Ni, 87.9% Mn
IA, OP (Zhu et al., 2012)	2 M H$_2$SO$_4$ at 60°C	2 vol.% H$_2$O$_2$ as reducing agent, (NH$_4$)$_2$C$_2$O$_4$ as precursor	87.5% Li, 96.3% Co
IA, OP (Aktas et al., 2006)	4 M H$_2$SO$_4$, 80°C, 4 h	H$_2$O$_2$ as reducing agent, C$_2$H$_5$OH as precipitant	-
IA, OP (Pinna et al., 2017)	H$_3$PO$_4$, H$_2$O$_2$	Oxalic acid precursor	99% Co, 88% Li

(Continued)

Table 5.1 (Continued)

Method	Leaching conditions	Additive	Recovery
IA, OP (Lee & Rhee, 2003)	1 M HNO_3, 75°C, 1 h	1.7 vol.% H_2O_2 as reducing agent, citric acid precursor	–
IA, OP (Lee & Rhee, 2002)	1 M HNO_3, 75°C, 1 h	1.7 vol.% H_2O_2 as reducing agent, citric acid precursor	–
IA, IP (Li et al., 2017)	0.3 M H_2SO_4, H_2O_2, 60°C, 2 h	Na_3PO_4 as precipitant	95.56% Li
IA, IP (Barik et al., 2017)	1.7 M HCl, 50°C for 1.5 h	NaClO as precipitant	90% Co, 95% Mn
IA, IP, SX (Chen et al., 2011)	5 wt.% NaOH (Al), 4 M H_2SO_4, L/S = 10/1, 85°C for 2 h	10 vol.% H_2O_2 as reducing agent, P507 and oxalate precursor	95% Co, 96% Li
IA and OA (Li et al., 2014)	H_2SO_4/HCl/citric acid, 60°C, 5 h	0.055 M H_2O_2 as reducing agent	100% Li, 96% Co
H (Kim et al., 2004)	Regeneration at 200°C	LiOH	–
H, E (Ra & Han, 2006)	Regeneration at 40°C−100°C	LiOH, KOH	–
Bio (Bahaloo-Horeh & Mousavi, 2017)	–	*Aspergillus niger*	100% Cu, 100% Li, 77% Mn, 75% Al, 64% Co, and 54% Ni
Bio (Mishra et al., 2008)	–	*Acidithiobacillus ferrooxidans*	10% Li, 65% Co
Bio (Horeh et al., 2016)	1% pulp density	*Aspergillus niger*	100% Cu, 95% Li, 70% Mn, 65% Al, 45% Co, and 38% Ni
B (Ku et al., 2016)	NH_3 + $(NH_4)_2SO_3$ + $(NH_4)_2CO_3$, 80°C, 1 h	–	80% Co, 100% Cu 2% Al, 1% Mn, and 25% Ni
B (Chen et al., 2018)	NH_3−$(NH_4)_2SO_4$− $(NH_4)_2SO_3$	–	98% Ni, 81% Co, 92% Mn, and 98% Li
B (Zheng et al., 2017)	NH_3−$(NH_4)_2SO_4$, 200−700 rpm, L/S = 20−100:1, 50°C−80°C for 0−8 h	Na_2SO_3 as reductant	94.8% Co, 88.4% Mn, and 6.34% Li
B, IA (Nayl et al., 2017)	NH_4OH, H_2SO_4	H_2O_2 as reducing agent	<97% of Al, Mn, Ni, Co, and Li, about 65% Cu
SS (Paulino et al., 2008)	Fusion at 500°C for 5 h	$KHSO_4$, KF	–
SS (Fouad et al., 2007)	Al sheets and powder at 800°C−900°C for 2 h	–	γ-$LiAlO_2$ formation

B, Basic leaching; *Bio*, bioleaching; *E*, electrochemical method; *H*, hydrothermal regeneration method; *IA*, inorganic acid; *IP*, inorganic precursor/precipitant; *OA*, organic acid; *OP*, organic precursor; *SS*, solid state; *SX*, solvent extraction.

Contestabile et al. (1999) reported on a laboratory-scale hydrometallurgical route to recycle LIBs and then validated the commercial usability of the process by reconstructing a battery cell with the recycled products. The process included mechanical separation of the batteries and NMP treatment to separate the black powder from metallic Al and Cu current collectors. The black powder was then leached with HCl (4 M, $L/S = 10/1$, 80°C for 1 h) and consecutively treated with 4 M NaOH to recover Co at pH 8 as $Co(OH)_2$. The Co salt was than treated and $LiCoO_2$ reproduced. A functioning battery cell was then produced with it and the electrochemical properties were studied. The authors did not report the yield and recovered only Co via the chemical treatment.

Myoung et al. (2002) proposed an electrodeposition method to recover Co from $LiCoO_2$ leach liquor as $Co(OH)_2$. Virgin and waste LIB were leached with hot 1 M HNO_3 to leach the $LiCoO_2$ and the solution subjected to electrodeposition at -1.0 V on a Ti electrode. The resultant film was characterized as $Co(OH)_2$ which was then heat treated at 400°C for 2 h to produce Co_3O_4. It was reported that the waste LIBs yielded less Li and more Co than the virgin variant possibly due to changes in $LiCoO_2$ spinel through the service history. The authors did not report the Li recovery.

Freitas et al. (2009) reported work on the electrodeposition of metallic Co on Al electrodes from a waste LIB leach solution. Dismantled LIB electrodes were heated at 80°C for 2 h to remove organics (e.g., ethylene carbonate and propylene carbonate) and subsequently washed at 40°C for 1 h to remove $LiPF_6$ and $LiCl_4$. Then 3 M HCl and 6% v H_2O_2 was used to leach the $LiCoO_2$ from the treated electrodes (at 80°C for 2 h). Electrodeposition of Co was carried out at 1.0 V at two different pH levels (2.7 and 5.4) using NaOH and 0.1 M H_3BO_4 buffer. It was reported that the lower pH yielded finer grained 2D (110 plane) growth and the higher pH yielded a coarser grained 3D growth (002 plane).

Li et al. (2009) published a technical note on their work on the improvements of mechanical separation of crushed LIB portions by a combination of ultrasound and agitation. It was claimed that the underflow of 2 mm screen containing the most $LiCoO_2$ and Cu, Fe, and Al as impurities (2% in total) was completely separated by the following treatment process which was also claimed to be energy-saving and environment-friendly. The separated black powder was leached with HCl (4 M for 2 h, at 80°C) and sequentially precipitated with 40% NaOH at pH 4.5−6 to remove the Al, Fe, and Cu impurities. The dependence of pH, temperature, and leach time of the process was also determined. The authors reported the efficiency of the leaching process to be 89% of the total Co content of LIB.

Zou et al. (2013) described a hydrometallurgical route to process mixed types of LIBs containing $LiCoO_2$, $LiMn_2O_4$, $LiNi_{0.33}Mn_{0.33}Co_{0.33}O_2$, and $LiFePO_4$, even though the leaching efficiency of $LiFePO_4$ was reported to be very low. To avoid possible Cl_2 emissions from the HCl, H_2SO_4 was used to leach the mixed active salts (4 M + 30 wt.% H_2O_2 at 70°C–80°C for 2–3 h). NaOH was added to raise the solution pH to >3 (to remove Fe), and later to pH 11 to coprecipitate Ni, Co, and Mn as respective hydroxides after adjusting the molar ratio to Mn:Co:Ni = 1:1:1. By adding Na_2CO_3 at 40°C, Li was extracted (80%) as Li_2CO_3. Resynthesis of mixed lithium oxides was achieved by mixing the reclamation products [Li: (Ni + Mn + Co) = 1.1:1] at 900°C for 15 h. A laboratory cell was constructed, and cyclic voltammetry done to compare the electrochemical properties. The authors compared the recycled versus virgin active cathode material and suggested the process to be financially (cost differential US$10,440/ton) and environmentally beneficial.

Meshram et al. (2016) compared the effect of $NaHSO_3$ (0.075 M) and H_2O_2 (5%) as reducing agents in a H_2SO_4 based leaching system (1 M, L/S = 20/1, 95°C for 4 h at 500 rpm) and studied the mass balance as well as thermodynamic kinetics. The scope of the work was limited only to leaching (not reclaiming) the respective elements, the relevant Eh–pH diagram analysis, a thermodynamic study using HSC Chemistry 7.14 software, and a kinetic study of different parameters using the shrinking core empirical model. Notably, they used lower concentrations (1 M) of H_2SO_4 and reported higher yields with $NaHSO_3$ (96.7% Li, 91.6% Co, 96.4% Ni, and 87.9% Mn) rather than with H_2O_2 (Co 79.2%, Mn 84.6%).

Li et al., 2008) described a hydrometallurgical process to recycle cylindrical 18650 cell-type LIBs using mechanical shredding and NMP dissolution at 40°C for 15 min followed by H_2SO_4 leaching (3 M + 1.5 M H_2O_2 at 70°C for 1 h). Cobalt was reclaimed (99.4%) as oxalate by the addition of oxalic acid (1 M $Na_2C_2O_4$ at 50°C) and Li (94.5%) as carbonate by Na_2CO_3 addition (100% excess saturated). The reclamation salts were used to resynthesize $LiCoO_2$ by mixing (L/S 1.05:1 Li/Co) and then successive heating (600°C for 6 h and 900°C for 10 h). The authors reported the cyclic voltammetry results (specific capacity of 153 mAh/g and good cyclability) of a laboratory cell constructed with a Li metal electrode utilizing the resynthesized $LiCoO_2$.

2. SX

Certain organic solvents have the ability to impregnate themselves with selective metallic ions by the interaction of an aqueous–organic phase emulsion. This emulsion could be modified or saponified by specific additives acting as buffers. The liquid–liquid separation process involves transportation of metal cations, M^{n+}, metalate anions, MX_x^{n-}, or metal salts, MX_x into a water-immiscible solvent while interacting with metal-rich phases, or more commonly leach liquors. Aqueous and organic mixtures can be easily separated by various mixer-settlers and elements of interest can be extracted from the organic phase by stripping, often achieved with pH modification. Table 5.2 summarizes a few common reagents used in LIB SX. The process has

Table 5.2 Common solvent extraction solvents and target metal elements for extraction.

Abbreviation	Chemical name	Target metal
D$_2$EHPA	Di-(2-ethylhexyl) phosphoric acid	Cu, Co, Ni, Mn, Al, and Li (mostly as aqueous raffinate)
DEHPA	Diethylhexyl phosphoric acid	
PC-88A or P507	2-Ethylhexyl phosphonic acid mono-2-ethylhexyl ester	
Acorga M5640	2-Hydroxy-5-nonylbenzaldehyde oxime (ester)	
Cyanex 272	Bis-(2,4,4-tri-methyl-pentyl) phosphinic acid	
TOA	Trioctylamine	
TBP	Tributyl phosphate	

excellent selectivity, ease of operational conditions, low energy consumption, and high product purity (Wilson et al., 2013). Common drawbacks include high extractant price and treatment costs for scaled-up liquid handling (Xu et al., 2008).

In one of the earliest works on LIBs using a SX route, Zhang, Yokoyama, Itabashi, Wakui, et al. (1998) compared the efficiencies of sulfurous acid, hydroxylamine hydrochloride, and hydrochloric acid as leachants considering the effects of acid concentration, temperature, reaction time, and solid-to-liquid ratio. It was reported that HCl gave the best leaching performance under the given conditions (4 M, L/S = 10/1, 80°C for 1 h) leaching 99% of the Li and Co combined. In the SX process comparing PC-88A and D$_2$EHPA solvents, the former showed superior performance (0.9 M in kerosene, O:A = 0.85/1 at pH 6.7) selecting almost 99% Co and 13% Li. The Li was then separated out by scrubbing with a CoCl$_2$ + HCl solution (30 g/L of Co^{2+}, pH = 1.0, O:A = 10:1) and then reclaiming Co in a 2 M H$_2$SO$_4$ solution at an O:A ratio of 5:1. The remaining Li was recovered as Li$_2$CO$_3$ by treatment with a saturated Na$_2$CO$_3$ solution at 100°C. The authors reported the yield of Co to be 99.99% (at 99.99% purity) and that of Li to be 80% (at 99.93% purity).

Dorella & Mansur (2007) published a short communication describing a method of separating Co from LIBs using acid leaching and SX. Sorted and dismantled electrodes from LIBs were cut and leached with H$_2$SO$_4$ (6% v/v + 1% v/v H$_2$O$_2$, L/S = 30/1 at 65°C for 1 h) and the different leaching parameters investigated. The leach solution was treated with NH$_4$OH to reach a pH value of 5 to separate out around 80% of the Al from the leach solution with an associated loss of 10% of the Li and 7% of the Co. The Al-bearing solution was then treated by SX using extractant Cyanex 272 dissolved in kerosene to recover the Co. A maximum tenor of 63 g/L Co was achieved in the H$_2$SO$_4$ stripping stage, reportedly adequate for subsequent electrowinning.

5.2.2 Pyrometallurgical processes

Metal extraction from ores or concentrates is often performed at high temperatures, typically $500°C-2000°C$, with complex interactions between gases and solids or between gases and molten materials. This branch of extractive metallurgy is called pyrometallurgy and has widespread commercial popularity and operation, mainly for base metal smelting (e.g., Cu, Pb, Fe). Pyrometallurgic techniques include drying, roasting, incineration, smelting, sintering, refining, calcination, and alloying, and similarly, e-waste can also be treated in existing primary or secondary base metal smelters. Specifically developed LIB pyrometallurgical set-ups are also in operation where the furnaces are designed to handle the often more corrosive feed materials. Furthermore, the off-gas treatment facilities need to be able to capture the high amounts of halogenic and organic compounds, which may also contain decomposition products of organics and volatile metal products such as Pb, Hg, and Zn if present. At present, processing of LIBs is usually deemed a high-cost operation.

At high temperatures, pyrometallurgical reactions are usually kinetically fast. Spent LIBs are decomposed in high-temperature, oxygen-rich furnaces, with temperatures exceeding $1200°C$. The different thermochemical reactions occurring in the pyrometallurgical processing of LIBs can be segmented into three main categories. Firstly, *evaporation* and *pyrolysis* will occur at relatively low temperatures ($300°C-380°C$). Binders (PVDF) and solvents (ethylene carbonate, propylene carbonate, dimethyl carbonate, etc.) generally have low boiling points ($250°C-300°C$) and these compounds will start evaporating in this temperature range whereas at higher temperatures ($> 300°C$), electrolytes ($LiBF_6$, $LiPF_6$, $LiAsF_6$, $LiClO_4$, etc.) will begin to break down through pyrolysis to lower hydrocarbons producing emissions such as LiF and HF. Initiation of carbon oxidation is also achieved at the same time. The second mechanism involves *calcination* and *roasting* which operates at a relatively higher temperature range ($\geq 700°C$). Generally, calcination can be described as a thermal treatment process for solid materials that involves thermal decomposition, phase transitions, and/or removal of volatile substances in the absence or limited supply of air or oxygen to produce an upgraded solid substance, for example, the reduction of carbonates to oxides or the reduction of $LiPF_6$ to LiF and PF_5. Roasting, on the other hand, relates to the heating of substances, usually in air or oxygen, resulting in oxidation, burning, and/or high temperature solid–solid reactions. At this temperature range, electrolyte salts and other compounds will break down to various solid/gaseous products. Carbon and other carbonaceous products (including gases) would completely burn-off supplying energy to the system and active salts would react with other solids (such as $NaHSO_4$ if present). The metallic components on the other hand would remain stable in metallic or oxidized form. The third stage is *oxide reduction* and *smelting* and is achieved at very high temperatures ($1200°C-1500°C$). At these temperatures it is reported that vacuum pyrolysis or other carbothermic reactions can reduce lithium metal oxides (LMO) to form metals or their oxides and can also be used to convert Li to a carbonate (Xiao et al., 2017). Theoretically Fe, Ni, and Co can be reduced by C, CO, or H_2 (produced as pyrolysis products) at this stage according to the Ellingham diagram (An, 2019). Furthermore, in smelting, all constituents of the charge are in a molten state and elements can exist in two or more phases such as slags, matte, speiss, and metal. Reducing agents and fluxes are common additives for such processes to convert the oxides to metallic alloys and produce a low melting—point slag phase. To assist the carbothermal reduction of metals coke (as reductant) and lime (as a slag former) are often added mimicking classical ore reduction or iron making. Aluminum in the melt also serves as a reductant, reducing the fuel demand. Many metals are reduced and deport to the melt/metal phase while others having a lower affinity for oxygen report to the slag phase. The temperature at this stage is kept relatively high to help lower the viscosity of the slag which can contain significant amounts of Al, Si, Ca, Fe, Mn, Li, and rare earth elements (REEs). Metals such as Co, Ni, and Cu are recovered with the greatest efficiency as mixed metallic melts or as Fe-based alloys. The alloy outputs can be further refined or separated into constituent elements via hydrometallurgy, or they can be sold commercially as is. High silica levels in the slag poses process difficulties in subsequent hydrometallurgical steps for metal recovery.

Unwanted or nuisance components (F, Cl, C, P, etc.) can also be removed by pyrometallurgy and hence produce a more uniform intermediate product for subsequent hydrometallurgical processing, noting again the sensitivity of hydrometallurgical processes to feedstock composition variation. It makes sense to use pyrometallurgical conditions that promote the deportment of low-value, nonprecious metals such as iron, manganese, or aluminum to the slag phase to reduce the processing load and duty in the hydrometallurgical steps that follow. Conceivably, the greatest advantages of pyrometallurgy are the scalable simplicity of the operation and its ability to cope with a wide range of feedstock compositions and physical parameters, for example, size, shape, and chemistry. However, pyrometallurgical processes are constrained by their immense power usage and the propensity for them to produce greenhouse gas emissions.

5.2.3 Bioleaching processes

Bioleaching processes utilize the acid produced during metabolization of microorganisms; hence they are considered environmentally friendly methods. Generally, bacteria form inorganic acids, while fungi form organic acids, and both can be utilized to leach spent batteries. Common organic acids like formic acid, oxalic acid, acetic acid, and citric acids are known to form in certain cultures. The organic acids have several advantages including lower gaseous emissions, relatively longer stability (than H_2O_2), biodegradability, and reduced corrosion of equipment and infrastructure.

King et al. (2018) conducted a rigorous evaluation of a novel zero waste, closed-loop processing technologies to treat LIBs by bio-electrochemical routes. The evaluation included technoeconomic analyses. Using a 0.5 M NaCl solution as the electrolyte, a laboratory-scale electrochemical cell was used for acidic (HOCl + HCl), H_2, and basic (OH^- rich) leachant production. The best leaching conditions were reported to be sequential leaching (L/S = 10/1, 25°C, 200 rpm for 1–4 h) with an 8 V anolyte, equivalent to 0.135 M HCl, with extractions reaching almost 80% efficiency for Li and Co from shredded LIBs fines. Mixed cultures (*Desulfobacterium* spp., *Desulforhopalus* spp., and *Desulfovibrio* spp.) cultured in seawater were used in a laboratory-scale bioreactor, consuming the H_2 produced in the electrochemical cell to produce H_2S for metal recovery (as sulfides) from the leach solutions. It was recommended that long-term batches be conducted for actual leaching. The technoeconomic evaluation was favorable.

Calvert et al. (2019) evaluated the metal recovery using hydrogen sulfide in a lactate fed fluidized bed reactor (FBR) from a LIB waste leachate. The H_2S was generated by a consortium of sulfate-reducing bacteria, containing *Desulfovibrio* genus. Ground and processed LIB powder (<500 μm) was leached with 2 N HCl at 25°C at a pulp density of 10% (w/v) for 24 h at 200 rpm and subsequently filtered. The FBR was prepared using a Postgate medium with an adjusted pH of 7 using NaOH or H_2SO_4, keeping a sulfate-to-lactate ratio of 1:0.35 and an operating temperature of 35°C. Granulated activated carbon was used to carry the biomass as bio-film former on the surface. Under computer assisted pH control, the biomass produced H_2S under N_2 sparging, which was utilized to precipitate metals. Aluminum, Ni, Co, and Cu were simultaneously precipitated with an efficiency over 99% using the biogenic sulfide and NaOH.

5.2.4 Alternative technologies

Recent developments in LIB recycling technologies are trending toward alternative compact options targeting the recovery of all the battery materials. Methods proposed include: cathode-to-cathode, mechanical, electrochemical, and "healing" technologies. In principle, such routes are termed as "direct recycling," and are designed to recover the functional cathode material without decomposition into substituent elements as practiced with classical hydrometallurgy. Such processes may use solvents in supercritical conditions (H_2O, CO_2, etc.) to reclaim the solvents and some extend to physically liberate the constituent phases such as Cu, Al, polymer, and black powder. Efforts have been made to also regenerate the active cathode mix in Li-rich solutions electronically, thereby "rejuvenating" the charge storage for reuse. The downside of the existing development of alternative technologies is that the recovered material may not perform as well as virgin material and thereby losing cycle life and resulting in poorer charge attenuation and discharge intensity. Additionally, mixing cathode materials could also reduce the value of recycled products.

Fujiwara et al. (2002) reported a low-energy and low-waste electrodeposition method to directly synthesize crystalline (R3c) $LiCoO_2$. A laboratory cell was built with polytetrafluoroethylene construction utilizing 5 mol/L $LiOH \cdot H_2O$ and 0.5–0.1 mol/L $CoSO_4 0.7H_2O$ assisted by $Na_2S_2O_3$ (0.1 mol/L) as a Co^{+3} stabilizing agent, a flow rate = 100 mL/h, and a current density = 1 mA/cm^2. It was shown that the process was particularly suitable at or above 120°C and resulted in a geometry specific (patterned) synthesis of the ceramic.

5.3 Commercial operations

Commercially, re-manufacturing and repurposing batteries is an attractive option as the required processing mostly relates to already built or finished units. According to the United States Advanced Battery Consortium (Hesselbach & Herrmann, 2011), batteries are generally not suitable for EV use when the delivered capacity or power of a cell, module, or pack is less than 80% of its original rated value. This presents an opportunity to examine the battery assembly and carefully replace the depleted cells or modules to remanufacture the battery pack for reuse in EVs, potentially saving up to 40% of the cost of a new unit (Foster et al., 2014). Spiers New Technology and Global Battery Solutions (formerly, Sybesma's Electronics) run operations in the US and Europe remanufacturing EoL EV batteries. Depending on the condition of the battery units, they can be repaired, remanufactured, or repurposed for other application after a comprehensive diagnosis. Such repurposing often involves less severe servicing conditions relative to initial requirements and can ultimately extend the service life of the LIB unit before eventual destructive recycling options kick in. Examples of this repurposing are seen with the Nissan Leaf, Chevrolet Volt, and Renault EoL EV batteries which are now being used to power households, stadiums, data centers, and highways providing peak shaving, frequency regulation, renewable power storage, or EV charging (Automotiveworld, n.d.; DeRousseau et al., 2017; Mobilityhouse, n.d.; Olsson et al., 2018).

Even after reuse, remanufacture, or repurposing, every battery meets its EoL eventually and destructive recycling via pyrometallurgical or hydrometallurgical routes (as discussed earlier) are then required. Many facilities combine both routes to extract values much more efficiently. A list of the major LIB recycling operations around the world is provided in Table 5.3. Precise details of the processing conditions at each plant are not always easy to obtain with many manufacturers keeping such details secret.

One of the forerunners, if not the world leader in LIBs recycling, is Umicore of Belgium, who recycles LIBs using an ultra-high-temperature pyrometallurgical process

Table 5.3 Major worldwide industrial lithium-ion battery recycling operations (Dahllöf et al., 2019; Ellis & Mirza, n.d.; Fan et al., 2020; Heelan et al., 2016; Khaliq et al., 2014; Shareef, 2019).

S. No.	Country	Company	Process classification	Annual capacity (tons)
1.	Australia	CMA Ecocycle	Collection, mechanical treatment and separation, export	N/A
2.	Australia	Envirostream		N/A
3.	Australia	MRI (Aust) Pty Ltd		N/A
4.	Australia	PF Metals		N/A
5.	Australia	Powercell (Australia) Trading Pty Ltd		N/A
6.	Australia	Schlumberger Australia Pty Ltd		N/A
7.	Australia	SIMs Recycling Solutions		3300
8.	Australia	Tes-Amm Australia Pty Ltd		N/A
9.	Austria	Neometals	Hydrometallurgy	N/A
10.	Belgium	Revatech	N/A	3000
11.	Belgium	VAL'EAS, Umicore Solutions Recycling	Pyro–hydrometallurgy	7000
12.	China	Brunp	Hydrometallurgy	30,000
13.	China	Fuoshan Bangpu Ni/Co High Tech Co.	N/A	3600
14.	China	GEM	Hydrometallurgy	N/A
15.	China	Shenzhen Green Eco-Manufacturer Hi-Tech Co.	N/A	20,000
16.	Finland	AkkuSer Oy Ltd	N/A	4000
17.	France	AFE Group (Valdi)	Pyrometallurgy	N/A
18.	France	Citron	Pyrometallurgy	N/A
19.	France, Singapore, China, India	TES-AMM/ Recupyl SA (IPGNA Ent.)	Hydrometallurgy	3000
20.	France	SNAM	Pyrometallurgy	300
21.	France	SARP/Euro Dieuze Industrie	N/A	220
22.	Germany	Accurec Recycling GmbH	Pyrometallurgy	6000
23.	Germany	DK Recycling und Roheisen GmbH	Pyrometallurgy	N/A
24.	Germany	Duesenfeld GmbH	Hydrometallurgy	3000
25.	Germany	IME-GmbH	Pyro–hydrometallurgy	N/A
26.	Germany	Redux Recycling GmbH	Pyrometallurgy	N/A
27.	Germany	Rockwood Lithium GmbH	Hydrometallurgy	N/A

#	Country	Company	Technology	Capacity
28.	Germany	Stiftung Gemeinsames Rucknahmesystems Batterien	N/A	340
29.	Israel	Metal-Tech Ltd	N/A	N/A
30.	Japan	Dowa Eco System Co Ltd	Pyrometallurgy	1000
31.	Japan	Japan Recycle Center	Pyrometallurgy	N/A
32.	Japan	JX Nippon Mining and Metals Co.	Pyro-hydrometallurgy	5000
33.	Japan	Mitsubishi	Pyrometallurgy	N/A
34.	Japan	Sony Corp & Sumitomo Metals & Mining Co, Japan	Pyrometallurgy	150
35.	Netherlands	Stichting Batterijen (StiBat)	N/A	N/A
36.	South Korea	Kobar Limited	Pyro-hydrometallurgy	1000
37.	South Korea	SungEel Hi-Tech	Hydrometallurgy	8000
38.	Sweden	uRecycle	Mechanical	N/A
39.	Switzerland	Batrec Industries AG (Sarpi/Veolia)	Mechanical treatment and hydrometallurgy	1000
40.	Switzerland	Glencore (Formerly Xstrata Nickel)	Pyrometallurgy	3000–7000
41.	Switzerland	SAFT AB (SAFT-NIFE)	Pyrometallurgy	N/A
42.	United Kingdom	AEA Technology	Shredding, electrolyte extraction, electrode dissolution, and Co reduction	N/A
43.	United Kingdom	G&P Batteries	N/A	145
44.	United States/Canada	Toxco Process (McLaughlin)	Cryogenic, mechanical shredding, hydrolysis	4500
45.	United States	AERC (now Clean Earth)	Pyrometallurgy	N/A
46.	United States	Retriev Technologies	Hydrometallurgy	3,500
47.	United States	BDT	N/A	350
48.	United States	INMETCO (CVRD)	Pyrometallurgy	6000
49.	United States	OnTo Technology	Hydrometallurgy	N/A
50.	United States	Salesco Systems	Pyrometallurgy	N/A

N/A, Not available.

where plastics, organic solvents, and graphite in the batteries provide heat during combustion, while the Co, Ni, and Cu components are reduced and converted to a high purity alloy. Lithium and other elements (Al, Si, Ca, Fe, Mn, and REEs) form a slag phase (Georgi-Maschler et al., 2012; Zheng et al., 2018). The three-zone shaft furnace used in this process facilitates electrolyte evaporation (300°C), plastic pyrolysis (700°C), and reductive smelting (1200°C−1450°C) (Cheret & Santen, 2007). The Al−Li-containing slag can be further treated using standard Li recovery techniques. The alloy is subjected to sulfuric acid leaching and SX to produce cobalt oxide and nickel hydroxide which are used to resynthesize active cathode materials (Meshram et al., 2014). Importantly, this process (also known as VAL'EAS process) deliberately controls gaseous effluents by recirculation, cooling, and filtration.

The German-based company Accurec Recycling GmbH utilizes comprehensive pretreatments including comminution, sorting, sieving, magnetic and air separation to recover Al, Cu, and steel (Georgi-Maschler et al., 2012; Zhang et al., 2018) and extracts the electrolyte, solvents, and hydrocarbons using a proprietary vacuum thermal treatment at 250°C. Separated active cathode materials are pelletized with binders and subjected to a two-stage Co-based smelting process where the Li deports to the slag phase for subsequent hydrometallurgical recovery (Espinosa et al., 2004). The LithoRec process uses battery discharging and warm (120°C) shredding of automotive LIBs under a nitrogen atmosphere. Using zig-zag air separation Al, Fe, Cu, and plastics are separated and fines processed by repeated shredding and separation. After the second stage of separation hydrometallurgical treatments are used to recover Co, Ni, and Mn as oxides (Velázquez-Martínez et al., 2019). The Swiss company, Batrec Industrie AG utilizes CO_2 in an inert atmosphere to comminute sorted LIB units and moist air to reduce any released Li. The CO_2 is scrubbed before exhaustion and the fines are treated hydrometallurgically. AkkuSer Oy of Finland utilizes cyclone gaseous filtration during its two-stage mechanical shredding process to produce <6 mm particles at moderate temperatures from presorted LIBs. The final metal recovery is via hydrometallurgical or pyrometallurgical processes (Pudas et al., 2015).

One well-established commercial process is the Valibat process, developed by the French company Recupyl (later known as TES-AMM). It comprises the physical separation of Fe, Cu, and polymers after mechanical treatment under an inert CO_2 atmosphere followed by leaching in two stages—first a basic hydrolyzing treatment to separate out the Li followed by separation of other impurities and Co (Meshram et al., 2014). Lithium is recovered as Li_2CO_3 using the recirculated CO_2 from the pretreatment stage or as Li_3PO_4, and Co is recovered as hydroxide after H_2SO_4 leaching. AEA Technology in the United Kingdom utilizes a N_2 atmosphere during mechanical treatment after firstly removing the casing. The fines ($\sim 1 \text{ cm}^2$) are treated with acetonitrile to leach the electrolyte and solvent which are reclaimed after evaporation.

Using NMP the PVDF binder is removed and the washed residue is electrolyzed to recover LiOH and CoO_2 (Lain, 2001).

In Japan Sony Corporation calcines LIBs at around 1000°C to burn off plastics and organics leaving the actives and metallics (Cu, Fe, and Al) behind to be separated magnetically. Sumitomo Corporation receives the actives and separates out the Cu by pyrometallurgical means and the Cu/Ni by hydrometallurgical methods (Zhang et al., 2018). Lithium is discarded in this process and metallic scraps like Cu and ferrous portions are recovered as by-products (Meshram et al., 2014).

Another popular processing route is the Toxco LIB recycling process in Canada, which involves cryogenic mechanical treatment. Using a caustic bath to neutralize the acidity and leach the Li salts, the LIB units are sheared while submerged. The processed solids are then treated with Li brine and the metallic portions are recovered subsequently (McLaughlin & Adams, 1999). The International Metals Reclamation Company (INMETCO) in the United States produces iron-based alloys recovering Co, Ni, and Cu from LIBs. A direct reduction process is used. It was originally designed to treat electric arc furnace dust but can now accept battery shreds in the feed. The actives are mixed with carbon-rich compounds, pelletized, and reduced at high temperature (Espinosa et al., 2004). Retriev Technologies Inc. dismantles LIBs and shreds under submerged conditions and separates the polymers, metallic solids, and Li-enriched liquids. The submerged milling has the advantage of reduced gaseous effluents and avoids electrolyte neutralization. Lithium is recovered after the treatment and the solid products rich in Co, Ni, Mn, Cu, and Al are on-sold to other ventures for metal refining and recovery (Retriev, n.d.). The OnTo Technology utilizes high-pressure super critical fluids consisting of liquid CO_2, alkyl ethers, ammonia, etc. to treat discharged and dismantled batteries before pulverizing and separating the values by physical means (Meshram et al., 2014). The cathode materials gained are regenerated by hydrothermal treatments. Battery Resourcers also synthesizes different mixed actives (Ni−Co−Mn) from LIBs in a closed loop from a variety of feed stocks (Chen et al., 2019) by mechanical treatment and a series of hydrometallurgical processes. Glencore, on the other hand, utilizes processed EoL batteries (including solids from Retriev Technologies) as a secondary feedstock from which the Co, Ni, and Cu are extracted (Zhang et al., 2018).

One of China's leading battery recyclers Brunp Recycling along with several others (GEM, GHTech, TES-AMM China, and Highpower Int.) utilizes sulfuric acid leaching assisted by hydrogen peroxide to produce metal hydroxides from LIBs.

South Korea-based SungEel Hi-Tech uses aqueous grinding of LIBs to avoid fire hazards, and applies magnetic and electrostatic separation to recover Fe, Cu, and Al. They also recover Cu, Ni, Co, Li, and Mn as sulfates and phosphates by hydrometallurgical processes from the primary processing residue and supply these products to LIB manufacturers (성일하이텍주, n.d.; Sojka et al., 2020).

5.4 Informal trends

Informal sectors of recycling e-waste in developing countries (i.e., Bangladesh, India, Pakistan, Nigeria, Brazil, regions of China, and many others) form an important part of the domestic economy. Although the developing countries generate a significant amount of e-waste themselves, volumes of e-wastes are also imported from developed countries as secondary destinations (Lundgren, 2012). Generally, e-waste recycling operations take place in large cities and often in slum areas where many poor, uneducated people including vulnerable women and children live. The crude rudimentary processes used in recycling expose them and others in their community to significant health risks. They generally collect from door to door or from repair shops and partially dismantle items before passing them on to the larger informal sector locations (Dutta & Goel, 2021).

5.4.1 Challenges with informal recycling

Unplanned urbanization in developing countries represents a substantial challenge for municipalities while collecting and managing wastes, due to the wide variety of waste composition, the accessibility to waste sources, weather conditions, management capacity, and lack of skilled manpower, not to mention the status of the technology itself (Matter et al., 2013). Municipal waste collection procedures in such localities are usually operated in terms of "collection and dumping," overlooking waste segregation, possible waste to value recovery, and safe disposal. As a consequence, the waste collection rate is quite low, burdening backyards, city streets, open stormwater, and wastewater drains and low-lying slums in particular. Usually, the informal e-waste recycling sector is a low-paid, labor-intensive, and low-technology amalgamation of people or family groups that may have migrated to cities from rural areas, have no trading license, pay no taxes, and are not involved in government insurance or social welfare schemes. They operate on small-scale, principally unregulated and unregistered. Generally unacknowledged by the formal sector or authorities, the informal sector recycles materials with some economic value, thus reducing the solid waste streams to be managed and feeding the manufacturing sector with cheaper raw materials for processing.

Informal recycling or "backyard recycling" of e-waste is often done by nonscientific methods avoiding any regulatory bodies or guiding entities and is conducted by underprivileged people with very limited knowledge about recycling processes, safety regulations, environmental effects, and associated risks. Such initiatives are dangerous and harmful in a multitude of circumstances. The primary effects are direct exposure to a variety of toxins, fumes, and particles, as well as fire, explosion, injuries, and many other forms of occupational hazards, due largely to a lack of structural guidelines about the nature and composition of the e-wastes. Local communities are also affected by the secondary emission of fumes, deposition of heavy metals in the soil, leachates in the water bodies, and the occasional radiation and sound pollutions. Tertiary effects include vegetation and livestock affected by the emissions which can spread country wide by means of the food chain. Additionally, incineration of halogenated components contributes to the greenhouse effect, microplastic sediments get entrapped in water bodies, and acid fumes form acid-rains in extreme cases. Overall, informal handling of batteries pose undeniable health hazards in modern society.

5.4.2 International settings

Among a significant few, India is one of the major countries producing massive amounts of e-wastes and handling about 95% of it informally (Kannan et al., 2016). Populous cities including New Delhi, Kolkata, Mumbai, Bangalore, and Chennai have both formal and informal e-waste collection practices destined for satellite recycling locations such as Moradabad (Uttar Pradesh) (CSE, 2015). Underprivileged men, women, and children slum-dwellers, migrants, and immigrants are reported to contribute to the e-waste recycling facilities operating at the back end of door-to-door collectors, formal contractors, and traders. In most of the cases, dismantling is done by wire cutters and pliers, incineration by torches and acid baths are used in open without any suitable safety garments and gears. After the elementary collection of scraps and fractions, residue is dumped in open waterways, yards, and landfills (Purushothaman et al., 2020). Although formal recycling facilities of LIBs are unknown, initiatives have been seen to cope with the anticipated ingress of EVs by battery manufacturers and handlers (Sharma, 2021).

Bangladesh is another country that is yet to formalize its hazardous and electronic waste collection infrastructure, although in recent years guidelines (e-waste rules) have been introduced (Enviliance, n.d.). Although there is a scarcity of established scientific data about the rate of e-waste generation, it is to be assumed that a country with such a vast population and rapid digitalization is producing significant amounts of electronic wastes. A few ventures (Foraji, 2019) have been operating formally with government offices, multinational companies, telecom companies, and defense units by means of tenders and agreements relating to the scope of collection, dismantling, sorting, and export. The majority of the e-wastes generated by the local corporations, unofficial bodies, and households are typically dumped with municipal wastes and any recycling is limited to urban communities which collect and treat the material informally (Rahman, 2017). Such crude recyclers are only interested in recovering resalable commodities like common metals scraped mechanically or via incineration and dumping the residue. For example, Azizu Trading Company works closely with telecom industries, corporate houses, and government agencies in addition to door-to-door collection of e-wastes. Methods include copper scraping from wires and recovery of metallic fractions including steel and Al by means of magnetic and electrostatic separation. Products are sold in the local scrap market. The company is also known for exports of crushed telecom PCBs to Tes-Amm Singapore. There are no known recyclers of LIBs in Bangladesh except for the limited informal reuse of cylindrical batteries by cottage industries to manufacture rechargeable lights, fans, mosquito swatters, and similar appliances. Informal ventures concentrated in the Nimtoli, Hatirpool, and Dholaikhal regions in the capital district are operating mostly using child and female labor to hammer, incinerate, and break batteries to extract Fe, Al, and Cu portions.

In South America, Chile and Uruguay are known to generate significant amounts of e-wastes per capita and are treating it informally in most cases (Savino et al., 2018). Studies show that due to the informal incineration of wires to manually scrap the metallic Cu, several regions in Uruguay are affected by lead poisoning which manifests itself often in children (Pascale et al., 2016). Formal recycling plants for LIBs are scarce in Latin America, though initiatives are slowly nucleating (Dialogochino, n.d.).

5.5 Advantages and disadvantages

Understandably, both formal and informal methods have their own advantages and disadvantages. The organization of affairs is much more structured for the former, making it "formal" per se. As a result, formal recycling of any product makes it heavily dependent on fixed costs like plant establishment, construction, and machineries. But in turn, this greatly increases the productivity and profit margins which are also lucrative given the current market scenario. Conversely, informal sectors perform like cottage industries.

For the formal sector, the initiation of the production process is rooted on solid scientifically proven processes derived from rigorous study and most often leasing patents. Additionally, in many types of e-wastes certain critical elements would alone produce the maximum of value addition yet require the most complex route of extraction making it the single most crucial aspect of the venture; for example, gold or precious metals from PCBs or cobalt/nickel from LIBs—without extracting these components the processes are usually not profitable. Many commercial bodies also offer turn-key plant set-ups and technical expertise for common processes. For electronic wastes, or in the case of LIBs, these could be an expensive feature and a difficult one too, given few countries have actually commercialized LIB recycling and the processes are multistaged depending on the required end product. For uncommon electronic products or newer alternatives of existing ones, the study for the recycling process development generally takes much longer than actually launching the commercial product. For example, commercialization of LIBs with alternative anodic material like graphene of graphene oxides might take very little time compared to the start of commercially recycling graphene or graphene oxide from LIBs. These processes, however, deliver maximum productivity, known outputs and parameters, better control, and most of all safer risk management. While in the case of informal sector, it is more of a "watch and learn" process vacant of all the advantages mentioned above at almost no cost involved.

Methodically recycling LIBs require plant designs as detailed and complex as any chemical plant or foundry. Additional set-ups like process accelerators (boilers, heat exchangers, pumps, and valves), by-product, and effluent treatment plants and most importantly safety arrangements (generators, fire and personnel safety) are incorporated to encounter the hazardous or sometimes toxic elements of electronic waste. Such plants have to maintain local and international regulatory compliances to keep the environment clean and the ecology safe. In comparison, the requirements of skilled manpower to address the operation as well as other demanding operating costs are minimal for the informal sector.

Another aspect of the LIB recycling is the geo-socio-economic scenario for any given region. While most developed countries have the budget, infrastructure, technology, and market for high valued specific chemical products derived from e-waste recycling, many densely populated developing countries lack those amenities. Attracting massive investment for such establishments in these countries could prove to be very difficult especially where there is no local or steady demand of the engineered recycled products. Such regions generally move toward informal recycling practices to keep the operation cheap enough to profit from the crude and less attractive products. For example, in Bangladesh, being one of the most densely populated countries, LIB waste production from mobile devices is huge, but the local requirements for $LiCO_3$, Co, Co_2O_3, or $LiCoO_2$ are almost nil shaping the recycling segment to be focused only on informally scrapping the steel, copper, and aluminum and dumping the rest. Similar scenarios exist in many South American and African countries, especially where some of them have natural mineral resources that remain commercially more attractive than active recycling.

One of the very few advantages the informal sector has is the ability to nucleate with extreme locality, as they function with smaller scale and minimal budget making it one of the promising start-ups of the decade. Regions where formal recycling is yet to begin may observe such start-ups to popularize the idea of e-waste recycling and create a potential market, shaping the general consensus and raising awareness. Such initiatives can often form a cluster and convolute toward formal recycling as observed in some countries like China and India.

5.6 Conclusion

The publicly available literature on the recycling status of LIBs has been reviewed. The processing technologies generally fall under the categories of hydrometallurgical, pyrometallurgical, or a combination of these. The LIB usage and potential production are expected to grow due to the increasing market entry of the EV in various countries where there is a need to reduce carbon emissions from petrochemical fuels in the transport sector. The use of LIBs in everyday electronic devices is also showing an increasing trend. In the next decade or so a substantial amount of EoL LIBs will require recycling for recovery of the valuable components, particularly the metals. The recycling of LIBs is necessary to reduce the cost of the large giga grid scale battery electrical energy storage systems.

LIB recycling currently occurs in developed and developing countries both in a formal and informal capacity. There are advantages of formal recycling since it is undertaken under strict environmental protocols; however, the cost is higher compared to that of informal recycling. The informal recycling sector is generally segregated and small but usually undertaken by the local poor and uneducated group of people living with extremely low socio-economic status in developing countries. The processes used are crude and rudimentary and pose a significant health risk to the operators and workers, particularly vulnerable women and children in the slums of the large cities in those countries.

Efficient collection and sorting of LIBs for better utilization of resources and recycling is the first step that is lacking in many developing countries. Once collected, with proper identification tags, and stockpiled, sustainable throughput-based small mobile processing facilities and technologies can recover various grades of materials. Potentially it can be intermediate concentrates of metals that can be refined by further processing in conventional plants located in both developing and developed countries.

This hazardous waste management issue is becoming an international focus globally by many governments and the United Nations Environment Programs. Further research is required for developing cost-effective and environmentally acceptable solutions and technologies. New flowsheet development, evaluation, and testing using techno-economic and life-cycle assessment methodology can provide a list of indicators for selecting such technologies.

References

Aktas, S., Fray, D. J., Burheim, O., Fenstad, J., & Açma, E. (2006). Recovery of metallic values from spent Li ion secondary batteries. *Transactions of the Institutions of Mining and Metallurgy, Section C: Mineral Processing and Extractive Metallurgy, 115*(2), 95—100. Available from https://doi.org/10.1179/174328506X109040.

An, L. (2019). Recycling of spent lithium-ion batteries: Processing methods and environmental impacts. *Recycling of spent lithium-ion batteries: Processing methods and environmental impacts.* Springer International Publishing. Available from https://doi.org/10.1007/978-3-030-31834-5.

Automotiveworld. (n.d.). *From plug-in cars to plug-in homes — EV batteries get a second life.* Retrieved August 15, 2021, from https://www.automotiveworld.com/articles/plug-cars-plug-homes-ev-batteries-get-second-life/.

Bahaloo-Horeh, N., & Mousavi, S. M. (2017). Enhanced recovery of valuable metals from spent lithium-ion batteries through optimization of organic acids produced by Aspergillus niger. *Waste Management*, *60*, 666−679. Available from https://doi.org/10.1016/J.WASMAN.2016.10.034.

Barik, S. P., Prabaharan, G., & Kumar, L. (2017). Leaching and separation of Co and Mn from electrode materials of spent lithium-ion batteries using hydrochloric acid: Laboratory and pilot scale study. *Journal of Cleaner Production*, *147*, 37−43. Available from https://doi.org/10.1016/J.JCLEPRO.2017.01.095.

Bok, J. S., Lee, J. H., Lee, B. K., Kim, D. P., Rho, J. S., Yang, H. S., & Han, K. S. (2004). Effects of synthetic conditions on electrochemical activity of $LiCoO_2$ prepared from recycled cobalt compounds. *Solid State Ionics*, *169*(1−4), 139−144. Available from https://doi.org/10.1016/J.SSI.2003.07.003.

Calvert, G., Kaksonen, A. H., Cheng, K. Y., Van Yken, J., Chang, B., & Boxall, N. J. (2019). Recovery of metals from waste lithium ion battery leachates using biogenic hydrogen sulfide. *Minerals*, *9*(9), 1−19. Available from https://doi.org/10.3390/min9090563.

Castillo, S., Ansart, F., Laberty-Robert, C., & Portal, J. (2002). Advances in the recovering of spent lithium battery compounds. *Journal of Power Sources*, *112*(1), 247−254. Available from https://doi.org/10.1016/S0378-7753(02)00361-0.

Chen, L., Tang, X., Zhang, Y., Li, L., Zeng, Z., & Zhang, Y. (2011). Process for the recovery of cobalt oxalate from spent lithium-ion batteries. *Hydrometallurgy*, *108*(1−2), 80−86. Available from https://doi.org/10.1016/J.HYDROMET.2011.02.010.

Chen, M., Ma, X., Chen, B., Arsenault, R., Karlson, P., Simon, N., & Wang, Y. (2019). Recycling end-of-life electric vehicle lithium-ion batteries. *Joule*, *3*(11), 2622−2646. Available from https://doi.org/10.1016/J.JOULE.2019.09.014.

Chen, Y., Liu, N., Hu, F., Ye, L., Xi, Y., & Yang, S. (2018). Thermal treatment and ammoniacal leaching for the recovery of valuable metals from spent lithium-ion batteries. *Waste Management*, *75*, 469−476. Available from https://doi.org/10.1016/J.WASMAN.2018.02.024.

Cheret, D., & Santen, S. (2007). *US7169206B2 Battery recycling* (Patent No. US7169206B2). https://patentimages.storage.googleapis.com/ab/96/56/b1960bc0741bc3/US7169206.pdf.

Contestabile, M., Panero, S., & Scrosati, B. (1999). A laboratory-scale lithium battery recycling process. *Journal of Power Sources*, *83*(1−2), 75. Available from https://doi.org/10.1016/S0378-7753(99)00261-X.

Contestabile, M., Panero, S., & Scrosati, B. (2001). Laboratory-scale lithium-ion battery recycling process. *Journal of Power Sources*, *92*(1−2), 65−69. Available from https://doi.org/10.1016/S0378-7753(00)00523-1.

CSE. (2015). *Recommendations to address the issues of informal sector involved in E-waste handling.* <https://cdn.downtoearth.org.in/pdf/moradabad-e-waste.pdf>.

Dahllöf, L., Romare, M., & Wu, A. (2019). *Mapping of lithium-ion batteries for vehicles.* Nordic Council of Ministers, https://doi.org/10.6027/TN2019-548.

Dai, Q., Spangenberger, J., Ahmed, S., Gaines, L., Kelly, J. C., & Wang, M. (2019). *EverBatt: A closed-loop battery recycling cost and environmental impacts model.* Argonne National Laboratory. Available from https://publications.anl.gov/anlpubs/2019/07/153050.pdf.

DeRousseau, M., Gully, B., Taylor, C., Apelian, D., & Wang, Y. (2017). Repurposing Used electric car batteries: A review of options. *JOM*, *69*(9), 1575−1582. Available from https://doi.org/10.1007/S11837-017-2368-9.

Dialogochino. (n.d.). *Ganfeng announces lithium battery recycling plant in Mexico.* Retrieved September 12, 2021, from https://dialogochino.net/en/climate-energy/38594-ganfeng-announces-lithium-battery-recycling-plant-in-mexico/?__cf_chl_captcha_tk__ = pmd_o9hMCopaQqDGVS2h7MTaXA6WyMGM0_M2HmS0FUFHKZs-1631438952-0-gqNtZGzNAyWjcnBszQel.

Dorella, G., & Mansur, M. B. (2007). A study of the separation of cobalt from spent Li-ion battery residues. *Journal of Power Sources*, *170*(1), 210−215. Available from https://doi.org/10.1016/j.jpowsour.2007.04.025.

Dutta, D., & Goel, S. (2021). Understanding the gap between formal and informal e-waste recycling facilities in India. *Waste Management*, *125*, 163−171. Available from https://doi.org/10.1016/J.WASMAN.2021.02.045.

Ellis, T. W., & Mirza, A. H. (n.d.). *Battery Recycling: defining the market and identifying the technology required to keep high value materials in the economy and out of the waste dump*. Retrieved August 16, 2021, from http://www.theiet.org.

Enviliance. (n.d.). *Bangladesh publishes E-waste management rule*. Retrieved October 17, 2021, from https://enviliance.com/regions/south-asia/bd/report_2900?__cf_chl_managed_tk__ = pmd_Vk0K.F.0Amer8Fiq7fb0N.PQW0StjGOWFJ4n.UM26rk-1634440239-0-gqNtZGzNAvujcnBszQi9.

Espinosa, D. C. R., Bernardes, A. M., & Tenório, J. A. S. (2004). An overview on the current processes for the recycling of batteries. *Journal of Power Sources*, *135*(1—2), 311—319. Available from https://doi.org/10.1016/j.jpowsour.2004.03.083.

Fan, E., Li, L., Wang, Z., Lin, J., Huang, Y., Yao, Y., Chen, R., & Wu, F. (2020). Sustainable recycling technology for Li-ion batteries and beyond: Challenges and future prospects. *Chemical Reviews*, *120*(14), 7020—7063. Available from https://doi.org/10.1021/ACS.CHEMREV.9B00535.

Ferreira, D. A., Prados, L. M. Z., Majuste, D., & Mansur, M. B. (2009). Hydrometallurgical separation of aluminium, cobalt, copper and lithium from spent Li-ion batteries. *Journal of Power Sources*, *187*(1), 238—246. Available from https://doi.org/10.1016/J.JPOWSOUR.2008.10.077.

Foraji, M. K. H. (2019). *e-Waste management policy & practices in Bangladesh* (pp. 1—15). BTRC. https://www.itu.int/en/ITU-D/Regional-Presence/AsiaPacific/SiteAssets/Pages/Events/2019/Policy-awareness-workshop-on-E-waste/E-WasteManagement Policy in Bangladesh.pdf.

Foster, M., Isely, P., Standridge, C. R., & Hasan, M. M. (2014). Feasibility assessment of remanufacturing, repurposing, and recycling of end of vehicle application lithium-ion batteries. *Journal of Industrial Engineering and Management*, *7*(3), 698—715. Available from https://doi.org/10.3926/JIEM.939.

Fouad, O. A., Farghaly, F. I., & Bahgat, M. (2007). A novel approach for synthesis of nanocrystalline γ-LiAlO$_2$ from spent lithium-ion batteries. *Journal of Analytical and Applied Pyrolysis*, *78*(1), 65—69. Available from https://doi.org/10.1016/J.JAAP.2006.04.002.

Freitas, M. B. J. G., Garcia, E. M., & Celante, V. G. (2009). Electrochemical and structural characterization of cobalt recycled from cathodes of spent Li-ion batteries. *Journal of Applied Electrochemistry*, *39*(5), 601—607. Available from https://doi.org/10.1007/s10800-008-9698-9.

Fujiwara, T., Nakagaw, Y., Nakaue, T., Song, S. W., Watanabe, T., Teranishi, R., & Yoshimura, M. (2002). Direct fabrication of crystallized LiCoO$_2$ films on paper by artificial biomineralization with electrochemically activated interfacial reactions. *Chemical Physics Letters*, *365*(5—6), 369—373. Available from https://doi.org/10.1016/S0009-2614(02)01444-6.

Gao, W., Zhang, X., Zheng, X., Lin, X., Cao, H., Zhang, Y., & Sun, Z. (2017). Lithium Carbonate recovery from cathode scrap of spent lithium-ion battery: A closed-loop process. *Environmental Science and Technology*, *51*(3), 1662—1669. Available from https://doi.org/10.1021/ACS.EST.6B03320.

Georgi-Maschler, T., Friedrich, B., Weyhe, R., Heegn, H., & Rutz, M. (2012). Development of a recycling process for Li-ion batteries. *Journal of Power Sources*, *207*, 173—182. Available from https://doi.org/10.1016/j.jpowsour.2012.01.152.

Globaldata. (2020). Global lithium demand to more than double by 2024, as electric vehicle battery production quadruples. https://www.globaldata.com/global-lithium-demand-double-2024-electric-vehicle-battery-production-quadruples/.

Golmohammadzadeh, R., Rashchi, F., & Vahidi, E. (2017). Recovery of lithium and cobalt from spent lithium-ion batteries using organic acids: Process optimization and kinetic aspects. *Waste Management*, *64*, 244—254. Available from https://doi.org/10.1016/j.wasman.2017.03.037.

Guo, X., Cao, X., Huang, G., Tian, Q., & Sun, H. (2017). Recovery of lithium from the effluent obtained in the process of spent lithium-ion batteries recycling. *Journal of Environmental Management*, *198*, 84—89. Available from https://doi.org/10.1016/J.JENVMAN.2017.04.062.

He, L.-P., Sun, S.-Y., Mu, Y.-Y., Song, X.-F., & Yu, J.-G. (2016). Recovery of lithium, nickel, cobalt, and manganese from spent lithium-ion batteries using L-tartaric acid as a leachant. *ACS Sustainable Chemistry and Engineering*, *5*(1), 714—721. Available from https://doi.org/10.1021/ACSSUSCHEMENG.6B02056.

He, L. P., Sun, S. Y., Song, X. F., & Yu, J. G. (2017). Leaching process for recovering valuable metals from the LiNi1/3Co1/3Mn1/3O$_2$ cathode of lithium-ion batteries. *Waste Management*, *64*, 171—181. Available from https://doi.org/10.1016/J.WASMAN.2017.02.011.

Heelan, J., Gratz, E., Zheng, Z., Wang, Q., Chen, M., Apelian, D., & Wang, Y. (2016). Current and prospective Li-ion battery recycling and recovery processes. JOM 68(10), 2632–2638. Available from https://doi.org/10.1007/S11837-016-1994-Y.

Hesselbach, J., & Herrmann, C. (2011). Glocalized solutions for sustainability in manufacturing. In *Proceedings of the 18th CIRP international conference on life cycle engineering*. Technische. https://books.google.com/books? hl = en&lr = &id = ZE9_BXSVoTwC&oi = fnd&pg = PR3&dq = Hesselbach, + J., + and + Herrmann, + C. + . + Glocalized + solutions + for + sustainability + in + manufacturing. + Proceedings + of + the + 18th + CIRP + International + Conference + on + Life + Cycle + Engineering + .&ots = xP8vs EbwmA&sig = qYawXjHAzIExjzHGroMNzvVBM5Q.

Horeh, N. B., Mousavi, S. M., & Shojaosadati, S. A. (2016). Bioleaching of valuable metals from spent lithium-ion mobile phone batteries using *Aspergillus niger*. *Journal of Power Sources*, *320*, 257–266. Available from https://doi.org/10.1016/J.JPOWSOUR.2016.04.104.

Joulié, M., Laucournet, R., & Billy, E. (2014). Hydrometallurgical process for the recovery of high value metals from spent lithium nickel cobalt aluminum oxide based lithium–ion batteries. *Journal of Power Sources*, *247*, 551–555. Available from https://doi.org/10.1016/J.JPOWSOUR.2013.08.128.

Kang, J., Senanayake, G., Sohn, J., & Shin, S. M. (2010). Recovery of cobalt sulfate from spent lithium ion batteries by reductive leaching and solvent extraction with Cyanex 272. *Hydrometallurgy*, *100*(3–4), 168–171. Available from https://doi.org/10.1016/J.HYDROMET.2009.10.010.

Kannan, D., Govindan, K., & Shankar, M. (2016). Formalize recycling of electronic waste. Nature 530 (7590), 281. Available from https://doi.org/10.1038/530281b.

Khaliq, A., Rhamdhani, M. A., Brooks, G., & Masood, S. (2014). Metal extraction processes for electronic waste and existing industrial routes: A review and Australian perspective. *Resources*, 152–179. Available from https://doi.org/10.3390/resources3010152.

Kim, D. S., Sohn, J. S., Lee, C. K., Lee, J. H., Han, K. S., & Lee, Y.-I. (2004). Simultaneous separation and renovation of lithium cobalt oxide from the cathode of spent lithium ion rechargeable batteries. Journal of Power Sources, 132(1–2), 145–149. Available from https://doi.org/10.1016/j.jpowsour.2003.09.046.

King, S., Boxall, N. J., & Bhatt, A. I. (2018). Lithium battery recycling in Australia. (pp. 1–74). CSIRO, Australia. https://www.csiro.au/en/Research/EF/Areas/Energy-storage/Battery-recycling.

Ku, H., Jung, Y., Jo, M., Park, S., Kim, S., Yang, D., Rhee, K., An, E. M., Sohn, J., & Kwon, K. (2016). Recycling of spent lithium-ion battery cathode materials by ammoniacal leaching. *Journal of Hazardous Materials*, *313*, 138–146. Available from https://doi.org/10.1016/J.JHAZMAT.2016.03.062.

Lain, M. J. (2001). Recycling of lithium ion cells and batteries. *Journal of Power Sources*, *97–98*, 736–738. Available from https://doi.org/10.1016/S0378-7753(01)00600-0.

Lee, C.-K., & Rhee, K. (2002). Preparation of $LiCoO_2$ from spent lithium-ion batteries. *Journal of Power Sources*, *109*, 17–21.

Lee, C. K., & Rhee, K. I. (2003). Reductive leaching of cathodic active materials from lithium ion battery wastes. *Hydrometallurgy*, *68*(1–3), 5–10. Available from https://doi.org/10.1016/S0304-386X (02)00167-6.

Li, H., Xing, S., Liu, Y., Li, F., Guo, H., & Kuang, G. (2017). Recovery of lithium, iron, and phosphorus from spent $LiFePO_4$ batteries using stoichiometric sulfuric acid leaching system. *ACS Sustainable Chemistry and Engineering*, *5*(9), 8017–8024. Available from https://doi.org/10.1021/ACSSUSCHEMENG.7B01594.

Li, J., Shi, P., Wang, Z., Chen, Y., & Chang, C. C. (2009). A combined recovery process of metals in spent lithium-ion batteries. *Chemosphere*, 77(8), 1132–1136. Available from https://doi.org/10.1016/ j.chemosphere.2009.08.040.

Li, J., Zhao, R., He, X., & Liu, H. (2008). Preparation of $LiCoO_2$ cathode materials from spent lithium-ion batteries. *Ionics*, *15*, 111–113. Available from https://doi.org/10.1007/s11581-008-0238-8.

Li, L., Lu, J., Ren, Y., Zhang, X. X., Chen, R. J., Wu, F., & Amine, K. (2012). Ascorbic-acid-assisted recovery of cobalt and lithium from spent Li-ion batteries. *Journal of Power Sources*, *218*, 21–27. Available from https://doi.org/10.1016/j.jpowsour.2012.06.068.

Li, L., Qu, W., Zhang, X., Lu, J., Chen, R., Wu, F., & Amine, K. (2015). Succinic acid-based leaching system: A sustainable process for recovery of valuable metals from spent Li-ion batteries. *Journal of Power Sources*, *282*, 544–551. Available from https://doi.org/10.1016/j.jpowsour.2015.02.073.

Li, L., Zhai, L., Zhang, X., Lu, J., Chen, R., Wu, F., & Amine, K. (2014). Recovery of valuable metals from spent lithium-ion batteries by ultrasonic-assisted leaching process. *Journal of Power Sources*, *262*, 380−385. Available from https://doi.org/10.1016/J.JPOWSOUR.2014.04.013.

Lundgren, K. (2012). *The global impact of e-waste: Addressing the challenge*. International Labour Office. Available from http://www.ilo.org/sector/Resources/publications/WCMS_196105/lang-en/index.htm.

Lupi, C., & Pasquali, M. (2003). Electrolytic nickel recovery from lithium-ion batteries. *Minerals Engineering*, *16*(6), 537−542. Available from https://doi.org/10.1016/S0892-6875(03)00080-3.

Matter, A., Dietschi, M., & Zurbrügg, C. (2013). Improving the informal recycling sector through segregation of waste in the household − The case of Dhaka Bangladesh. *Habitat International*, *38*, 150−156. Available from https://doi.org/10.1016/J.HABITATINT.2012.06.001.

McLaughlin, W., & Adams, T. S. (1999). *Li reclamation process*. Toxco (Patent No. 5888463). Available from https://patents.google.com/patent/US5888463.

Meshram, P., Abhilash., Pandey, B. D., Mankhand, T. R., & Deveci, H. (2016). Comparision of different reductants in leaching of spent lithium ion batteries. *JOM*, *68*(10), 2613−2623. Available from https://doi.org/10.1007/s11837-016-2032-9.

Meshram, P., Pandey, B. D., & Mankhand, T. R. (2014). Extraction of lithium from primary and secondary sources by pre-treatment, leaching and separation: A comprehensive review. *Hydrometallurgy*, *150*, 192−208. Available from https://doi.org/10.1016/j.hydromet.2014.10.012.

Meshram, P., Pandey, B. D., & Mankhand, T. R. (2015). Hydrometallurgical processing of spent lithium ion batteries (LIBs) in the presence of a reducing agent with emphasis on kinetics of leaching. *Chemical Engineering Journal*, *281*, 418−427. Available from https://doi.org/10.1016/J.CEJ.2015.06.071.

Mishra, D., Kim, D. J., Ralph, D. E., Ahn, J. G., & Rhee, Y. H. (2008). Bioleaching of metals from spent lithium ion secondary batteries using *Acidithiobacillus ferrooxidans*. *Waste Management*, *28*(2), 333−338. Available from https://doi.org/10.1016/j.wasman.2007.01.010.

Mobilityhouse. (n.d.). *Nissan and Eaton provide efficient battery storage for Amsterdam ArenA*. Retrieved August 15, 2021, from https://www.mobilityhouse.com/int_en/magazine/press-releases/the-mobility-house-provides-energy-solution-in-amsterdam-arena-project.html.

Myoung, J., Jung, Y., Lee, J., & Tak, Y. (2002). Cobalt oxide preparation from waste $LiCoO_2$ by electrochemical − hydrothermal method. *Journal of Power Sources*, *112*, 639−642.

Nan, J., Han, D., Yang, M., & Cui, M. (2005). Dismantling, recovery, and reuse of spent nickel−metal hydride batteries. *Journal of the Electrochemical Society*, *153*(1), A101. Available from https://doi.org/10.1149/1.2133721.

Nayaka, G. P., Pai, K. V., Manjanna, J., & Keny, S. J. (2016). Use of mild organic acid reagents to recover the Co and Li from spent Li-ion batteries. *Waste Management*, *51*, 234−238. Available from https://doi.org/10.1016/J.WASMAN.2015.12.008.

Nayaka, G. P., Pai, K. V., Santhosh, G., & Manjanna, J. (2016). Recovery of cobalt as cobalt oxalate from spent lithium ion batteries by using glycine as leaching agent. *Journal of Environmental Chemical Engineering*, *4*(2), 2378−2383. Available from https://doi.org/10.1016/j.jece.2016.04.016.

Nayl, A. A., Elkhashab, R. A., Badawy, S. M., & El-Khateeb, M. A. (2017). Acid leaching of mixed spent Li-ion batteries. *Arabian Journal of Chemistry*, *10*, S3632−S3639. Available from https://doi.org/10.1016/j.arabjc.2014.04.001.

Olsson, L., Fallahi, S., Schnurr, M., Diener, D., & van Loon, P. (2018). Circular business models for extended EV battery life. *Batteries*, *4*(4). Available from https://doi.org/10.3390/BATTERIES4040057.

Pagnanelli, F., Moscardini, E., Altimari, P., Abo Atia, T., & Toro, L. (2016). Cobalt products from real waste fractions of end of life lithium ion batteries. *Waste Management*, *51*, 214−221. Available from https://doi.org/10.1016/J.WASMAN.2015.11.003.

Pascale, A., Sosa, A., Bares, C., Battocletti, A., Moll, M. J., Pose, D., Laborde, A., González, H., & Feola, G. (2016). E-waste informal recycling: An emerging source of lead exposure in South America. *Annals of Global Health*, *82*(1), 197−201. Available from https://doi.org/10.1016/j.aogh.2016.01.016.

Paulino, J. F., Busnardo, N. G., & Afonso, J. C. (2008). Recovery of valuable elements from spent Li-batteries. *Journal of Hazardous Materials*, *150*(3), 843−849. Available from https://doi.org/10.1016/J.JHAZMAT.2007.10.048.

Pinna, E. G., Ruiz, M. C., Ojeda, M. W., & Rodriguez, M. H. (2017). Cathodes of spent Li-ion batteries: Dissolution with phosphoric acid and recovery of lithium and cobalt from leach liquors. *Hydrometallurgy*, *167*, 66−71. Available from https://doi.org/10.1016/J.HYDROMET.2016.10.024.

Pudas, J., Erkkila, A., & Viljamaa, J. (2015). *Battery recycling method* (Patent No. US8979006B2). Available from https://patents.google.com/patent/US8979006B2/en.

Purushothaman, M., Gousuddin Inamdar, M., & Muthunarayanan, V. (2020). Socio-economic impact of the e-waste pollution in India. *Materials Today: Proceedings*, *37*(Part 2), 280−283. Available from https://doi.org/10.1016/j.matpr.2020.05.242.

Qiao, H., & Wei, Q. (2012). 10 − Functional nanofibers in lithium-ion batteries. In Q. Wei (Ed.), *Functional nanofibers and their applications* (pp. 197−208). Woodhead Publishing. Available from https://doi.org/10.1533/9780857095640.2.197.

Ra, D.-il, & Han, K. S. (2006). Used lithium ion rechargeable battery recycling using Etoile-Rebatt technology. *Journal of Power Sources*, *163*(1), 284−288. Available from https://doi.org/10.1016/j.jpowsour.2006.05.040.

Rahman, M. A. (2017). E-waste management: A study on legal framework and institutional preparedness in Bangladesh. *The Cost and Management*, *45*(1), 28−35. Available from http://www.icmab.org.bd/images/stories/journal/2017/Jan-Feb/5.E-waste.pdf.

Retriev. (n.d.). *Lithium ion*. Retriev Technologies. Retrieved August 17, 2021, from https://www.retrievtech.com/lithiumion.

Savino, A., Solorzano, G., Quispe, C., & Correal, M.C. (2018). Waste *management outlook for* Latin America and the Caribbean. https://wedocs.unep.org/handle/20.500.11822/26448.

Shareef, J. (2019). *New energy futures paper: Batteries & the circular economy* (pp. 1−71). <moz-extension://a120fea7-04a2-a746-bbfb-135ac8b68d8b/enhanced-reader.html?openApp&pdf = https%3A%2F%2Fblob-static.vector.co.nz%2Fblob%2Fvector%2Fmedia%2Fvector%2Fvector_new_energy_futures_paper_batteries.pdf>.

Sharma, T. (2021). *India's largest battery recycling company eyes electric vehicle battery business*. https://www.moneycontrol.com/news/business/companies/indias-largest-battery-recycling-company-eyes-electric-vehicle-battery-business-7096671.html.

Shin, S. M., Kim, N. H., Sohn, J. S., Yang, D. H., & Kim, Y. H. (2005). Development of a metal recovery process from Li-ion battery wastes. *Hydrometallurgy*, *79*(3−4), 172−181. Available from https://doi.org/10.1016/j.hydromet.2005.06.004.

Sojka, R., Pan, Q., & Billmann, L. (2020). Comparative study of Li-ion battery recycling processes. https://accurec.de/wp-content/uploads/2021/04/Accurec-Comparative-study.pdf.

Sun, L., & Qiu, K. (2012). Organic oxalate as leachant and precipitant for the recovery of valuable metals from spent lithium-ion batteries. *Waste Management*, *32*(8), 1575−1582. Available from https://doi.org/10.1016/J.WASMAN.2012.03.027.

Swain, B., Jeong, J., Lee, J.-c, Lee, G. H., & Sohn, J. S. (2007). Hydrometallurgical process for recovery of cobalt from waste cathodic active material generated during manufacturing of lithium ion batteries. *Journal of Power Sources*, *167*(2), 536−544. Available from https://doi.org/10.1016/J.JPOWSOUR.2007.02.046.

Velázquez-Martínez, O., Valio, J., Santasalo-Aarnio, A., Reuter, M., & Serna-Guerrero, R. (2019). A critical review of lithium-ion battery recycling processes from a circular economy perspective. *Batteries*, *5*(4), 5−7. Available from https://doi.org/10.3390/batteries5040068.

Virolainen, S., Fallah Fini, M., Laitinen, A., & Sainio, T. (2017). Solvent extraction fractionation of Li-ion battery leachate containing Li, Ni, and Co. *Separation and Purification Technology*, *179*, 274−282. Available from https://doi.org/10.1016/j.seppur.2017.02.010.

Wang, F., Sun, R., Xu, J., Chen, Z., & Kang, M. (2016). Recovery of cobalt from spent lithium ion batteries using sulphuric acid leaching followed by solid−liquid separation and solvent extraction. *RSC Advances*, *6*(88), 85303−85311. Available from https://doi.org/10.1039/C6RA16801A.

Wang, R. C., Lin, Y. C., & Wu, S. H. (2009). A novel recovery process of metal values from the cathode active materials of the lithium-ion secondary batteries. *Hydrometallurgy, 99*(3–4), 194–201. Available from https://doi.org/10.1016/j.hydromet.2009.08.005.

Wilson, A. M., Bailey, P. J., Tasker, P. A., Turkington, J. R., Grant, R. A., & Love, J. B. (2013). Solvent extraction: The coordination chemistry behind extractive metallurgy. *Chemical Society Reviews, 43*(1), 123–134. Available from https://doi.org/10.1039/C3CS60275C.

Xiao, J., Li, J., & Xu, Z. (2017). Novel approach for in situ recovery of lithium carbonate from spent lithium ion batteries using vacuum metallurgy. *Environmental Science & Technology*. Available from https://doi.org/10.1021/acs.est.7b02561.

Xu, J., Thomas, H. R., Francis, R. W., Lum, K. R., Wang, J., & Liang, B. (2008). A review of processes and technologies for the recycling of lithium-ion secondary batteries. *Journal of Power Sources, 177*(2), 512–527. Available from https://doi.org/10.1016/j.jpowsour.2007.11.074.

Yang, Y., Zheng, X., Cao, H., Zhao, C., Lin, X., Ning, P., Zhang, Y., Jin, W., & Sun, Z. (2017). A closed-loop process for selective metal recovery from spent lithium iron phosphate batteries through mechanochemical activation. *ACS Sustainable Chemistry and Engineering, 5*(11), 9972–9980. Available from https://doi.org/10.1021/ACSSUSCHEMENG.7B01914.

Zeng, X., Li, J., & Shen, B. (2015). Novel approach to recover cobalt and lithium from spent lithium-ion battery using oxalic acid. *Journal of Hazardous Materials, 295*, 112–118. Available from https://doi.org/10.1016/J.JHAZMAT.2015.02.064.

Zhang, P., Yokoyama, T., Itabashi, O., Suzuki, T. M., & Inoue, K. (1998). Hydrometallurgical process for recovery of metal values from spent lithium-ion secondary batteries. *Hydrometallurgy, 47*(2–3), 259–271. Available from https://doi.org/10.1016/S0304-386X(97)00050-9.

Zhang, P., Yokoyama, T., Itabashi, O., Wakui, Y., Suzuki, T. M., & Inoue, K. (1998). Hydrometallurgical process for recovery of metal values from spent nickel-metal hydride secondary batteries. *Hydrometallurgy, 50*(1), 61–75. Available from https://doi.org/10.1016/S0304-386X(98)00046-2.

Zhang, X., Li, L., Fan, E., Xue, Q., Bian, Y., Wu, F., & Chen, R. (2018). Toward sustainable and systematic recycling of spent rechargeable batteries. *Chemical Society Reviews, 47*(19), 7239–7302. Available from https://doi.org/10.1039/C8CS00297E.

Zheng, R., Zhao, L., Wang, W., Liu, Y., Ma, Q., Mu, D., Li, R., & Dai, C. (2016). Optimized Li and Fe recovery from spent lithium-ion batteries via a solution–precipitation method. *RSC Advances, 6*(49), 43613–43625. Available from https://doi.org/10.1039/C6RA05477C.

Zheng, X., Gao, W., Zhang, X., He, M., Lin, X., Cao, H., Zhang, Y., & Sun, Z. (2017). Spent lithium-ion battery recycling — Reductive ammonia leaching of metals from cathode scrap by sodium sulphite. *Waste Management, 60*, 680–688. Available from https://doi.org/10.1016/J.WASMAN.2016.12.007.

Zheng, X., Zhu, Z., Lin, X., Zhang, Y., He, Y., Cao, H., & Sun, Z. (2018). A mini-review on metal recycling from spent lithium ion batteries. *Engineering, 4*(3), 361–370. Available from https://doi.org/10.1016/J.ENG.2018.05.018.

Zhu, S. G., He, W. Z., Li, G. M., Zhou, X., Zhang, X. J., & Huang, J. W. (2012). Recovery of Co and Li from spent lithium-ion batteries by combination method of acid leaching and chemical precipitation. *Transactions of Nonferrous Metals Society of China, 22*(9), 2274–2281. Available from https://doi.org/10.1016/S1003-6326(11)61460-X.

Zou, H., Gratz, E., Apelian, D., & Wang, Y. (2013). A novel method to recycle mixed cathode materials for lithium ion batteries. *Green Chemistry, 15*(5), 1183–1191. Available from https://doi.org/10.1039/c3gc40182k.

성일하이텍(주). (n.d.). Retrieved August 18, 2021, from http://www.sungeelht.com/business.php.

CHAPTER 6

Hydrometallurgy: urban mining of E-waste and its future implications

Aneri Patel[1], Sunil Kumar[1] and Abhishek Awasthi[2]
[1]Waste Re-processing Division, CSIR-National Environmental Engineering Research Institute (CSIR-NEERI), Nagpur, Maharashtra, India
[2]Tsinghua University, Beijing, P.R. China

Abbreviations

Mt	million metric ton
EEE	electrical and electronic equipment
PCB	printed circuit board
CE	circular economy
CP	cleaner production
LM	lean manufacturing
EPR	extended producer responsibility
SDGs	sustainable development goals

6.1 Introduction

Electronic and electrical equipment (EEE) is innovated to satisfy advanced human needs world wide. Nowadays, humans have shifted to more comfortable and time-saving devices for their daily routine ranging from small to large electric gadgets for their convenience from leisure to hygiene. E-waste with its fast growth rate is expected to reach 74 million metric tons (Mt) by 2030. E-waste generation has risen by 8.9 Mt in just 3 years since 2016, as per E-Waste monitor 2020 data. The aforementioned number shows that the maximum amount of E-Waste generated was coming from the small electric gadget, whose end of life was accounted to be very limited (Forti et al., 2020). The legislative framework of E-Waste management is already been signed by 78 countries across the globe (Forti et al., 2020), but it still remains in an area that is unfocused and inappropriately financed (Forti et al., 2020; Kazançoglu et al., 2020). The extensively rising order of elite designs and productions aspects were observed under the umbrella of the E-Waste management framework which readily focuses on the circular economy (CE) guidelines as well (Forti et al., 2020). Control over demands falls under the acceptability of humans which takes higher stakes to

Global E-waste Management Strategies and Future Implications
DOI: https://doi.org/10.1016/B978-0-323-99919-9.00007-6
105

overcome the behavioral change toward sustainability (CNBC International, 2018). The major factor of this change tends to depend on profitability (Zeng et al., 2021). Profitability can be earned by mining the urban waste accessible in the surroundings as per the E-waste generated reflected in figures. Studies have noted that recycling the printed circuit board (PCB), which is the heart of EEE, can enable us to recover higher concentrations of precious metals in comparison with the ores being mined (Tuncuk et al., 2012). We need to accept the CE concept in our business opportunities for balancing the economy along with nature (Velenturf & Purnell, 2021). On the other hand, we are very well aware of the monetary load of the mining activities along with the resources, and energy required for them (Graedel, 2011; Zeng et al., 2021). Virgin metals also bring several environmental threats at the time of mining (Zeng et al., 2021).

Graedel mentions that prudence is in adapting recycled metals instead of wasting metal ores gifted us billion years ago by stars. Several studies are held on the cost–benefit analysis for understanding the economic loads of both conventional ore mining and urban mining (Zeng et al., 2018, 2021). A study suggests that urban mining will be accepted industrially in the future due to its better cost–benefit performance (Zeng et al., 2021). These statements encourage recycling and reusing from the waste and increase product life within a closed-loop, and this way the concept of CE can be well maintained (Zeng et al., 2021). The E-waste sector is not yet completely formalized (Arya et al., 2021; Sharma et al., 2021) thus the recycling via informal sectors uses displeasing techniques which have detrimental effects on the people closely dealing with E-waste recycling (Arya et al., 2021; Sharma et al., 2021). Instant resources availability and higher scopes of making cash from E-waste have indeed branched out the informal sectors.

Traditionally, formal and informal sectors extract the metals via two approaches: pyrometallurgy and hydrometallurgy (Arya et al., 2021). Bio-hydrometallurgy came along later. The driving force of the acceptability of these approaches is the financial ability of the countries; usually developed countries are using pyrometallurgy or hydrometallurgy as per their industrial scalability (Zeng et al., 2021). Developing countries are commonly accepting hydrometallurgical techniques due to their comparatively lower capital investments and effective results (Sharma et al., 2021). This book chapter's focal point is on urban mining via hydrometallurgical techniques and their eligibilities based on the CE concept.

6.2 Urban mining of E-waste

Urban mining is an idea that originated from landfill mining which was brought to light for recuperating the urban waste (Sharma et al., 2021). Virgin mining is a well-industrialized sector where the amount of advancement is now saturated and urban mining has a new scope that may bring value to the obsolete materials (Kazançoglu et al., 2020). Urban mining is initially conceptualized to create wealth from all the anthropogenic minerals (Zeng et al., 2021). Economic and ecological benefits of urban mining are more evidently observed as the anthropogenic minerals become rich in waste material, increasing its recycling potential (Zeng et al., 2021). E-waste is a metal-rich source; hence it is eligible for settling in the urban mining concept (Kazançoglu et al., 2020; Zeng et al., 2018). The metals present in the waste gadgets are already refined and are in pure form in most of the parts except from some complex composition parts which is an essential advantage of urban mining (Graedel, 2011). A study says that urban mining scope can be evaluated by finding answers to the following questions: amount of metal existing in nature from the E-waste, when the stock of that E-waste will be made available to us, and in what form it will be available (Graedel, 2011). Finding answers to these questions will guide us to assess the urban mining potential from E-waste. The challenge for urban mining lies where we demand the utmost performance in our products; meanwhile, the same product must have the decent potential of being recycled (Graedel, 2011). It will take quite a time to reach such a stage by product designers to go completely sustainable but the ready solution to urban mining is to look for more competent resulting treatment techniques. E-waste can be plausibly recycled via pyrometallurgy, hydrometallurgy, bio-hydrometallurgy, and hybrid leaching techniques. This chapter focuses on the chemical leaching techniques which are widely used methods with evident results.

6.2.1 Urban mining via hydrometallurgical techniques

The urban mining concept demands regeneration of maximum resources possibly by finding competent recycling solutions. Extraction of metals via hydrometallurgy is noted to be a more foreseeable technique using a set of chemicals. Hydrometallurgy involves chemical leaching methods, such as ligands, acid leaching, and etching methods (Iannicelli-zubiani et al., 2017; Pant et al., 2012). Hydrometallurgy has grounded its roots in the industrial market for several decades and is widely accepted by recyclers due to its effective results with low-cost operations. Hydrometallurgical treatments are usually water-intensive and it is dealing with different chemicals during the treatment, polluting a large amount of water which cannot be discharged into the natural streams without treating the wastewater generated (Iannicelli-zubiani et al., 2017; Tuncuk et al., 2012). Additionally, waste residue after the leaching process is an untouched area of research with very limited studies available. A study of toxicity characterization leaching procedures showed that the levels of copper and lead are expected to be hazardous if not properly disposed of in the environment even after applying the halide leaching process (Arya et al., 2021). Details of the various leaching techniques

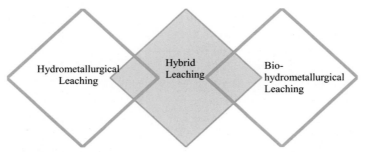

Figure 6.1 Hydrometallurgical techniques.

of hydrometallurgy are shown in Fig. 6.1, and Fig. 6.2 which shows different chemicals used in these leaching techniques which are practiced on the ground for decades. Presently, there is a constant emergence of new leaching techniques, of which some of them are as follows:

1. Cyanide leaching
2. Halide leaching
3. Thiourea leaching
4. Thiosulfate leaching
5. Bio-hydrometallurgical leaching
6. Hybrid leaching

6.2.1.1 Cyanide leaching

Traditionally in hydrometallurgy, cyanide leaching is the oldest amongst halide, thiourea, thiosulfate leaching, or any of the acid leaching techniques. Cyanide leaching is a nearly 1.18 years older technique for the extraction of metals such as gold, silver, platinum, and palladium (Cui & Zhang, 2008), simultaneously being cheaper than its alternatives (Hilson & Monhemius, 2006). Cyanide leaching is said to be known as a toxic leaching procedure for which all these effective alternative leaching procedures were taken into considerations like thiourea and thiosulfate leaching. Most of the serious accidental cases of contamination are caused due to cyanide leaching; hence this leaching technique should be archived from the conventional leaching techniques (Cui & Zhang, 2008; Hilson & Monhemius, 2006).

6.2.1.2 Halide leaching

Metal dissolutions of valuable and precious metals by halide leaching process provide good results using chlorine, bromine, fluorine, iodine, and astatine, out of which

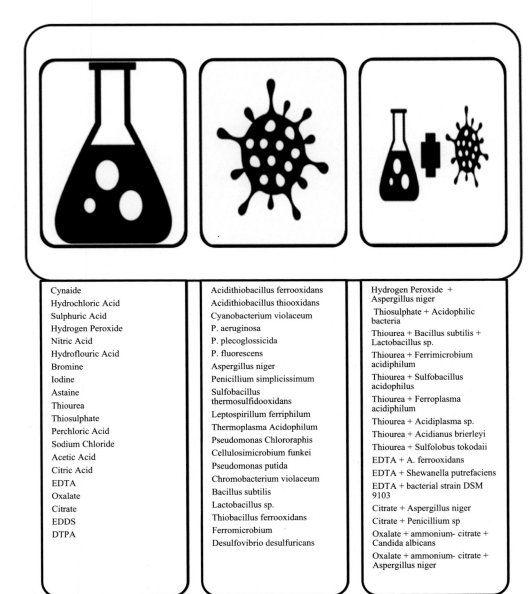

Figure 6.2 Different chemicals used in hydrometallurgical leaching technique (Arya & Kumar, 2020; Chauhan et al., 2018; Cui & Anderson, 2016; Cui & Zhang, 2008; Jadhav & Hocheng, 2012; Jing-ying et al., 2012; Pant et al., 2012; Rizki et al., 2019; Sheel & Pant, 2021).

chlorine leaching is extensively used in industries (Cui & Zhang, 2008), selectively for Cu extraction (Kaya, 2018). Nitric acid and hydrochloric acid are the most commonly used halides in hydrometallurgical leaching which leaches valuable and precious metals. Aqua regia is a combination of two highly reactive acids, hydrochloric acid and nitric acid with the 3:1 ratio v/v solution (Zhong et al., 2014), whereas the informal sectors prefer using this acid more often for leaching. Usually, Aqua regia leaching is a nonselective metal leaching technique due to its aggressive digestion (Kaya, 2018). Extremely poisonous chlorine gas is released while leaching, giving the worker undesirable health effects (Cui & Zhang, 2008). Halogens are considered to be the noncyanide lixiviants for leaching precious metals. Furthermore, it is noted that halogen leaching have good chemical stability during the leaching process (Akcil et al., 2015; Tuncuk et al., 2012). In the meantime, the possibilities of halide leaching still are studied rigorously until a correct alternative is established in the field of E-waste recycling.

6.2.1.3 Thiourea leaching

Thiourea is a considerably costly technique for the extraction of precious metals (Jing-ying et al., 2012). Gold leaching is successfully caused by organosulfur compound (NH_2CSNH_2) which nearly gives a 99% effective result (Cui & Zhang, 2008; Hilson & Monhemius, 2006). Despite being an expensive technique, it is observed that it has received enormous consideration as an alternative to cyanide in the past half-century (Hilson & Monhemius, 2006). Studies explain that the rate of gold leaching improves as the ferric ion is added to the solution forming complexions with thiourea (Cui & Zhang, 2008; Ilankoon et al., 2018; Jing-ying et al., 2012). However, high thiourea leaching rates may possess carcinogenic effects in some cases caused due to lower chemical stability (Akcil et al., 2015). Thiourea to an additional extent has now experimented with several combinations with microbial leaching aiming to establish a cost-effective method without disturbing any ecological system (Sheel & Pant, 2021).

6.2.1.4 Thiosulfate leaching

Thiosulfate leaching does not have any issue related to corrosivity because of its alkaline nature (To Alternatives and Cyanide I N Gold, 2008). Au leaching via ammonium thiosulfate has faster reaction kinetics (Akcil et al., 2015). Thiosulfate leaching includes compounds of ammonium thiosulfate and ammonium sulfate which are potentially used as fertilizers, creating an advantage of being an eco-friendly technology (To Alternatives and Cyanide I N Gold, 2008). The bottleneck for thiosulfate leaching is its low chemical stability which causes difficulties in standing out in the commercial market; however, cost-effectiveness is found to be high (Akcil et al., 2015).

6.2.1.5 Biohydrometallurgical leaching

Bio-hydrometallurgy is a microbial leaching technique where the microbes convert the metals from insoluble into a soluble form (Isildar, 2018; Marra et al., 2018). It is an unindustrialized leaching technique to the urban mining sector which is rapidly emerging (Isildar, 2018; Jia et al., 2020; Sheel & Pant, 2021) unlike conventional approaches without wounding nature. Bioleaching is free from the huge material loads due to which it is a low-cost process (Jia et al., 2020). In times where the CE is promoted rigorously, bio-hydrometallurgical leaching can be a new bioeconomy for the future. Bioleaching is a greener approach for recovering metals from E-waste (Isildar, 2018; Jia et al., 2020) but it still remains infancy due to its demand for specific conditions required for the survival of microorganisms (Akcil et al., 2015; Chauhan et al., 2018).

6.2.1.6 Hybrid leaching

Hybrid technology involves a combination of chemical and biological leaching process which uses safer chemicals and supportive microbes to have efficient results and in trimming down the environmental hazards (Pant et al., 2012; Sheel & Pant, 2021). A variety of combinations were found efficient in the hybrid leaching technique such as a combination of Ethylenediaminetetraacetic acid (EDTA -A ligand) with *Acidithiobacillus ferrooxidans* bacteria or with bacterial strain DSM 9130 (Pant et al., 2012). Thiourea leaching is also effectively combined with the *Acidiplasma* sp. which can dissolve Au nearly by 98% (Rizki et al., 2019). Similarly, studies on leaching lithium-ion batteries via hybrid leaching combinations are also presented in research articles (Sheel & Pant, 2021). Investigations on hybrid combinations are extensively studied for reducing the time consumed by the bioleaching technique alone.

6.3 Circular economy and urban mining via hydrometallurgy

The fundamental of a CE is to sustain the economic balance, at the same time not exploiting nature. The elimination of the waste and pollution entailed by-products can be carefully taken as a raw material for other products which simultaneously maintains the balance of natural systems. This natural ecosystem is conveniently explained in Fig. 6.3, incorporating the CE concept. CE is highly dependent on environmental, social, and political factors to secure resources (Velenturf & Purnell, 2021). Governments and private parties should be encouraged to design business models/schemes, considering all the factors of CE (Velenturf & Purnell, 2021). Reusing and recycling the products are core tools to promote the CE. Urban mining is an ameliorate choice that assists us to dig out the supplies from discarded items that have already served their primary purpose. Waste management agencies and recyclers are majorly dependent on waste/discarded products for which they require an efficient business model (accepting CE concept) for building a profitable business. Such efficient

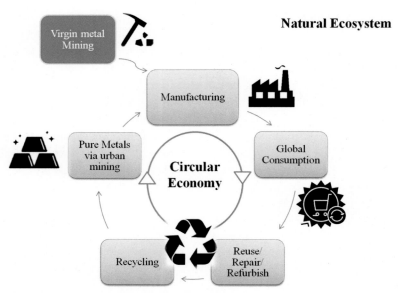

Figure 6.3 Circular economy concept in E-waste recycling.

business models for building profitable business need to set their actions based on the ruling criteria to upgrade their models as per the CE concept. Hydrometallurgy being used extensively in E-waste management brings out a bag of opportunities to improvise their sector by including CE aspects in the field. E-waste managament is a rapidly growing sector and on infiltrating CE concept in the sector, will help us expedite sustainability with a boost. The crux of including CE in hydrometallurgy is that it provides us the bandwidth to choose alternative chemicals, process alternative, based on its environmental, economic, and societal status.

6.3.1 Circular economy ruling criteria

The authors Velenturf and Purnell have thoughtfully encapsulated the detailed concept of CE whereas, "European Environment – An agency" has also released a short video explaining the concept of CE in 2014 (EU Environment, 2014). This chapter has articulated the essential ruling criteria which ladle out the CE concept. The concept of CE can be achieved when we assess the existing system through the ruling criteria which are as follows:

1. Cleaner Production (CP) and Lean Manufacturing (LM)
2. Ecodesign Initiation
3. Incorporating Industrial Ecology
4. Strengthening Reverse Logistics
5. Adapting Functional Economy

Applying these ruling criteria in the E-waste management stream can help us switch over from a linear economy to CE. Minimizing the resource inputs and energy consumptions are the two most evident solutions to move toward maintaining a sustainable offset in the existing business. These criteria are narrowed down from different research articles which particularly focused on the CE concept. A study says that we need to set planetary boundaries on the rates of renewable/nonrenewable inputs and biological/technical outputs to balance CE and the biosphere (Suárez-Eiroa et al., 2019). The ruling criteria here are aiming to help us set limits to human interventions, to eliminate the threats to nature. Studies explain that the baseline of the CE concept is to apply CP techniques in the life cycle of products or the existing process. Urban mining in E-waste management itself is aiming to recover the secondary raw materials from the E-waste which beautifully fits in all the factors defined to endorse the CE concept (Kazançoglu et al., 2020; Khayyam Nekouei et al., 2020). Since transitioning to CE will take quite time to convert the contemporary practices of the well-established industrial sector.

6.3.1.1 Ruling criteria 1: cleaner production and lean manufacturing

CE concept can be incorporated into the existing industrial processes, allowing us to make mindful decisions to move toward LM. CP and LM concepts are two approaches through which industrial sectors can design an efficient process that will not be a barrier for nature and society for the longer run. LM is where we keep our center of attention on the cost and time reductions with abrupt impacts on the manufactured goods and their market status (Gomes da Silva, José & Gouveia, 2019). Such lean productions are rewarded only after a thorough understanding of the process line or the product requirement. The industrial sector needs to upgrade its process and product lines after completely understanding its product requirement and applying the essential mitigate measures to make its existing system efficient and profitable. CP aims to continuously reduce risks on humans and the environment, building the strategies which are applied to the processes, products, and services altogether, along with increasing the eco-efficiency (Gomes da Silva, José & Gouveia, 2019). Several decision-making tools can help pull off the LM aspects and strategize CP aspects as well. Tools such as Material Flow Analysis, Life Cycle Assessment, and Substance Flow Analysis are support builders to get to the CP and LM concept.

CP and LM approach in hydrometallurgical leaching can be applied by driving many different factors such as adjusting the dosage of the chemicals, monitoring the residues and their toxicity levels, and analyzing expenses involved during the leaching process. Extracting metals via chemical leaching does pollute the water stream but applying CP strategies bring out the greener substitutes for the chemicals. Based on the environmental risks and effective results of the chemical leaching technologies the feasibility of the CP and LM was observed. The alternatives chosen for cyanide leaching were thiourea and thiosulfate leaching which were greener substitutes (Cui & Zhang, 2008; Hilson & Monhemius, 2006). Cyanide leaching was found cheaper in cost and its results are highly effective for recovering metals but at the cost of damaging nature which is not feasible as per the CP and LM approach (Cui & Zhang, 2008; Hilson & Monhemius, 2006). Halide leaching is acceptable due to its accurate and predictable results (Arya et al., 2021). Depending upon the type of halogen compound used the degree of toxicity is defined. Halide leaching was observed to be aggressive based on its vulnerability. The Bio-hydrometallurgy leaching technique was a green technique that substitutes the use of harmful chemicals used for leaching but the constraints of using this technology are that it is still on a pilot scale (Priya & Hait, 2018). Hydrometallurgy is a water-intensive technique that pollutes a large amount of water after the chemical leaching process, whereas the application of biohydrometallurgy reduces the number of chemicals infiltrated in the water during the leaching process (Isildar, 2018). The bottlenecks faced by the microbial leaching technique are sensitivity of the microbes on the toxic heavy metals (Chauhan et al., 2018) and it is highly selective in the case of extracting metals (Marra et al., 2018). Hybrid technologies could become alternatives for conventional techniques. Considering the constraints from both the leaching techniques (chemicals and microbial), the hybrid leaching technique seems to be environmentally sound and effective in nature (Rizki et al., 2019).

6.3.1.2 Ruling criteria 2: ecodesign innovation

The author Graedel said we need to change the approach from the very beginning of the life of a product, designing in a way that the product remains in the system for a longer-term and can be easily reused or recycled. Innovations in the designs, material intake, and efficient performance with high metal recycling potential are explored. Most of the E-waste products are relying on PCBs for their functioning (Dutta et al., 2018). Essentially the driver of these supportive functions is signaled from the PCBs. PCB is a metal-rich component, innovative eco-friendly, and certainly needed to be addressed. Earlier the PCBs were manufactured in large size and consisted of a huge amount of metals in it; with time the designers have downsized the PCBs keeping the cost and the raw materials required in mind (Bhunia & Tehranipoor, 2019). Similarly, other curative ways of eco-designs need to be promoted by the industrial sectors. An eco-artist from Bengaluru named Vishwanath Mallabadi in his free time creates attractive home decors and art pieces with discarded E-waste components that are not harmful until it is not exposed to unfavorable conditions (The Hindu, 2020).

6.3.1.3 Ruling criteria 3: incorporating industrial ecology

The term industrial ecology is used when two or more companies depend on each other for their raw material intake but here the trash of one company can be a raw material of another company (Kangas & Gary, 2018; Mohammadi et al., 2021; Suárez-Eiroa et al., 2019). It is the engagement of multiple firms for resource sharing that can eliminate the demand for raw materials, balancing the material flow into the industry (Mohammadi et al., 2021). In E-waste management, recyclers who lack in having proper infrastructure and investments (Sharma et al., 2021; Shirodkar & Terkar, 2017) can start with a small-scale business. Practices of small-scale recyclers can efficiently take out the repairable parts and send them to the local vendors or service providers who repair or refurbish the parts dismantled from the E-waste. The waste consists of materials with different compositions and has different recycling techniques; this way it demands to bring on the industrial ecology alive. The waste can be collected altogether in a materials recovery facility and can be sent to suitable recycling centers. Companies like NEPRA Resource Management Private Limited, Saahas Zero Waste Private Limited, E–Coli Waste Management Private Limited, etc. are some top waste management companies in India which are involved in the collection, segregation, and sorting of solid waste; further the waste segregated and sorted waste are sent to authorized recyclers, in particular waste material recyclers. Another successful example of industrial symbiosis is noted in literature where residues of the hydrometallurgical processes such as nonmetallic fractions are used as raw materials in preparing bricks (Khayyam Nekouei et al., 2020; Mou et al., 2007; Wei et al., 2012) and also used with concrete to build the roads (Wei et al., 2012). Chinese researchers have claimed that a nonmetallic fraction of the E-waste can be used in the grates besides the walkways and it is also successfully used in making products for amusement parks of Beijing such as roller coasters, racing cars, boats, and trains for children (Mou et al., 2007).

6.3.1.4 Ruling criteria 4: strengthening reverse logistics

EEE are highly demanded products of the contemporary period. The expected annual doubling of the E-waste in the next three decades is a clear sign for us to strengthen the reverse logistics system from the EEE producers. The toughest milestone of E-waste management lies within the waste collection system. As the demand graph of EEE touches the skyline and the collection of obsolete EEE on the other hand still remains in its nascent stage. Extended Producer Responsibility is designed to enforce a fair system to allocate the E-waste generated to its right place (recycling centers) (Forti et al., 2020). The supply chain has to be closely managed in such a way that the products, regardless of losing their initial worth, can remain valuable in society. Storage habits and improper disposal of waste EEE into the municipal solid waste stream are the major reasons for low recycling practices around the globe. Awareness campaigns and accessible E-waste collection spots are required as a maximizing strategy for E-waste collection (Islam et al., 2021; Mohammadi et al., 2021). E-waste consists of various different materials which are supposed to be recycled in their own way depending upon the material type. If the recycling sequence is correctly followed, then the resources can be conveniently conserved into a closed-loop system (Graedel, 2011). Incorporating a systematic collection system via a reverse logistic approach can best serve the E-waste management framework along with sufficing the CE concept.

6.3.1.5 Ruling criteria 5: adapting functional economy

A functional economy is a service-based economy where one company lends its tools or services on a regular or a required basis so that the company taking the service does not have to invest in a tool or have to employ manpower for certain services. Such an aspect should certainly be promoted as on-ground practices that bilaterally suffice the needs of two different corporations, depriving any material loads on their business development.

The functional economy may vary based on various business models. A study says that a functional economy can attain better environmental performance through various business models adopted (Bisiaux et al., 2014). Different business implementations may lead to better environmental benefits or the business model can be designed in a way to gain environmental performance or can be such and business model where they never serve to disturb the natural stream (Bisiaux et al., 2014).

6.3.2 E-waste management system along with these ruling criteria

Systematic management of E-waste can turn out to be resourceful. Application of ruling criteria on the E-waste management system can enhance the chances of precious and valuable metals being extracted up to a certain limit. Metal extraction is the core step in the entire E-waste management system as it is where the metals are drawn out of the discarded metal and possibly can be recycled. Hydrometallurgical leaching techniques were focused on in this chapter; thus their eligibilities are judged based on the environmental risks and economic status of the leaching techniques as per their market establishments shown in Table 6.1. Potentially thiosulfate and thiourea leaching techniques should be taken under consideration by the researchers looking over the present profiles of parameters like environmental risk and economic status of the leaching technique. Since the bio-hydrometallurgy leaching tends to target low-grade ores. The low metal content in the E-waste component, that is, PCBs, resistors, transmitters, etc. could be easily targeted if we commercialize bio-hydrometallurgical techniques. Meanwhile, promoting hybrid leaching may help us reduce the time consumed by bio-hydrometallurgical leaching alone (Pant et al., 2012). This time is reduced in hybrid technology by adding some process-enhancing chemical reagents which are not harmful in nature (Pant et al., 2012). Thus CE should be applied from the beginning and needs to be followed esthetically throughout the E-waste management system. The leaching technique processes should also be selected based on their

Table 6.1 Hydrometallurgical techniques and their eligibility for circular economy (CE).

S. No.	Hydrometallurgical technique	Eligibility parameters		Eligibility for CE (judged on the two parameters)	Market penetration (market applicability: small scale/medium scale/wide scale)	References
		Environmental risks	Economic status (cost-effectively: low/medium/high)			
1.	Cyanide leaching	High	High	Not eligible due to its lethal pollutions exposed in nature	Wide scale	Cui and Zhang (2008), Kaya (2018, 2020), Rizki et al. (2019)
2.	Halide leaching	High	High	Partially eligible as the wastewater generated is acidic/alkaline in nature; however, the leaching potential is high as compared to other bioleaching techniques	Wide scale	Akcil et al. (2015)
3.	Thiourea leaching	Low	Low	Eligible based on the low environment risks; however, the chemical intake is high due to its low chemical stability	Medium scale	Akcil et al. (2015)
4.	Thiosulfate leaching	Low	High	Eligible based on its high cost-effectiveness and low environmental risks	Small scale	Akcil et al. (2015)
5.	Bio-hydrometallurgical leaching	Low	High	Eligible as it is an eco-friendly and low-investment technique which should be duly promoted	Small scale	Jia et al. (2020); Isildar (2018)
6.	Hybrid leaching	Low	High	Eligible as it will not harm the nature; however, it is still being studied	Small scale	Pant et al. (2012), Sheel and Pant (2021)

eligibility judged based on their cost-effectiveness and their environment friendliness represented in Table 6.1.

These ruling criteria individually or altogether can systematically help organize the E-waste management sector. The first ruling criteria, CP is all about installing a system where we re-fit a less harmful process in the system, without disrupting the product's quality and its market grade. The second one aims to redesign the product in its early phase to avoid the loss of resources at the end of the product life. Moreover, as we know industrial involvement has a huge role to play in bulk production products, and here's where our third criteria play its role by developing an organic industrial relationship between different industries dealing with each other in the same sector. This balanced relationship of industries can simplify the product database, modifies the product quality in an eco-friendly manner, and saves the monetary load of companies during the processing time of the product. The fourth criterion is extremely important as it connects the chain from the resource extraction to the end user and thus can eliminate a lot of carbon emissions during the entire life span of the product if the reverse logistics are strengthened. Furthermore, the last criterion is one which can help industries save their carbon credits because of their common efforts toward sustainability, as this sector builds a platform where the companies can provide their services in a way such that they gain profits and reduce emissions altogether. These criteria will direct the E-waste management sector toward a sustainable path.

6.4 Conclusion

This chapter focused on the factors promoting sustainability in a longer run that were explained by covering the urban mining and CE concepts. These concepts also help us achieve SDGs as per the current trends globally. Therefore the SDGs which can be achieved in E-waste management sectors in upcoming time are noted in further section.

6.4.1 Closing the loop of economy

Sustainability says that we borrow resources from the upcoming generations in the present period which requires strategic management to pay the debt back. It can only be achieved by limiting our substantial way of using resources, regardless of our present needs and demands. A linear economy is an enemy to the source security, bringing the products to their end of life due to its Take-Make-Dispose aspect. The CE is one where products brought to the system, remains for a long run, and is regenerative in nature (Suárez-Eiroa et al., 2019). Closing the loop of such a system conserves resources and balances the economy (Suárez-Eiroa et al., 2019). E-waste management is one of the best supporting fields for urban mining; it permits us to recycle most of the obsolete products which makes it a demanding field, promoting the CE concept. Despite such demand for E-waste in CE, a bibliographic study explaining the density map of CE concept in E-waste research hardly presented a spot for the hydrometallurgy section. Thus no major dependency was observed in the hydrometallurgical techniques for the core concept of the CE. Hydrometallurgy is a crucial method for metal extraction in the E-waste management sector. Apart from chemical leaching, metals are also extracted by burning them at high temperatures a.k.a. pyrometallurgy. Such extraction process releases dioxins and furans which are threats to the biosphere (Arya & Kumar, 2020), whereas the capital expenditure for the treatment is also very high (Cui & Zhang, 2008). These techniques will not have major changes in the field; possible challenges are a lack of

Table 6.2 Summary of benefits of urban mining through the circular economy and contribution toward sustainable development goals.

S. No.	Benefits for nature, society, and industrial sector by urban mining through circular economy model	References
1.	Resource security	Suárez-Eiroa et al. (2019)
2.	Eliminates the threats on water, land, and air	Suárez-Eiroa et al. (2019)
3.	Diverts the solid waste from reaching the landfills	Suárez-Eiroa et al. (2019)
4.	Formalize the E-waste recycling	Suárez-Eiroa et al. (2019)
5.	Reduces cost entailed by virgin mining	Suárez-Eiroa et al. (2019)
6.	Escalating product life and its potential to be recycled by eco-product designs	Suárez-Eiroa et al. (2019)
7.	Increasing the career opportunities	Arya et al. (2021), Mohammadi et al. (2021), Zeng et al. (2018)
8.	Builds business relationship	Suárez-Eiroa et al. (2019)
Contributing toward sustainable development goals		
Goal 3	Good Health and Well-being	Arya et al. (2021), Forti et al. (2020)
Goal 6	Clean Water and Sanitation	Forti et al. (2020), Ilankoon et al. (2018)
Goal 8	Decent Work and Economic Growth	Arya et al. (2021), Forti et al. (2020)
Goal 11	Sustainable Cities and Communities	Forti et al. (2020), Ilankoon et al. (2018), Sharma et al. (2021)
Goal 12	Responsible Consumption and Production	Forti et al. (2020), Ilankoon et al. (2018), Sharma et al. (2021)
Goal 14	Life Below Water	Ilankoon et al. (2018), The Global E-Waste Monitor (2017)

know-how in the treatment techniques and weak infrastructure (Kazançoglu et al., 2020). Extensive efforts are made by strengthening the policies and frameworks of E-waste management, CE, and simultaneously considering the sustainable development goals (SDGs) in mind (Suárez-Eiroa et al., 2019). Thus the criteria discussed in this chapter enable us to target the correct and efficient ways of closing the loop. Table 6.2 shows the aftermath results of these criteria being applied to urban mining for the E-waste management system which can bring substantial benefits the nature, society, and the industrial sector. Attaining all the paybacks should pave the road toward the SDGs (noted in Table 6.2) by urban mining of E-waste. CE in the E-waste sector will also bring opportunities of partnering multistakeholders to achieve a sustainable system altogether contributing toward the SDGs goals 17.

6.5 Future Implications of urban mining in hydrometallurgy

Sustaining in such a fast-moving society is quite a task for the present generation with the rapidly depleting resources on the Earth of confined quantity. Urban mining will facilitate us with the new possible future and opportunities which did not exist due to the linear economy system. EEE is a stream where the application of the urban mining concept can best serve. A series of different stages are executed during E-waste management (Kazançoglu et al., 2020), in which after achieving the collection of waste, it is carefully sent for leaching metals. The urban mining concept is a subset of the CE concept. Thus, once we organized a closed-loop system, we automatically start gaining advantages mentioned in Table 6.2. Urban mining, in general, will change the mindset of the business makers in the long run, and with understanding the resources and severity of environmental threats, the future industrial economy can be balanced sustainably.

Urban mining via hydrometallurgy is an essential step of recovering valuable and precious strives to find out eco-friendlier leaching techniques. The chapter explained all the chemical leaching techniques briefly along with their economic and ecological status in the present time. These factors were even considered to understand their eligibility with the CE concept, which is expected to be adopted by all the micro to macro-scale businesses. Halide leaching is expected to be used even when the waste water released at the end is highly corrosive and toxic until we find an efficient and environmentally friendly treatment. Hydrometallurgy has stronger hold over the industrial market since a longer period of time so it will be harder to replace this traditionally grounded technology so quickly. Bio-hydrometallurgy and hybrid technologies are outgrown from hydrometallurgy itself so they are expected to be more focused area in future research, in anticipation of leaching techniques with higher acceptance. These techniques have the potential to be implemented on a large scale in the E-waste management sector in upcoming time.

References

Akcil, A., Erust, C., Gahan, C. S., Ozgun, M., Sahin, M., & Tuncuk, A. (2015). Precious metal recovery from waste printed circuit boards using cyanide and non-cyanide lixiviants — A review. *Waste Management (New York, N.Y.), 45*, 258−271. Available from https://doi.org/10.1016/j.wasman.2015.01.017.

Arya, S., Aneri Patel, S. K., & Pau-Loke, S. (2021). Urban mining of obsolete computers by manual dismantling and waste printed circuit boards by chemical leaching and toxicity assessment of its waste residues. *Environmental Pollution, 283*, 117033. Available from https://doi.org/10.1016/j.envpol.2021.117033.

Arya, S., & Kumar, S. (2020). E-Waste in India at a glance: Current trends, regulations, challenges and management strategies. *Journal of Cleaner Production, 271*, 122707. Available from https://doi.org/10.1016/j.jclepro.2020.122707.

Bhunia, S., & Tehranipoor, M. (2019). Printed circuit board (PCB): Design and test. *Hardware Security*, 81−105. Available from https://doi.org/10.1016/b978-0-12-812477-2.00009-5.

Bisiaux, J., Thierry Gidel, F. H., & Millet, D. (2014). How functional economy would be an environmental economy? Mode of endogenization of environmental issues in functional economy. In: *2014 international conference on engineering, technology and innovation: engineering responsible innovation in products and services (ICE 2014)*. <https://doi.org/10.1109/ICE.2014.6871617>.

Chauhan, G., Jadhao, P. R., Pant, K. K., & Nigam, K. D. P. (2018). Novel technologies and conventional processes for recovery of metals from waste electrical and electronic equipment: challenges & opportunities — A review. *Journal of Environmental Chemical Engineering, 6*(1), 1288−1304. Available from https://doi.org/10.1016/j.jece.2018.01.032.

CNBC International. (2018, January 19). *What is circular economy? CNBC Explains* [Video file]. YouTube. <https://www.youtube.com/watch?v = 0Spwj8DkM&t = 1s>.

Cui, H., & Anderson, C. G. (2016). Literature review of hydrometallurgical recycling of printed circuit boards (PCBs). *Journal of Advanced Chemical Engineering, 6*(1), 11, 20904568. Available from https://doi.org/10.4172/2090-4568.1000142.

Cui, J., & Zhang, L. (2008). Metallurgical recovery of metals from electronic waste: A review. *Journal of Hazardous Materials, 158*(2−3), 228−256. Available from https://doi.org/10.1016/j.jhazmat.2008.02.001.

Dutta, D., Panda, R., Kumari, A., Goel, S., & Jha, M. K. (2018). Sustainable recycling process for metals recovery from used printed circuit boards (PCBs). *Sustainable Materials and Technologies*, e00066. Available from https://doi.org/10.1016/j.susmat.2018.e00066.

EU Environment. (2014, November 26). *How to become a green SME in a circular economy* [Video file]. YouTube. <https://www.youtube.com/watch?v = V1Tszs48xCI>.

Forti, V., Baldé, C. P., Kuehr, R., & Bel, G. (2020). *The global E-Waste monitor 2020: Quantities, flows, and the circular economy potential*. Bonn/Geneva/Rotterdam: United Nations University (UNU)/United Nations Institute for Training and Research (UNITAR) — Co-Hosted SCYCLE Programme, International Telecommunication Union (ITU) & International Solid Waste Association (ISWA).

Gomes da Silva, J. F., & Gouveia, R. M. (2019). Cleaner production definition and evolution. *Cleaner Production: Toward a Better Future*, 1−13.

Graedel, T. (2011). The prospects for urban mining. *Bridge, 41*(1), 43−50.

Hilson, G., & Monhemius, A. J. (2006). Alternatives to cyanide in the gold mining industry: What prospects for the future? *Journal of Cleaner Production, 14*(12−13), 1158−1167. Available from https://doi.org/10.1016/j.jclepro.2004.09.005.

Iannicelli-zubiani, E. M., Irene, M., Recanati, F., Dotelli, G., Puricelli, S., & Cristiani, C. (2017). Environmental impacts of a hydrometallurgical process for electronic waste treatment: A life cycle assessment case study. *Journal of Cleaner Production, 140*, 1204−1216. Available from https://doi.org/10.1016/j.jclepro.2016.10.040.

Ilankoon, I. M. S. K., Ghorbani, Y., Nan, M., Herath, G., & Moyo, T. (2018). E-Waste in the international context — A review of trade flows, regulations, hazards, waste management strategies and technologies for value recovery. *Waste Management, 82*, 258−275. Available from https://doi.org/10.1016/j.wasman.2018.10.018.

Isildar, A. (2018). *Biological versus chemical leaching of electronic waste for copper and gold recovery.*

Islam, M. T., Huda, N., Baumber, A., Shumon, R., Zaman, A., Ali, F., Hossain, R., & Sahajwalla, V. (2021). A global review of consumer behavior towards E-Waste and implications for the circular economy. *Journal of Cleaner Production, 316*, 128297. Available from https://doi.org/10.1016/j.jclepro.2021.128297.

Jadhav, U. U., & Hocheng, H. (2012). A review of recovery of metals from industrial waste. *Journal of Achievements in Materials and Manufacturing Engineering, 54*(2), 159–167.

Jia, L., Huang, J. J., Ma, Z.-l., Liu, X.-h., Chen, X.-y., Li, J.-t., He, Li.-h., & Zhao, Z.-w. (2020). Research and development trends of hydrometallurgy: An overview based on hydrometallurgy literature from 1975 to 2019. *Transactions of Nonferrous Metals Society of China (English Edition), 30*(11), 3147–3160. Available from https://doi.org/10.1016/S1003-6326(20)65450-4.

Jing-ying, L., Xiu-li, X., & Wen-quan, L. (2012). Thiourea leaching gold and silver from the printed circuit boards of waste mobile phones. *Waste Management, 32*(6), 1209–1212. Available from https://doi.org/10.1016/j.wasman.2012.01.026.

Kangas, P., & Gary, E. S. (2018). An industrial ecology teaching exercise on cycling E-waste. *Ecological Modelling, 371*, 119–122. Available from https://doi.org/10.1016/j.ecolmodel.2017.12.008.

Kaya, M. (2018). *Current WEEE recycling solutions. Waste* electrical and electronic equipment recycling: aqueous recovery methods. Elsevier Ltd.. Available from https://doi.org/10.1016/B978-0-08-102057-9.00003-2.

Kazançoglu, Y., Ada, E., Ozturkoglu, Y., & Ozbiltekin, M. (2020). Analysis of the barriers to urban mining for resource melioration in emerging economies. *Resources Policy, 68*, 101768. Available from https://doi.org/10.1016/j.resourpol.2020.101768.

Khayyam Nekouei, R., Tudela, I., Pahlevani, F., & Sahajwalla, V. (2020). Current trends in direct transformation of waste printed circuit boards (WPCBs) into value-added materials and products. *Current Opinion in Green and Sustainable Chemistry, 14*(20). Available from https://doi.org/10.1016/j.cogsc.2020.01.003.

Marra, A., Cesaro, A., Rene, E. R., Belgiorno, V., & Lens, P. N. L. (2018). Bioleaching of metals from WEEE shredding dust. *Journal of Environmental Management, 210*, 180–190. Available from https://doi.org/10.1016/j.jenvman.2017.12.066.

Mohammadi, E., Singh, S. J., & Habib, K. (2021). How big is circular economy potential on Caribbean Islands considering E-waste? *Journal of Cleaner Production, 317*, 128457. Available from https://doi.org/10.1016/j.jclepro.2021.128457.

Mou, P., Xiang, D., & Duan, G. (2007). Products made from nonmetallic materials reclaimed from waste printed circuit boards. *Tsinghua Science and Technology, 12*(3), 276–283. Available from https://doi.org/10.1016/S1007-0214(07)70041-X.

Pant, D., Joshi, D., Upreti, M. K., & Kotnala, R. K. (2012). Chemical and biological extraction of metals present in E waste: A hybrid technology. *Waste Management, 32*(5), 979–990. Available from https://doi.org/10.1016/j.wasman.2011.12.002.

Priya, A., & Hait, S. (2018). Extraction of metals from high grade waste printed circuit board by conventional and hybrid bioleaching using *Acidithiobacillus ferrooxidans. Hydrometallurgy, 177*, 132–139. Available from https://doi.org/10.1016/j.hydromet.2018.03.005.

Rizki, I. N., Yu, T., & Okibe, N. (2019). Thiourea bioleaching for gold recycling from E-waste. *Waste Management, 84*, 158–165. Available from https://doi.org/10.1016/j.wasman.2018.11.021.

Sharma, M., Joshi, S., & Govindan, K. (2021). Issues and solutions of electronic waste urban mining for circular economy transition: An Indian context. *Journal of Environmental Management, 290*, 112373. Available from https://doi.org/10.1016/j.jenvman.2021.112373.

Sheel, A., & Pant, D. (2021). Thiourea Bacillus combination for gold leaching from waste lithium-ion batteries. *Bioresource Technology Reports, 15*, 100789. Available from https://doi.org/10.1016/j.biteb.2021.100789.

Shirodkar, N., & Terkar, R. (2017). Stepped recycling: The solution for E-waste management and sustainable manufacturing in India. *Materials Today: Proceedings, 4*(8), 8911–8917. Available from https://doi.org/10.1016/j.matpr.2017.07.242.

Suárez-Eiroa, B., Fernández, E., Méndez-Martínez, G., & Soto-Oñate, D. (2019). Operational principles of circular economy for sustainable development: Linking theory and practice. *Journal of Cleaner Production, 214*, 952–961. Available from https://doi.org/10.1016/j.jclepro.2018.12.271.

The Global E-Waste Monitor 2017. (2017).

The Hindu. (2020, February 18). *Creating amazing art from e-waste* [Video file]. YouTube. <https://www.youtube.com/watch?v = v8JJCbfllws>.

To, Alternatives, and Cyanide I N Gold. (2008). Thiosulphate leaching — An alternative to cyanidation in gold processing alternatives to cyanide in gold, 2–3.

Tuncuk, A., Stazi, V., Akcil, A., Yazici, E. Y., & Deveci, H. (2012). Aqueous metal recovery techniques from E-scrap: Hydrometallurgy in recycling. *Minerals Engineering, 25*(1), 28–37. Available from https://doi.org/10.1016/j.mineng.2011.09.019.

Velenturf, A. P. M., & Purnell, P. (2021). Principles for a sustainable circular economy. *Principles for a sustainable circular economy, 27*, 1437–1457. Available from https://doi.org/10.1016/j.spc.2021.02.018.

Wei, B., Li, J., Liu, L., & Dong, Q. (2012). Progress in Research of Comprehensive Utilization of Nonmetallic Materials from Waste Printed Circuit Boards. *Procedia Environmental Sciences, 16*, 500–505. Available from https://doi.org/10.1016/j.proenv.2012.10.069.

Zeng, X., John, A. M., & Li, J. (2018). Urban mining of E-waste is becoming more cost-effective than virgin mining. *Environmental Science and Technology, 52*(8), 4835–4841. Available from https://doi.org/10.1021/acs.est.7b04909.

Zeng, X., Xiao, T., Xu, G., Albalghiti, E., Shan, G., & Li, J. (2021). Comparing the costs and benefits of virgin and urban mining. *Journal of Management Science and Engineering*. Available from https://doi.org/10.1016/j.jmse.2021.05.002.

Zhong, Y., Dan, L., Mao, Z., Huang, W., Peng, P., Chen, P., & Mei, J. (2014). Kinetics of tetrabromobisphenol A (TBBPA) reactions with H2SO4, HNO3 and HCl: Implication for hydrometallurgy of electronic wastes. *Journal of Hazardous Materials, 270*, 196–201. Available from https://doi.org/10.1016/j.jhazmat.2014.01.032.

CHAPTER 7

Pyrometallurgy: urban mining and its future implications

Rumi Narzari[1], Biswajit Gogoi[2] and Sachin Rameshrao Geed[2]
[1]Department of Energy, Tezpur University, Tezpur, Assam, India
[2]Engineering Science and Technology Division, CSIR-NEIST, Jorhat, Assam, India

7.1 Introduction

With global rise in population, the percentage of people residing in urban areas is also rising. As a result, everyday pile up of urban wastes being generated is also increasing throughout the world as shown in Fig. 7.1 (Forti et al., 2020). Such an increase in piles of waste is a threat to the environment as well as fauna and flora around the waste disposal sites. A large portion of urban waste comprises metals, and some of which contains even precious metals and rare earth metals like gold, silver, etc. Most of the waste containing such precious metals is from either electrical or electronic gadgets (E-wastes) in some form or the other. Disposal of such precious metals without use is a huge loss to the metal industry. Apart from the precious metals, urban wastes also contain hazardous materials (lead, mercury, chromium, cadmium, etc.) for which proper treatment is required before disposal. Inconsiderate and unchecked disposal of E-waste in the open without undergoing proper recycling has serious consequences on environment and human health (Arya, Rautela, et al., 2021; Dutta et al., 2021; Rautela et al., 2021). The impact of various metals on human health is listed in Table 7.1. Thus urban waste contains a vast mixture of glass, organic materials, and metals such as ferrous, nonferrous, and precious metals. With growing demand of metals and undue pressure on exploration and mining to fulfill the needs of the demand, there is an utter necessity of recycling these metals from urban wastes. Such process is known as urban mining where valuable metals are being recovered from wastes generated in urban areas (Arya & Kumar, 2020a). The process of recycling also reduces the carbon footprint, as less energy per unit of metals recovered is used, as compared to mining (Ebin & Isik, 2016). In addition, recycling process reduces both the amount of waste stockpile as well as the waste generated during mining.

A study conducted by United Nations University reported that a single smartphone could provide approximately 40 various critical metals. The metals thus obtained are much pure as compared to its ore. As reported the gold obtained from such waste is

Global E-waste Management Strategies and Future Implications
DOI: https://doi.org/10.1016/B978-0-323-99919-9.00011-8

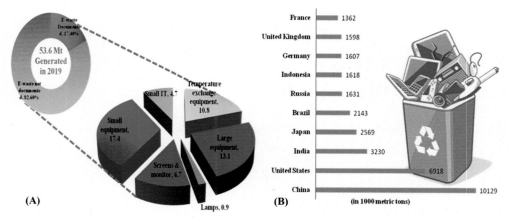

Figure 7.1 Representation of E-waste generation worldwide: (A) documented and its source and (B) top leading countries with reference to E-waste generated (Forti et al., 2020).

Table 7.1 Impact of contaminants on human health (Kiddee et al., 2013).

Contaminants	Application	Impact on health
Antimony (Sb)	• Melting agent in cathode ray tube (CRT) glass, plastic computer housings • Solder alloy in cabling	• Carcinogenic • Leads to stomach pain, vomiting, diarrhea, and stomach ulcers
Arsenic (As)	• Used in light emitting diodes	• Skin disease and lung cancer • Impaired nerve signaling
Barium (Ba)	• Sparkplugs, fluorescent lamps, and CRT gutters in vacuum tubes	• Brain swelling • Muscle weakness • Damage to the heart, liver, and spleen
Beryllium (Be)	• Power supply boxes, motherboards, relays, and finger clips	• Berylliosis • Lung cancer and skin disease
Brominated flame retardants	• Used to reduce flammability in printed circuit boards (PCBs), and plastic housings, keyboards, and cable insulation	• Combustion of PCBs and plastic causes hormonal disorders
Cadmium (Cd)	• Rechargeable NiCd batteries, semiconductor chips, infrared detectors, printer inks, and toners	• Irreversible impacts kidneys
Chlorofluorocarbons	• Cooling units and insulation foam	• These substances impact on the ozone layer which can lead to greater incidence of skin cancer
Hexavalent chromium (Cr VI)	• Plastic computer housing, cabling, hard disks, and as a colorant in pigments	• Is extremely toxic in the environment, causing DNA damage and permanent eye impairment

(Continued)

Table 7.1 (Continued)

Contaminants	Application	Impact on health
Lead (Pb)	• Solder, lead-acid batteries, CRTs, cabling, PCBs, and fluorescent tubes	• Damage the brain, nervous system, kidney, and reproductive system • Blood disorders • Both acute and chronic effects on human health
Mercury (Hg)	• Batteries, backlight bulbs or lamps, flat panel displays, switches, and thermostats	• Damage the brain, kidneys, and fetuses
Nickel (Ni)	• Batteries, computer housing, CRT, and PCBs	• Allergic reaction • Bronchitis • Reduced lung function and lung cancers
Polychlorinated biphenyls	• Condensers, transformers, and heat transfer fluids	• Cancer in animals • Liver damage in humans
Polyvinyl chloride	• Monitors, keyboards, cabling, and plastic computer housing	• Incomplete combustion leads to formation of hydrochloric acid after which can cause respiratory inconvenience
Selenium (Se)	• Older photocopy machines	• Selenosis

Source: Adapted from Kiddee, P., Naidu, R., & Wong, M. H. (2013). Electronic waste management approaches: An overview. *Waste management, 33* (5), 1237−1250.

25−30 times purer than its ore. United Nations General Assembly states that Sustainable Development Goals (SDGs) 2030 Agenda is directly associated with the success of waste management. Hence, in order to maximize the resource utilization and conservation, it was urged to implement sustainable approach toward waste management. In this regard various technologies and approaches have been deployed and one of the mostly accepted processes is the pyrometallurgy (Ayres & Peiro, 2013; Arya & Kumar, 2020b).

Pyrometallurgy is the process of using elevated temperatures to extract metals from various types of wastes. However, the energy used per unit of metal recovered or extracted is less during pyrometallurgy as compared to that used during mining. This is because a higher concentration of metals could be found per unit volume of urban waste as compared to mining. Pyrometallurgy along with electrometallurgy and hydrometallurgy are extensively used in urban waste recycling or recovering metals. Some of the common pyrometallurgical methods are (Ebin & Isik, 2016; Gurgul et al., 2017; Khaliq et al., 2014):

1. Smelting
2. Incineration

3. Pyrolysis

4. Molten salt

7.1.1 Smelting

Smelting is one of the common pyrometallurgical processes which are used to recover nonferrous metals from urban wastes (Ebin & Isik, 2016; Hsu et al., 2019; Rene et al., 2021). Amongst all the smelting processes copper and lead smelting techniques are common methods for metal recovering. The process starts with physical separation of metals from other constituents of the wastes. Once separated the metallic constituents mainly consists of iron, aluminum, copper, lead, tin, antimony, zinc, and precious metals. During copper smelting the metallic constituents are fed to the furnace where they are heated to their molten state. Metals such as lead, copper, and precious metals remain in the molten state while the other constituents form the slag. The molten metal is collected in parent phase of copper which is further casted to form anode. Later electrometallurgy is used to refine the anode and yield a copper cathode with high purity (Ebin & Isik, 2016). The anode once dissolved leaves anodic slime as residue which also contains other metallic fractions. The residue is further processed to recover valuable metals from it. However, iron and aluminum portions are oxidized and mixed with slag. Some of the major smelting companies are Boliden (Sweden), Dowa (Japan), Umicore (Belgium), Noranda (Canada), and Aurubis (Germany). The Dowa smelting and refining system can recover 18 different metallic constituents. In addition to application of copper and lead smelters as single units, they are also used as integrated units.

Integrated copper and lead smelters are used for recovering metals such as copper, lead, zinc, and precious metals. One of the important parts of such integrated system is the Kaldo furnace. Kaldo furnace is a cylindrical rotating furnace with an oxygen lance and an oxygen-oil burner. The use of Kaldo furnace removes the requirement for segregating the plastic components from the E-waste and is fed directly along with metals to it. The plastics thus fed are burned along with the metal and act as a source of energy during melting process. Scraps with high copper content are fed directly to the copper smelting line, whereas scraps with lower copper content are fed to the Kaldo furnace at first. The Kaldo furnace yields a molten copper alloy consisting of gold, copper, silver, selenium, palladium, nickel, and zinc. This molten alloy is also known as black copper. In addition, dusts are also formed in the Kaldo furnace consisting of zinc, lead, antimony, indium, and cadmium, which are sent for refining process. The molten alloy is further connected with the copper smelting line for metal recovery and refining. Refining is done in a converter where oxygen-enriched air or air is forced to remove impurities such as zinc, lead, tin, and iron.

In another attempt Umicore (Hoboken, Belgium) has also developed an integrated pyrometallurgical process to recover precious and base metals (Ebin & Isik, 2016). The urban waste is fed to a copper smelter which produces slag, copper bullion, and gas. The produced slag is fed to a lead blast furnace and gases are fed to sulfuric acid plant. The copper bullion produced is further leached and undergoes electrowinning operation. The lead bullion from lead blast furnace is further refined in a lead refinery. A portion fed to the lead refinery recovers lead, tin, antimony, and bismuth. Precious metals thus formed as residue from leaching, electrowinning, and lead refinery are sent to cupellation for precious metal refining such as palladium, silver, gold, platinum, ruthenium, rhodium, and iridium. Special metals such as indium, tellurium, and selenium are recovered using a special metal refining unit.

In another attempt roasted gold concentrates having Fe_tO and SiO_2 are used as flux material during smelting of printed circuit boards (PCBs) at high temperature. As a result, slag composed of oxide impurities such as Al_2O_3, Cao, and SiO_2 as well as Fe_tO and SiO_2 is formed and is easily removed for copper extraction from PCBs. Further treatment of leaching and extraction is done to purify the recovered material (Park & Kim, 2019).

Even though a lot of improvements have been done, smelting still has the drawbacks of generating new wastes, polluted exhaust gases, and loss of iron and aluminum due to oxidation. Moreover, organic and oxide fractions are also lost and could not be recovered. Exhaust emissions includes hazardous dioxins and halogenic compounds. Further improvements are required in minimizing recovery loss as dust and recovering from complex products such as cathode ray tube (CRT) and PCB. The smelting process still requires steps such as hydrometallurgical and electrometallurgical for increased recovery of precious metals. With increase in complimentary steps for increased refining efficiency and reduced emissions a higher number of investments are required.

7.1.2 Incineration

Incineration is the process of combusting organic materials along with solid waste for high volume reduction efficiency, energy generation, and sterilization. Urban wastes contain a considerable amount of organic material which occupies a large amount of volume as well as high energy content. Thus their combustion could reduce the volume by about 70% and can also act as source of energy. However, combustion of waste produces slag and flue gas with high concentration of heavy metals.

In the developing economies open-space combustion is practiced mostly to recover valuable metals due to their low-cost involvement. It is usually done to remove the outer coating or other unwanted materials in wastes. This causes a lot of environmental issue and causes health hazards. However, incineration is a controlled combustion method where emissions are monitored. The process is usually carried out at higher temperatures of around 1200°C to reduce carbon monoxide formation. However, issues of heavy metals in ashes arise, and it has a lower metal recovery rate.

7.1.3 Pyrolysis

Pyrolysis is the process of decomposing any organic material in the absence of oxygen using heat. During pyrolysis reaction temperature is varied from 450°C to 1100°C. This results in thermal degradation of urban waste and produces low molecular—weight products such as liquid as well as gas and char. The products thus obtained are of high value and used for various chemical processes or as fuels. The metals thus recovered remains in the char and could be easily separated. Commercially available reactors used in pyrolysis are of fixed bed, moving bed, and fluidized bed type.

There are various factors which affect the pyrolysis process such as temperature, heating rate, reactor type, time, plastic in waste, catalyst, and pressure. The temperature and heating rate affect the bond breaking, as smaller molecules are formed at higher temperature and heating rates. The time of pyrolysis process or residence time helps in production of more stable products as longer time of residence results in higher yields. The residence period of the waste depends on the organic content, reactor type, and operating temperature. The type of plastic of the waste determines the type of products obtained both in terms of structure and yield. The formation of heavier condensed products and coke is affected by the operating pressure, whereas catalysts change the type of products obtained by changing the kinetics and mechanism. Similarly, selection of the reactor type influences the heat transfer process, residence time, and mixing of waste.

As compared to combustion the number of emissions during pyrolysis is far less and metals could be easily recovered from the residue thus formed. Moreover, appropriate condensation process along with pyrolysis allows higher recovery rate of valuable fuels in form of liquid and gas. The entire process of pyrolysis is governed by the pyrolysis kinetics which in turn controls the efficiency. The process of pyrolysis could further be divided into vacuum pyrolysis, low-temperature pyrolysis, and high-temperature pyrolysis (Arya & Kumar, 2020b). A detailed discussion about these processes is discussed in the following sections.

7.1.3.1 Low-temperature pyrolysis

Low-temperature pyrolysis refers to the operating temperature lower than 500°C. Application of low temperature helps in recovering the organic content of E-waste in the form of liquid. Low-temperature pyrolysis provides the advantages of emitting lower emissions by dehalogenating the plastics present in wastes and converting them into liquids. Another advantage is the removal of chlorine in the form of hydrogen chloride from the waste during low operating temperature. This reduces the chances of chlorine contamination in the oil formed later during high temperature application. Similarly, with increase in residence time the organic bromine is converted to nontoxic inorganic bromine and could be further processed using alkali solution.

7.1.3.2 High-temperature pyrolysis

High-temperature pyrolysis process is undertaken at a temperature range from 600°C to 1000°C. At this temperature the organic parts are decomposed to lower fragments of low molecular weights and thus the yield of gases increases. This adds to the advantages of recovering low boiling—point metals such as cadmium, zinc, lead, and mercury by condensing the exhaust gases. Further increasing the operating temperature above the melting point could be used to recover the major metals such as lead, copper, and antimony. On the other hand, precious metals such as gold, palladium, silver, and platinum remain in trace amounts in metallic form with the residue.

7.1.3.3 Vacuum pyrolysis

Vacuum pyrolysis is a relatively newer concept in which a lower operating temperature is used with a shorter residence time (Ebin & Isik, 2016; Hsu et al., 2019). The advantages are the prevention of other secondary reactions occurring during elevated temperatures and production of residues with lower interfacial bonds resulting in easier separation and refining. Vacuum pyrolysis is economical when the copper-rich waste is used and has the added advantages of recovering hazardous metals such as cadmium and lead. Even though separation of carbon and glass constituents from the metallic content becomes easier, separation of metals from the metallic constituent needs further treatment. The metallic content could further be separated using the difference in vapor pressure of the metallic constituent to create metallic vapors along with suitable condensing units. The relation between vapor pressure and temperature could be determined using Clausius—Clapeyron and Antoine equations as shown in Eqs. (7.1) and (7.2), respectively (Ebin & Isik, 2016).

$$\frac{d\,(\ln P)}{dT} = \frac{\Delta H}{RT^2} \tag{7.1}$$

$$\ln P = A - \frac{B}{C + T} \tag{7.2}$$

where P is the vapor pressure, T is the temperature, ΔH is the standard molar enthalpies (evaporation or sublimation), and R is the ideal gas constant. A, B, and C are specific constants for metals.

7.1.4 Molten salt process

During molten salt process inorganic salts are melted along with E-wastes in an inert atmosphere at the desired operating temperature (Hsu et al., 2019). Thus the organic components and halogens are captured by the molten salt in the form of carbonates, silicates, and alkali metal halides. The flue gas thus formed is enriched with hydrogen and the metallic components are separated from the residue. Metals with low melting points are collected as alloy and the molten salt containing the metals is further treated to separate the metals.

The type of salt used is dependent on type of molten salt process used for recovery, such as molten salt oxidation and molten salt pyrolysis. The temperature usually varies between 300°C and 1100°C. Some of the commonly used salt mixtures are potassium hydroxide (KOH)—sodium hydroxide (NaOH) eutectic mixture, sodium sulfate (NaSO4)—sodium carbonate (NaCO$_3$) mixture, and lithium chloride (LiCl)—potassium chloride (KCl) eutectic mixture. The recovering temperature could be further reduced using eutectic salt compositions as the separation of metallic components is performed without melting, making it economical (Ebin & Isik, 2016). The major advantages of molten salt process are the reduction in emissions, low cost, and ease in processing.

7.1.5 Pyrochemical process

Pyrochemical or thermochemical process is the application of chemical reactions at elevated temperatures to separate metals from E-wastes. It is mainly preferred for recovery of valuable metals from its oxide form as could be found in liquid crystal display (LCD) and CRT components. Two most common pyrochemical processes are carbothermic and chlorination reactions which are discussed in the following sections (Ebin & Isik, 2016).

7.1.5.1 Carbothermic reactions

Lead monoxide is one of the prime components of CRT funnel glass and is recovered using smelting process and is not an economical process. Moreover, silica present in CRT funnel glass is also lost during smelting resulting in the formation of slag (Vishwakarma et al., 2022). Thus lead monoxide is reduced to lead using carbothermic reactions at high temperature (1000°C—1100°C) and low pressure, with higher residence period in inert atmosphere and vacuum conditions. These separate lead from the glass which could be further used after treatment.

Similarly, carbothermic reactions are also used to recover indium (a rare earth metal) and tin from LCD. Indium is one of the critically endangered metals as it is not mined directly and is rather found along with zinc ore. The extensive use of indium in creating indium tin oxide (ITO) electrodes for LCDs is depleting the metal at a much faster rate due to limited methods of recycling. However, in recent years, carbothermic reactions emerged as an important pyrochemical method to recover this rare metal. Indium and tin oxide from ITO are reduced to its metallic form using carbon or carbon monoxide using vacuum pyrometallurgy. Carbothermic reaction could recover 90% of indium used in LCD using 30% at temperature of 950°C and 1 Pa vacuum.

7.1.5.2 Chlorination reactions

Chlorination reaction is another method of recovering indium from LCD using hydrochloride and ammonium chloride at high temperatures (Krishnan et al., 2021). Indium oxide is reduced and chlorinated to indium chloride at a comparatively lower temperature of 350°C. It was reported that 99.48% recovery of indium from LCD could be achieved at 400°C and 0.09 MPa vacuum using ammonium chloride (Takahashi et al., 2009). The rates of indium recovery using chlorination process depend on reaction temperature, residence period, dosage, and atmosphere (Ma et al., 2012; Park et al., 2009; Takahashi et al., 2009). Similarly, recovery of precious metals using chlorination depends on the pressure, temperature, affinity with chlorine, and water solubility (Rautela et al., 2021).

7.2 Types of metal recovery

Urban waste management is one of the biggest issues causing environmental issues related to its disposal. The recovery techniques not only minimize wastes but also help in recovering important valuable metals. Such wastes contain about 95% metals and a large section of the mined metallic resources are used in electrical and electronic gadgets. Moreover, recovery rates of the metals are higher than mining. Thus utilization of recovered metals from e-waste saves a lot of energy as well as investments as compared to mining. Precious metals and other valuable metals are favored to be recovered from electronic wastes (Arya & Kumar, 2020b; Chatterjee & Kumar, 2009). Since long, many strategies were used for extraction of essential metals from urban waste such as magnetic, gravity, and electrostatic separation (Gollakota et al., 2020). Some of the recent techniques are pyrometallurgical, thermochemical processes, smelting, mechanical recycling, pyrolysis, combustion, bio-metallurgical processes, and hydrometallurgical (Gollakota et al., 2020). However, for industrial-scale recovery of metals from urban waste technology, sustainability of the process plays a greater role. Recovery techniques could recover metals like Ta (tantalum), Au (gold), Fe (iron), Cu (copper), Pt (platinum), Pd (palladium), Pb (lead), Al (aluminum), Fe (iron), Ag (silver), and Ni (Nickel) from electronic wastes (Chatterjee & Kumar, 2009).

Debnath et al. (2016) have reported that the metal composition of electronic waste varies according to operation and size of part. For example, in a typical computer 95% polymers are found in keyboards, while the waste PCBs are loaded with metals containing approximately 20% Cu and 250 mg/ton Au. Hageluken (2006) reported that the Cu contribution goes downward to 13% in PCBs from mobiles with 350 mg/ton of Au. Table 7.2 presents the general metal composition of electronic PCB.

The recycling of urban waste items is done to recover metallic components from cables, LCD, PCBs, capacitors, CRTs, etc. (Arya, Patel, et al., 2021). Metals recovery

Table 7.2 General metal distribution of electronic printed circuit board (Debnath et al., 2018).

Basic metals

Name	Al	Cu	Fe	Ni	Pb	Sn	Zn
Metal concentration in wt.%	14.17	6.92	20.47	0.85	6.29	1.01	2.20

Rare earth and valuable metals

Name	Ag	Au	Cd	Co	Ga	Ge	Ni	Pd	Se	Sb	Ta	Ti
Metal concentration in wt.% ($\times 10^{-3}$)	18.9	1.6	9.4	15.7	1.3	1.6	0.2	0.3	1.6	9.4	15.7	15.7

using pyrometallurgical methods such as pyrolysis, incineration, combustion, smelting, and molten salt are used for recovery of metals. In metallurgical methods high energy input is required as elevated temperatures are used to reduce or extract metals (Salhofer & Tesar, 2011). However, during pyrometallurgical processes lower energy is used and lower CO_2 is emitted compared to metal ore mining and refining processes. Smelting is an important technique used for extraction of metals from its ores. Nowadays, smelting is applied for extraction of nonferrous metals from urban waste. Copper smelting is mainly used for urban waste recycling. In general metal recovery processes starts with substantial separation. Initial separation of waste scrap consists of Cu, Al, Sb, Fe, Pb, Zn, Sn, and other valuable metals. These mixtures are used in Cu and Pb smelters (Hylander & Herbert, 2008).

Incineration is a widely used disposable technique involving combustion of waste solid waste in order to reduce volume fraction. Urban waste suitable for mining mostly comprises electronic waste having high calorific value mixed organic polymers and inorganic resources, which makes incineration feasible. However, recycling by open burning focuses only on recovery of valuable metals (Banerjee et al., 2019; Grigorescu et al., 2019). Copper recovery from cables is obtained by burning of plastic insulation or chips from circuit boards. Few industries have used open burning to obtain Cu and essential metals elements, as one of less effective processes of metal recovery (Osibanjo et al., 2016).

Pyrolysis of urban metal containing waste is also an important pyrometallurgical process where bromine and chlorine gas are obtained as pyrolysis outlet gas. The recovery of Cl and Br are controlled by capturing capability of calcium carbonate. Pyrolysis is process of thermal decaying of electronic waste by means of heat at prominent temperatures in O_2-free environment (Hsu et al., 2019). Pyrolysis has been carried out at temperatures above the melting point of metals Pb, Cu, and Sb present in the electronic waste. Then molten metals and nonmetallic portions are easily separated. The metallic part was made up of droplets (red and white) that segregated naturally at the end of pyrolysis of electronic waste above 1000°C temperature. Red droplets formed are Cu, whereas the white droplets are predominantly Sn with small amounts of Pb (Cayumil et al., 2014, 2018; Hsu et al., 2019).

The molten salt recycling is a pyrometallurgical method based on the extraction of metals from electronic waste. Molten salt recycling process has lower environmental pollution problems compared to other pyrometallurgical processes for recycling electronic waste like conventional pyrolysis, combustion, and smelting. Molten mixtures of inorganic salts are used for extraction of valuable metals from urban metal containing waste such as $NaSO_4$ and $NaCO_3$ mixture, KOH and NaOH mixture, or LiCl and KCl mixture. Salts are applied at melting temperatures of valuable metals ranging between 300°C and 1100°C (Kaya, 2016). The valuable metals are recovered using various processes reported by different researchers.

Sivakumar et al. (2018) have successfully recovered Cu and Pb metals from waste PCBs. The electrowinning technologies are effectively used for recycling Cu and Pb from PCBs. To simplify the process in a less expensive way the simpler method of ammonia leach extraction was employed. Selected electronic waste was placed with oil for 1 h at 500°C in a pyrolyzer. Subsequently, ash powder was made using the acid leaching process to obtain metals. Electrolysis solution of Pb $(NO_3)_2$ and $CuSO_4$ was used for the extraction of metals from electronic waste. The amount of metals Pb and Cu extracted was 73.29% and 82.17%, respectively.

Sun et al. (2015) reported that hydrometallurgical methods have been used to recover precious metals from industrial waste and urban metal containing waste. The two-step leaching techniques were adopted to extract Cu and enriched valuable metals. It has been found that extraction efficiency and Cu selectivity for recovery are more than 95% through the use of ammonia leach solutions. Sun et al. (2015) used electrodeposition with an efficiency of 90% at the time of copper recovery, while the purity of copper can reach 99.8 wt.%. The remains from the first phase were sorted into rough and small pieces. The solid part has been restored to extract the material for further copper detection. A good portion of the treatment in the second phase of immersion is the use of H_2SO_4 for concentration of precious metals that can achieve 100% increase in their concentration in the residue with minimal loss.

Barragan et al. (2020) used a combined process of electrochemical process and hydrometallurgy to obtain copper and antimony as key constituents in a leaching solution. Antimony extraction was developed as a refinement footstep for maximum benefit to get the pure Cu. For the reduction of size of waste circuit boards mechanical methods were used in order to improve effectiveness of Sb and Cu lixiviation with $FeCl_3$, followed by electrowinning process. Further, a reactor with rotating electrode cylinders was used to test the copper concentrations, special power consumption, cathodic current efficiency, and mass transfer coefficient. Again Sb was obtained by precipitation with changing pH. Under this approach, 96 wt.% Cu deposit and 81 wt.% Sb were obtained from electronic waste.

Hong et al. (2020) reported that the porphyrin polymer is used to extract the valuable metals from PCBs leachate. The nano-porous porphyrin polymer is synthesized from widely available monomers through reductive mechanism where Au extracted reaches 1.62 g/g. Hong et al. (2020) also studied the density functional theory calculations, which shows that the multinuclear Au obligatory enhances the adsorption, while platinum capture remains at single sites of porphyrin.

Vermeşan et al. (2020) have taken efforts to recycle waste PCB as it contains precious metals and earth metals. Waste obtained from information technology and telecommunication equipment contains around 80% of recoverable metals. Vermeşan et al.'s (2020) study reported on recycling techniques of waste PCBs with the application of mechanical separation, sieving, washing, and shredding with the combination of separation techniques, dismantling, and mechanical recycling processes. Vermeşan et al. (2020) also reported the thermal, chemical, and electrochemical processes for the extraction of metals from composite items such as waste circuit board. Table 7.3 presents the different types of metals recovered from urban metal containing waste.

Urban metal containing waste is an important source for pure metals recovery. Nowadays, extraction of precious metals is focused due to economic benefits related to these metals. It also increases the commercial value and basis of occupation, which eventually achieve the plan for SDGs. Arya and Kumar (2020b) reported strategic interventions for management of a sustainable e-waste and reduced environmental impacts. Arya and Kumar (2020b) have also reported relevant strategies like resource management, eco-product design, producer responsibility, life cycle assessment, 4R principle (reduce, reuse, recycle, and recover) and bio-leaching.

Table 7.3 Different types of metals recovered from urban wastes (Debnath et al., 2018).

S. No.	Metal recovered	Processes	Equipment required	References
1.	Cu, Fe	Pyrolysis	Pyrolyzer	Chuangzhi and Yong (2006)
2.	Hg and other metals	Laser cutting	Laser cutter	Ling and Poon (2012)
3.	Ag and other metals	Leaching	Reactor	Hageluken (2006)
4.	Ag, Au, and other metals	Bioleaching	Bioreactor	Norton et al. (2005)
5.	Base metals Cu, Fe, etc. (Ag, Au, Pt)	Electrochemical and smelting refining	Smelting device	Khaliq et al. (2014)
6.	Cu, Fe, and other base metals	Plasma process	Plasma torch chamber	Tippayawong and Khongkrapan (2009)

7.3 Challenges and impacts of pyrometallurgical process

Pyrometallurgy is one of the appropriate techniques for urban mining of metal containing waste and is being adopted throughout the world. However, it still poses some challenges for the technology to be highly sustainable. Some of the challenges are as follows:

1. One of the major challenges is the collection and segregation of metal containing urban waste, as it is essential to collect precious metals in fractions, for high recovery level, and feasibility (Ramanayaka et al., 2020).
2. Precious metals are found as an integral part of many urban wastes. However, these wastes also contain additional materials which are organic and inorganic in nature. This therefore increases the labor and energy requirement.
3. The PCBs found in electronic wastes are one of the richest sources of precious metals. However, it contains additional materials which are toxic and cause health hazards. Moreover, use of brominated flame, polychlorinated biphenyls, heavy metals, etc. can cause serious environmental problems.
4. Gold leaching extraction process is performed through use of cyanide, thiourea, and thiosulfate, etc., for recovery. The cyanide process is relatively simple in operation and is a low-cost process, which makes it one of the extensively used processes. However, cyanide possesses harmful toxic effect and therefore needs substitute (Ramanayaka et al., 2020).
5. In the pyrometallurgical processes elevated temperatures are used for recovery of some fractions of Zn, Pb, Sn, Cd, and Hg metals. Thus recovery of such metals is usually energy-intensive (Ramanayaka et al., 2020).
6. Further improvements in terms of efficiency and emissions are required during recovery of these metals.
7. Current pyrometallurgical processes have limited capability regarding processing of complex products such as CRTs and PCBs which contain various other components and materials.
8. Pyrometallurgical process has limitations in terms of partial separation of metals and needs additional secondary processes of hydrometallurgical and electrometallurgical operations for efficient metal recovery.
9. One of the major limitations during pyrometallurgical processes is the release of toxic gasses.
10. In smelters the organic components are not recovered and burned.
11. Similarly, oxide constituents of scrap are also lost due to slag formation.

However, apart from the challenges, adoption of pyrometallurgy process for mining of precious metals from urban waste has impacted both the environment as well as industrial sectors.

1. Pyrometallurgy has shown its impact on various environmental factors, such as climate change, greenhouse gas emissions, freshwater eutrophication, freshwater ecotoxicity, human toxicity, particulate matter formation, ozone depletion, terrestrial acidification, photochemical oxidant formation, total land occupation, fossil depletion, metal depletion, natural land transformation, and water depletion (Kulczycka et al., 2016). Thus application of pyrometallurgical processes directly to mining of urban wastes reduces the effects of each factor as mentioned above.

2. One of the prime impacts of adopting pyrometallurgy process for urban mining is the effective urban waste management as most of the components of such waste are nonbiodegradable and would take many years for decomposition.

3. It has been found that use of improved technology has the potential to reduce the environmental impact; for example, flash-based technology has lower environmental impacts as compared to shaft furnace during copper production (Kulczycka et al., 2016).

4. It has been reported that the energy consumption in pyrometallurgical processes of ores could be divided into mining (36%), smelting and refining (34%), and processing (30%) (Kulczycka et al., 2016). Thus the application of pyrometallurgical processes during urban metal mining has the potential to reduce the overall energy consumption by reducing the energy spent in mining.

5. The energy consumption during mining and processing stages is very high as compared to smelting and refining for low-grade ores which in turn depletes the fossil fuel resources (Kulczycka et al., 2016). Therefore direct use of urban waste for mining the precious metals reduces the energy wasted during mining and purification of low-grade ores.

6. Precious metal mining directly from the urban waste also helps in preserving as well as recycling the depleting rare earth metals such as indium (Takahashi et al., 2009).

7. One of the promising solutions for further reduction in environmental impact is through adoption of biohydrometallurgy process. Biohydrometallurgy uses microorganisms for metal recovery. It can be done through heterotrophic bacteria bioleaching, autotrophic bacteria bioleaching, and heterotrophic fungi bioleaching for leaching of metals (Jadhao et al., 2021). However, selection of right species and its culture is an essential task (Priya & Hait, 2018; Ramanayaka et al., 2020).

8. In another improvement, flash joule heating is reported as one of the fastest and safest techniques for urban mining. Temperature as high as 3400K is used for ultrafast separation and has the potential with recovery yield of more than 80% for Rh, Pd, and Ag, and 60% for Au. As compared to smelting and leaching time consumed is much lower and is a solvent-free process (Deng et al., 2021).

7.4 Ongoing projects

Noranda Inc. is an integrated mining and metals company located in Toronto, Ontario, Canada. Noranda is principally engaged in the production of copper and nickel, zinc, primary and fabricated aluminum, lead, silver, gold, sulfuric acid, and cobalt. In addition, it also recycles secondary copper, nickel, and precious metals. However, products such as copper, nickel, zinc, and aluminum accounts for 72% of its revenues. Its markets are steel, refinery and foundry, construction, telecommunications, automotive, agricultural and chemical industries (Cui & Roven, 2011; Noranda Inc.: https://www.sec.gov/Archives/edgar/data/889211/000104746904017738/a2136914zex-1.htm). Another metals company is Boliden with works in the fields of exploration, mining, and smelting. Boliden Rönnskär is one of the smelters which recycle electronics to produce copper, gold, and silver. The plant uses Kaldo technology (Boliden Group: https://www.boliden.com/operations/smelters/boliden-ronnskar; Cui & Roven, 2011). Umicore also operates a precious metal refining plant. It recycles wastes containing both ferrous and nonferrous metals. They recover metals like silver, gold, platinum, palladium, rhodium, iridium, ruthenium, indium, selenium, tellurium, antimony, tin, bismuth, lead, copper, and nickel (Cui & Roven, 2011; Umicore: https://pmr.umicore.com/).

7.5 Conclusion and recommendation

Any sustainable measure adopted to curb the menace of urban waste should focus on the exhaustive nature of primary resources, reduced energy consumption coupled with environmental management due to release of hazardous substances. Due to high abundance of precious and valuable metal coupled with its exponentially growing quantity, "urban mining" has become a vital or indispensable part of waste management. In this regard, pyrometallurgy technology has emerged as one of the viable options for extraction of metals and is continuously evolving with each passing year. It is considered as economical and eco-efficient, chiefly due to its ability to recover precious metals and utilization of nonmetallic fractions as fuels and reducing agents. However, certain limitations such as complicated Al and Fe recovery and emission of toxic compounds are associated with this method. It thus becomes very important to choose the method depending on its feed materials and desired products.

Therefore it is recommended to further expand our understanding to reduce the amount of pollutants generated through urban mining process. In order to overcome the inefficiency of pyrometallurgy while recovering some metals, it becomes extremely important to adopt new methods of pyrometallurgy such as vacuum metallurgy. Apart from this, application of pyrolysis prior to the pyrometallurgical processes helps in recovering metals like In and Ga. Process of volatilization of halides and hazardous organics in such wastes leads to enrichment of metallic parts along with recovery of energy.

References

Arya, S., & Kumar, S. (2020a). Bioleaching: Urban mining option to curb the menace of E-waste challenge. *Bioengineered*. Available from https://doi.org/10.1080/21655979.2020.1775988.

Arya, S., & Kumar, S. (2020b). E-waste in India at a glance: current trends, regulations, challenges and waste management strategies. *Journal of Cleaner Production, 271*. Available from https://doi.org/10.1016/j.jclepro.2020.122707.

Arya, S., Patel, A., Kumar, S., & Loke, S. P. (2021). Urban mining of obsolete computers by manual dismantling and waste printed circuit boards by chemical leaching and toxicity assessment of its waste residues. *Environmental Pollution*. Available from https://doi.org/10.1016/j.envpol.2021.117033.

Arya, S., Rautela, R., Chavan, D., & Kumar, S. (2021). Evaluation of soil contamination due to crude E-waste recycling activities in the capital city of India. *Process Safety and Environmental Protection*. Available from https://doi.org/10.1016/j.psep.2021.07.001.

Ayres, R. U., & Peiro, L. T. (2013). Material efficiency: Rare and critical metals. *Philosophical Transactions of the Royal Society A: Mathematical, Physical and Engineering Sciences, 371*(1986), 20110563.

Banerjee, P., Hazra, A., Ghosh, P., Ganguly, A., Murmu, N. C., & Chatterjee, P. K. (2019). *Solid waste management in India: A brief review. Waste management and resource efficiency* (pp. 1027−1049). Springer.

Barragan, J. A., Ponce de León, C., Alemán Castro, J. R., Peregrina-Lucano, A., Gómez-Zamudio, F., & Larios-Durán, E. R. (2020). Copper and antimony recovery from electronic waste by hydrometallurgical and electrochemical techniques. *ACS Omega, 5*(21), 12355−12363.

Boliden Group, Klarabergsviadukten 90 P.O. Box 44, SE-101 20 Stockholm, <https://www.boliden.com/operations/smelters/boliden-ronnskar> Accessed 13.07.22.

Cayumil, R., Ikram-Ul-Haq, M., Khanna, R., Saini, R., Mukherjee, P. S., Mishra, B. K., & Sahajwalla, V. (2018). High temperature investigations on optimising the recovery of copper from waste printed circuit boards. *Waste Management, 73*, 556−565.

Cayumil, R., Khanna, R., Ikram-Ul-Haq, M., Rajarao, R., Hill, A., & Sahajwalla, V. (2014). Generation of copper rich metallic phases from waste printed circuit boards. *Waste Management, 34*(10), 1783−1792.

Chatterjee, S., & Kumar, K. (2009). Effective electronic waste management and recycling process involving formal and non-formal sectors. *International Journal of Physical Sciences, 4*(13), 893−905.

Chuangzhi, X. Z. L. H. W., & Yong, C. (2006). A study on pyrolysis and kinetics of printed circuit boards wastes. *Techniques and Equipment for Environmental Pollution Control, 10*(010).

Cui, J., & Roven, H. J. (2011). *Electronic waste. Waste* (pp. 281−296). Academic Press.

Debnath, B., Chowdhury, R., & Ghosh, S. K. (2018). Sustainability of metal recovery from E-waste. *Frontiers of Environmental Science & Engineering, 12*(6), 1−12.

Debnath, B., Roychowdhury, P., & Kundu, R. (2016). Electronic components (EC) reuse and recycling − A new approach towards WEEE management. *Procedia Environmental Sciences, 35*, 656−668.

Deng, B., Luong, D. X., Wang, Z., Kittrell, C., McHugh, E. A., & Tour, J. M. (2021). Urban mining by flash Joule heating. *Nature Communications, 12*(1), 1−8.

Dutta, D., Arya, S., Kumar, S., & Lichtfouse, E. (2021). Electronic waste pollution and the COVID-19 pandemic. *Environmental Chemistry Letters*. Available from https://doi.org/10.1007/s10311-021-01286-9.

Ebin, B., & Isik, M. I. (2016). Pyrometallurgical processes for the recovery of metals from WEEE. *WEEE recycling* (pp. 107−137). Elsevier.

Forti, V., Balde, C. P., Kuehr, R., & Bel, G. (2020). *The global E-waste monitor 2020: Quantities, flows and the circular economy potential*. Bonn, Geneva, and Rotterdam: United Nations University/United Nations Institute for Training and Research, International Telecommunication Union, and International Solid Waste Association.

Gollakota, A. R., Gautam, S., & Shu, C. M. (2020). Inconsistencies of e-waste management in developing nations − Facts and plausible solutions. *Journal of Environmental Management, 261*, 110234.

Grigorescu, R. M., Grigore, M. E., Iancu, L., Ghioca, P., & Ion, R. M. (2019). Waste electrical and electronic equipment: A review on the identification methods for polymeric materials. *Recycling, 4*(3), 32.

Gurgul, A., Szczepaniak, W., & Zabłocka-Malicka, M. (2017). Incineration, pyrolysis and gasification of electronic waste. *E3S Web of Conferences, 22*, 00060.

Hageluken, C. (2006). Recycling of electronic scrap at Umicore's integrated metals smelter and refinery. *Erzmetall*, *59*(3), 152−161.

Hong, Y., Thirion, D., Subramanian, S., Yoo, M., Choi, H., Kim, H. Y., Stoddart, J. F., & Yavuz, C. T. (2020). Precious metal recovery from electronic waste by a porous porphyrin polymer. *Proceedings of the National Academy of Sciences*, *117*(28), 16174−16180.

Hsu, E., Barmak, K., West, A. C., & Park, A. H. A. (2019). Advancements in the treatment and processing of electronic waste with sustainability: A review of metal extraction and recovery technologies. *Green Chemistry*, *21*(5), 919−936.

Hylander, L. D., & Herbert, R. B. (2008). Global emission and production of mercury during the pyrometallurgical extraction of nonferrous sulfide ores. *Environmental Science & Technology*, *42*(16), 5971−5977.

Jadhao, P. R., Mishra, S., Pandey, A., Pant, K. K., & Nigam, K. D. P. (2021). Biohydrometallurgy: A sustainable approach for urban mining of metals and metal refining. *Catalysis for clean energy and environmental sustainability* (pp. 865−892). Cham: Springer.

Kaya, M. (2016). Recovery of metals and nonmetals from electronic waste by physical and chemical recycling processes. *Waste Management*, *57*, 64−90.

Khaliq, A., Rhamdhani, M. A., Brooks, G., & Masood, S. (2014). Metal extraction processes for electronic waste and existing industrial routes: A review and Australian perspective. *Resources*, *3*(1), 152−179.

Kiddee, P., Naidu, R., & Wong, M. H. (2013). Electronic waste management approaches: An overview. *Waste Management*, *33*(5), 1237−1250.

Krishnan, S., Zulkapli, N. S., Kamyab, H., Taib, S. M., Din, M. F. B. M., Abd Majid, Z., Chaiprapat, S., Kenzo, I., Ichikawa, Y., Nasrullah, M., & Chelliapan, S. (2021). Current technologies for recovery of metals from industrial wastes: An overview. *Environmental Technology & Innovation*, *22*, 101525.

Kulczycka, J., Lelek, è., Lewandowska, A., Wirth, H., & Bergesen, J. D. (2016). Environmental impacts of energy-efficient pyrometallurgical copper smelting technologies: The consequences of technological changes from 2010 to 2050. *Journal of Industrial Ecology*, *20*(2), 304−316.

Ling, T.-C., & Poon, C.-S. (2012). Development of a method for recycling of CRT funnel glass. *Environmental Technology*, *33*(22), 2531−2537.

Ma, E., Lu, R., & Xu, Z. (2012). An efficient rough vacuum-chlorinated separation method for the recovery of indium from waste liquid crystal display panels. *Green Chemistry*, *14*(12), 3395−3401.

Noranda Inc., BCE Place, 181 Bay Street, Suite 200, Toronto, Ontario, Canada M5J 2T3. <https://www.sec.gov/Archives/edgar/data/889211/000104746904017738/a2136914zex-1.htm> Accessed 13.07.22.

Norton, A., de Klerk Batty, J., Dew, D.W., & Basson, P., Billiton Intellectual Property BV, (2005). *Recovery of precious metal from sulphide minerals by bioleaching* (U.S. Patent No. 6860919).

Osibanjo, O., Nnorom, I. C., Adie, G. U., Ogundiran, M. B., & Adeyi, A. A. (2016). Global Management of Electronic Wastes: Challenges facing developing and economy-in-transition countries. *Metal Sustainability: Global Challenges, Consequences, and Prospects*, *33*(8), 51.

Park, H. S., & Kim, Y. J. (2019). A novel process of extracting precious metals from waste printed circuit boards: Utilization of gold concentrate as a fluxing material. *Journal of Hazardous Materials*, *365*, 659−664.

Park, K. S., Sato, W., Grause, G., Kameda, T., & Yoshioka, T. (2009). Recovery of indium from In_2O_3 and liquid crystal display powder via a chloride volatilization process using polyvinyl chloride. *Thermochimica Acta*, *493*(1−2), 105−108.

Priya, A., & Hait, S. (2018). Extraction of metals from high grade waste printed circuit board by conventional and hybrid bioleaching using *Acidithiobacillus ferrooxidans*. *Hydrometallurgy*, *177*, 132−139.

Ramanayaka, S., Keerthanan, S., & Vithanage, M. (2020). *Urban mining of E-waste: Treasure hunting for precious nanometals*. *Handbook of electronic waste management* (pp. 19−54). Butterworth-Heinemann.

Rautela, R., Arya, S., Vishwakarma, S., Lee, J., Kim, K.-H., & Kumar, S. (2021). E-waste management and its effects on the environment and human health. *Science of The Total Environment*. Available from https://doi.org/10.1016/j.scitotenv.2021.145623.

Rene, E. R., Sethurajan, M., Ponnusamy, V. K., Kumar, G., Dung, T. N. B., Brindhadevi, K., & Pugazhendhi, A. (2021). Electronic waste generation, recycling and resource recovery: Technological perspectives and trends. *Journal of Hazardous Materials*, 125664.

Salhofer, S., & Tesar, M. (2011). Assessment of removal of components containing hazardous substances from small WEEE in Austria. *Journal of Hazardous Materials, 186*(2–3), 1481–1488.

Sivakumar, P., Prabhakaran, D., & Thirumarimurugan, M. (2018). Optimization studies on recovery of metals from printed circuit board waste. *Bioinorganic Chemistry and Applications, 2018.*

Sun, Z., Xiao, Y., Sietsma, J., Agterhuis, H., & Yang, Y. (2015). A cleaner process for selective recovery of valuable metals from electronic waste of complex mixtures of end-of-life electronic products. *Environmental Science & Technology, 49*(13), 7981–7988.

Takahashi, K., Sasaki, A., Dodbiba, G., Sadaki, J., Sato, N., & Fujita, T. (2009). Recovering indium from the liquid crystal display of discarded cellular phones by means of chloride-induced vaporization at relatively low temperature. *Metallurgical and Materials Transactions A, 40*(4), 891–900.

Tippayawong, N., & Khongkrapan, P. (2009). Development of a laboratory scale air plasma torch and its application to electronic waste treatment. *International Journal of Environmental Science & Technology, 6*(3), 407–414.

Umicore. <https://pmr.umicore.com/> Accessed 13.07.22.

Vermęan, H., Tiuc, A. E., & Purcar, M. (2020). Advanced recovery techniques for waste materials from IT and telecommunication equipment printed circuit boards. *Sustainability, 12*(1), 74.

Vishwakarma, S., Kumar, V., Arya, S., Tembhare, M., Rahul., Dutta, D., & Kumar, S. (2022). E-waste in Information and Communication Technology Sector: Existing scenario, management schemes and initiatives. *Environmental Technology and Innovation*, 102797. Available from https://doi.org/10.1016/j.eti.2022.102797.

CHAPTER 8

Bioleaching: urban mining of E-waste and its future implications

Sartaj Ahmad Bhat[1,2], Guangyu Cui[3], Fuad Ameen[4], Fusheng Li[1] and Sunil Kumar[2]

[1]River Basin Research Center, Gifu University, Gifu, Japan
[2]Waste Re-processing Division, CSIR-National Environmental Engineering Research Institute (CSIR-NEERI), Nagpur, Maharashtra, India
[3]State Key Laboratory of Pollution Control and Resource Reuse, Tongji University, Shanghai, P.R. China
[4]Department of Botany and Microbiology, College of Science, King Saud University, Riyadh, Saudi Arabia

8.1 Introduction

Electronic waste, or E-waste, refers to the waste produced from electrical and electronic equipment (EEE) products without the plan of use again. E-waste consists of a broad variety of products such as any house or commercial thing with circuitry or electrical apparatus with power or battery supply (Baldé et al., 2017). According to Perkins et al. (2014) there are six categories that are to be considered as waste from EEE, such as screens and monitors (television monitors, laptops, and tablets), temperature exchange equipment (refrigerators, freezers, and heat pumps), lamps (LED lamps, fluorescent lamps, high-intensity discharge lamps, and tungsten bulbs), small information technology and telecommunication equipment (pocket calculators, mobile phones, global positioning systems, e-book reader, printers, and telephones), small (vacuum cleaners, electric kettles, small electrical and electronic tools and toys, microwaves, toasters, and electric shavers) and large equipment (washing machines, dish-washing machines, large printing machines, photovoltaic panels, and electric stoves). It is examined that globally waste from EEE disposal has been doubled about 41.8 million tons per year between 2009 and 2014, and in 2016, it attained about 44.7 million tons per year. Projections have been prepared and accounted that the overall quantity of waste from EEE disposal per year could surpass 50 million tons by 2021 (Hsu et al., 2019) and by 2030, computer waste will attain 1000 million tons (Tiwary et al., 2017). In recent years, urban mining of E-waste has established considerable interest due to its beneficial prospects, comprehensive business prospect, source of living and eventually attaining agenda for Sustainable Development Goals (SDGs) 2030. E-waste buildup in the environment is a severe environmental challenge. The call for heavy metal recovery, jointly with the abundance of valuable and base metals are strong incentives for investigators to uncover a sustainable process for recovery of metal from E-waste (Jagannath et al., 2017). The researchers are trying to advance the effectiveness of recovery of metals from E-wastes

Global E-waste Management Strategies and Future Implications
DOI: https://doi.org/10.1016/B978-0-323-99919-9.00003-9

by means of bioleaching, a new sustainable process in comparison to traditional means. Biological metal recovery from solid waste utilizes microorganisms for creation of reagents to remove metals (Brierley, 2016). Specificity, commercial, and environmental adequacy are key rewards for recovery of metals during microbial processes (Natarajan, 2018). Bioleaching method is easy and has advantages such as high effectiveness and protection, minor working expenses and energy usage, easier organization, accomplishment of operating environment at atmospheric pressure and room temperature, environment-friendly, and no require for trained recruits (Vakilchap et al., 2016). The aim of this chapter is to describe the fundamentals and application of bioleaching of E-wastes. The methods that deal with resource recovery, limitations, the challenges, and future prospects were also addressed.

8.2 E-waste generations and scope for bioleaching

Latest estimations accounted that the quantity of E-waste produced from Asia was 18.2 million tons, after that Europe (12.3 million tons), the Americas (11.3 million tons), Africa (2.2 million tons), and Oceania (0.7 million tons) (Ramanayaka et al., 2020). About 44.7 million metric tons of E-waste were produced globally in 2016 and is expected to grow up to 52.2 million metric tons in 2021 with annual growth rate of 3%–4%. As per the 2016 E-waste global trend results, China is the highest E-waste generating country with 7.2 Mt/annum, followed by 6.3 Mt in the United States, 2.1 Mt in Japan, 2.0 Mt in India, 1.9 Mt in Germany, 1.5 Mt in Brazil, 1.4 Mt in Russia and France, 1.3 Mt in Indonesia, and 1.2 Mt in Italy. The per capita E-waste production is maximum in Germany, that is, 22.8 kg/capita, followed by France (21.3 kg/capita), United States (19.4 kg/capita), Italy (18.9 kg/capita), Japan (16.9 kg/capita), Russia (9.7 kg/capita), Brazil (7.4 kg/capita), China (5.2 kg/capita), Indonesia (4.9 kg/capita), and India (1.5 kg/capita) (Associated Chambers of Commerce of India (ASSOCHAM), India, 2018; Baldé et al., 2017). In India E-waste generation was about 470,000 million tons as of 2011, out of which the yearly production from cities like Mumbai is approximately 11,000 million tons, Delhi 9000 million tons, Bengaluru 8000 million tons, and Chennai 5000–6000 million tons (Arya & Kumar, 2020a; Vidyadhar, 2016). Domestic households, community, and private sectors are the possible sources for E-waste production in India (GTZ-MAIT, 2007; Mundada et al., 2007). The numerical breakdown of the state-wise E-waste generation in India is presented in Fig. 8.1.

Maharashtra ranks the first for producing 19.8% of the total E-waste, followed by Tamil Nadu (13%), Uttar Pradesh (10.1%), West Bengal (9.8%), Delhi (9.5%), Karnataka (8.9%), Gujarat (8.8%), and Madhya Pradesh (7.6%). The per capita was maximum in Delhi (11.3), followed by Tamil Nadu (3.6), Maharashtra (3.52), Andhra Pradesh (2.95), Karnataka (2.91), Gujarat (2091), West Bengal (2.14), Madhya Pradesh (2.09), and Uttar Pradesh (1.01).

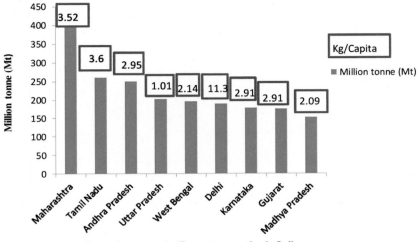

Figure 8.1 The numerical breakdown of the state-wise E-waste generation in India.

8.3 Urban mining of E-waste for metals

Urban mining is an idea associated to resource competence which widens landfill mining to the course of resource recovery (Cossu & Williams, 2015). It also ensures the development of resource preservation through 3R (reuse, recycling, and recovery) of costly and valuable resources from waste (Arora et al., 2017). E-waste is regarded as the backbone stream for urban mining amongst a range of categories of waste. Urban mining actions engage organized management of anthropogenic supply stocks and waste in line to defend the environment surroundings, preserve resources, and gain economic profit (Baccini & Brunner, 2012). Urban mining is a compilation of various processes. The first stage in the urban mining process is the collection and consolidation of E-waste (Fig. 8.2).

Cost minimization has been the main worry in the collection of E-waste collection (Nowakowski, 2017). Awareness plays a most important function to improve E-waste collections and recycling. According to the study reports, efficient urban mining has the possible potential in the direction of saving the waste substance value approximately 21 USD billion (Business World Online Bureau, 2018). The United Nations General Assembly in New York in 2015 reported that the E-waste management is directly connected with the achievement of the SDGs 2030 Agenda. Therefore, in the assembly gathering, it was advocated by the all countries to preserve a sustainable approach in the direction of E-waste management with a center of attention on decreasing the waste generation (United Nation Environment Management Group (UNEMG), 2018). The planned achievement of urban mining of E-waste shows the

Figure 8.2 Schematic diagram for urban mining in a perfect circular economy model.

ways in the direction of maximizing the resource maintenance and growing the economic cost, preparation and designing the infrastructure of cities, constructive advance in the direction of the health and safety and service opportunities as well as decreasing greenhouse gases and eventually contributing for achieving the SDGs 2030.

8.4 Bioleaching

Bioleaching is an easy practice for the recovery of metal particles from E-waste, with the help of microbes including bacteria, fungi, and actinomycetes (Vakilchap et al., 2016). Bioleaching is considered as small power-consuming, greatly competent, ecologically pleasant, and a short operational expenditure process that can be passed out at room temperature and atmospheric pressure (Xiang et al., 2010). The general process of bioleaching is presented in Fig. 8.3.

Bioleaching of E-waste has expanded much consideration due to its scientific development in provisions of sustainability and as transitional technologies particularly in developed countries (Acevedo, 2002). Bioleaching method has preferred to be sustainable practice with lesser energy costs (Arya & Kumar, 2020b) and better environmental legacies as compared to traditional methods (Rasoulnia et al., 2021). Bioleaching employs microorganisms to make possible leaching of minerals and tenders' important economic profit greater than traditional metallurgy owing to its cheap infrastructure costs

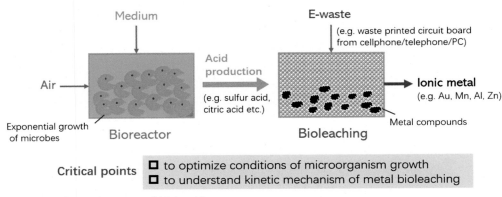

Figure 8.3 General process of bioleaching.

and small energy input during operation at ambient temperature and pressure (Agate, 1996; Kiddee & Naidu, 2013). Bioleaching works on the principle of the capability of the metals to get melted by the oxidizing microorganisms (Boon et al., 1998). Usually, the microbes leach the metal elements from the metal sulfide, which presents in E-waste by two kinds of mechanisms, which are well known as direct action and indirect action mechanisms. In direct action mechanism, microbes oxidize the metal sulfide directly, which denotes that the electron transfer takes place among the metal sulfides and on the surface of the exacting organism (Vera et al., 2013). In the indirect action mechanism, the microbes such as iron (II)-oxidizing bacteria regenerates the Fe^{3+} ions from Fe^{2+} ions by oxidization and after that Fe^{3+} ions oxidize the metal particles by itself. The indirect action mechanism of bioleaching can also be additionally classified into two groups: thiosulfate pathway and polysulfide pathway. According to Cui and Zhang (2008), thiosulfate can be formed as an intermediate in the thiosulfate pathway. In polysulfide pathway, acid-soluble metal sulfides go through oxidative assault by protons and/or Fe^{3+} ions and form chemically inert elemental sulfur (Vera et al., 2013). On the other hand, the elemental sulfur can lastly be oxidized by bacteria to sulfuric acid. The present mechanism can be catalyzed by sulfur-compound oxidizing bacteria such as *Acidithiobacillus ferrooxidans* and *Acidithiobacillus thiooxidans*. *A. ferrooxidans* and *A. thiooxidans* are the generally employed bacteria in the process of bioleaching and are regularly used in the metal recovery procedures from E-waste. Other Gram-negative bacteria like *Chromobacterium violaceum* and *Pseudomonas putida* are also being used in the bioleaching of metals (Liu et al., 2016). *C. violaceum* has the capability to generate cyanide ions, where Au particles can be leach out from the E-waste (Liu et al., 2016). The various metals recovered from various kinds of E-wastes using bioleaching method are summarized in Table 8.1. The bioleaching is an efficient method to recover the valuable metals from the E-waste. Hence, it is essential to learn the process in order to advance the leaching abilities and find out new species of microorganisms.

Table 8.1 Metals recovered from various kinds of E-waste using bioleaching method.

E-waste component	Microorganisms	Metals recovered	References
Dust from WEEE shredding	*Acidithiobacillus thiooxidans*	Ce, Eu, Nd, La, Y	Marra et al. (2018)
Liquid–crystal displays (LCDs)	*A. thiooxidans*	Ln, Sr	Jowkar et al. (2018)
Printed circuit boards (PCBs) of computer	*Acidithiobacillus ferrooxidans*	Cu, Zn, Pb, Ni	Priya and Hait (2018)
Mobile phone PCBs	*Acinetobacter* sp. Cr B2	Cu	Jagannath et al. (2017)
E-waste of PCBs	*Pseudomonas balearica* SAE1	Au, Ag	Kumar et al. (2018)
PCB powder	*Leptospiriuulm ferrriphilum* and *Sulfobacillus thermosulfidooxidans* (supernatant)	Cu	Wu et al. (2018)
PCBs (1 mm)	*A. ferrooxidans*	Cu	Annamalai and Gurumurthy (2019)
PCBs	*A. ferrooxidans + A. thiooxidans*	Cu	Isidar et al. (2016)
PCBs (4–10 mm)	*A. ferrooxidans*	Cu	Chen et al. (2015)
Personal computer (PC) PCBs	*A. ferrooxidans*	Cu, Ni	Arshadi and Mousavi (2014)
PCBs	*A. ferrooxidans*	Zn, Cu, Al	Yang et al. (2014)
Television (TV) PCBs	*A. ferrooxidans + Leptospirillum ferrooxidans + A. thiooxidans*	Cu	Bas et al. (2013)
PCBs	Mixed acidophilic bacteria	Cu, Al, Zn	Zhu et al. (2011)
PCBs	*Sulfobacillus thermosulfidooxidans + acidophilic isolate*	Cu, Al, Ni, Zn	Ilyas et al. (2007)
WEEE	*Pseudomonas putida + A. ferrooxidans + A. thiooxidans*	Al, Cu, Ni, Zn	Brandl et al. (2001)

8.5 Conclusions and future perspectives

The urban mining of valuable metals from E-waste is considered necessary since E-wastes are vastly rich in precious metals, and the buildup of E-waste in the environment is very common. The most important concern of urban mining is the environmental impacts. The bioleaching (hydrometallurgy) process produces fewer pollutants compared to the other processes. Nevertheless, pretreatment of contaminants prior to release to the environment is a necessity. Hence, additional studies are essential to decrease the contaminants produced after the urban mining process. The bioleaching process is a well-organized method as it is environment-friendly and is mediated by the secretion of extracellular polymeric substances by the microorganisms. Bioleaching process is also advantageous to industries, small units engaged in waste managing actions, pollution control boards, and, in addition can provide jobs to skilled and semi-skilled instructive youth. The wealth recovered from E-waste can be used as a circular financial system approach by the industries, thus decreasing the load of raw material for manufacturing new products. Resource resurgence by bioleaching process has significant profit on economic growth, public health, and sustainable surroundings. Bioleaching process has achieved much consideration due to its easy and cost-effective approach, compact energy requisite, and minor pollution. Nevertheless, the bioleaching process is very slow and takes a few days to leach the precious metals from the E-waste. Other limitations include minimum yielding productions, space obligation and appropriate infrastructure, and insufficiency for treating the difficult complex combination of precious metals. Additional studies are considered necessary to enhance the bioleaching process.

References

Acevedo, F. (2002). Present and future of bioleaching in developing countries. Biotechnology issues for developing countries. *Electronic Journal of Biotechnology*, *5*(2), 196–199.

Agate, A. D. (1996). Recent advances in microbial mining. *World Journal of Microbiology and Biotechnology*, *12*(5), 487–495.

Annamalai, M., & Gurumurthy, K. (2019). Enhanced bioleaching of copper from circuit boards of computer waste by *Acidithiobacillus ferrooxidans*. *Environmental Chemistry Letters*, *17*(4), 1873–1879.

Arora, R., Paterok, K., Banerjee, A., & Saluja, M. S. (2017). Potential and relevance of urban mining in the context of sustainable cities. *IIMB Management Review*, *29*(3), 210–224.

Arshadi, M., & Mousavi, S. M. (2014). Simultaneous recovery of Ni and Cu from computer-printed circuit boards using bioleaching: Statistical evaluation and optimization. *Bioresource Technology*, *174*, 233–242.

Arya, S., & Kumar, S. (2020a). E-waste in India at a glance: Current trends, regulations, challenges and management strategies. *Journal of Cleaner Production*, *271*, 122707. Available from https://doi.org/10.1016/j.jclepro.2020.122707.

Arya, S., & Kumar, S. (2020b). Bioleaching: urban mining option to curb the menace of E-waste challenge. *Bioengineered*, *11*(1), 640–660. Available from https://doi.org/10.1080/21655979.2020.1775988.

Associated Chambers of Commerce of India (ASSOCHAM), India (2018). *Electricals and electronics manufacturing in India* (pp. 1–56). NEC Technologies in India. <https://in.nec.com/en_IN/pdf/ElectricalsandElectronicsManufacturinginIndia2018.pdf> Accessed 28.12.19.

Baccini, P., & Brunner, P. H. (2012). *Metabolism of the anthroposphere*. MIT Press.

Baldé, C. P., Forti, V., Gray, V., Kuehr, R., & Stegmann, P. (2017). *The global E-waste monitor 2017: Quantities, flows and resources*. Bonn/Geneva/Vienna: United Nations University, International Telecommunication Union, and International Solid Waste Association (ISWA).

Bas, A. D., Deveci, H., & Yazici, E. Y. (2013). Bioleaching of copper from low grade scrap TV circuit boards using mesophilic bacteria. *Hydrometallurgy*, *138*, 65–70.

Boon, M., Heijnen, J. J., & Hansford, G. S. (1998). The mechanism and kinetics of bioleaching sulphide minerals. *Mineral Processing and Extractive Metallurgy Review*, *19*(1), 107–115.

Brandl, H., Bosshard, R., & Wegman, M. (2001). Computer-munching microbes: Metal leaching from electronic scrap by bacteria and fungi. *Hydrometallurgy*, *59*, 319–326.

Brierley, C. L. (2016). Biological processing: Biological processing of sulfidic ores and concentrates—Integrating innovations. *Innovative process development in metallurgical industry* (pp. 109—135). Cham: Springer.

Business World Online Bureau (2018). *Urban mining has the potential to unlock India's' E-waste economy.*

Chen, S., Yang, Y., Liu, C., Dong, F., & Liu, B. (2015). Column bioleaching copper and its kinetics of waste printed circuit boards (WPCBs) by *Acidithiobacillus ferrooxidans. Chemosphere, 141,* 162—168.

Cossu, R., & Williams, I. D. (2015). Urban mining: Concepts, terminology, challenges. *Waste Management (New York, N.Y.), 45,* 1—3.

Cui, J., & Zhang, L. (2008). Metallurgical recovery of metals from electronic waste: A review. *Journal of Hazardous Materials, 158,* 228—256.

GTZ-MAIT (2007). *A study on e-waste assessment in the country.* The German Technical Cooperation Agency (GTZ) and Manufacturer's Association for Information Technology Industry (MAIT). <http://www.mait.com/admin/press_images/press77-try.htm>.

Hsu, E., Barmak, K., West, A. C., & Park, A. H. A. (2019). Advancements in the treatment and processing of electronic waste with sustainability: A review of metal extraction and recovery technologies. *Green Chemistry: An International Journal and Green Chemistry Resource, 21*(5), 919—936.

Ilyas, S., Anwar, M. A., Niazi, S. B., & Ghauri, M. A. (2007). Bioleaching of metals from electronic scrap by moderately thermophilic acidophilic bacteria. *Hydrometallurgy, 88,* 180—188.

Isidar, A., van de Vossenberg, J., Rene, E. R., van Hullebusch, E. D., & Lens, P. N. (2016). Two-step bioleaching of copper and gold from discarded printed circuit boards (PCB). *Waste Management, 57,* 149—157.

Jagannath, A., Shetty, V., & Saidutta, M. B. (2017). Bioleaching of copper from electronic waste using *Acinetobacter* sp. Cr B2 in a pulsed plate column operated in batch and sequential batch mode. *Journal of Environmental Chemical Engineering, 5*(2), 1599—1607.

Jowkar, M. J., Bahaloo-Horeh, N., Mousavi, S. M., & Pourhossein, F. (2018). Bioleaching of indium from discarded liquid crystal displays. *Journal of Cleaner Production, 180,* 417—429.

Kiddee, P., & Naidu, R. (2013). Electronic waste management approaches: An overview. *Waste Management, 33*(5), 1237—1250.

Kumar, A., Saini, H. S., & Kumar, S. (2018). Bioleaching of gold and silver from waste printed circuit boards by *Pseudomonas balearica* SAE1 isolated from an E-waste recycling facility. *Current Microbiology, 75*(2), 194—201.

Liu, R., Li, J., & Ge, Z. (2016). Review on *Chromobacterium violaceum* for gold bioleaching from E-waste. *Procedia Environmental Sciences: An International Journal of Environmental Physiology and Toxicology, 31,* 947—953.

Marra, A., Cesaro, A., Rene, E. R., Belgiorno, V., & Lens, P. N. L. (2018). Bioleaching of metals from WEEE shredding dust. *Journal of Environmental Management, 210,* 180—190.

Mundada, M. N., Kumar, S., & Shekdar, A. V. (2007). E-waste: A new challenge for waste management in India. *International Journal of Environmental Studies, 61,* 265—279. Available from https://doi.org/10.1080/0020723042000176060.

Natarajan, K. A. (2018). *Biotechnology of metals: Principles, recovery methods and environmental concerns.* Elsevier.

Nowakowski, P. (2017). A proposal to improve E-waste collection efficiency in urban mining: Container loading and vehicle routing problems—A case study of Poland. *Waste Management, 60,* 494—504.

Perkins, D. N., Drisse, M.-N. B., Nxele, T., & Sly, P. D. (2014). E-waste: A global hazard. *Annals of Global Health, 80*(4), 286—295.

Priya, A., & Hait, S. (2018). Extraction of metals from high grade waste printed circuit board by conventional and hybrid bioleaching using *Acidithiobacillus ferrooxidans. Hydrometallurgy, 177,* 132—139.

Ramanayaka, S., Keerthanan, S., & Vithanage, M. (2020). Urban mining of E-waste: Treasure hunting for precious nanometals. *Handbook of electronic waste management* (pp. 19—54). Butterworth-Heinemann.

Rasoulnia, P., Barthen, R., & Lakaniemi, A. M. (2021). A critical review of bioleaching of rare earth elements: The mechanisms and effect of process parameters. *Critical Reviews in Environmental Science and Technology, 51*(4), 378—427.

Tiwary, C. S., Kishore, S., Vasireddi, R., Mahapatra, D. R., Ajayan, P. M., & Chattopadhyay, K. (2017). Electronic waste recycling via cryo-milling and nanoparticle beneficiation. *Materials Today, 20*(2), 67—73.

United Nation Environment Management Group (UNEMG). (2018). *The United Nations and E-waste System wide action on addressing the full life cycle of electrical and electronic equipment.* (pp. 1–40). Available from: <https://unemg.org/wp-content/uploads/2018/11/INF1.pdf>.

Vakilchap, F., Mousavi, S. M., & Shojaosadati, S. A. (2016). Role of *Aspergillus niger* in recovery enhancement of valuable metals from produced red mud in Bayer process. *Bioresource Technology, 218,* 991–998.

Vera, M., Schippers, A., & Sand, W. (2013). Progress in bioleaching: fundamentals and mechanisms of bacterial metal sulfide oxidation—Part A. *Applied Microbiology and Biotechnology, 97*(17), 7529–7541.

Vidyadhar, A. (2016). Chapter 6: A review of technology of metal recovery from electronic waste. In *E-waste in transition—From pollution to resource.* Rijeka: IntechOpen. <https://doi.org/10.5772/61569>.

Wu, W., Liu, X., Zhang, X., Zhu, M., & Tan, W. (2018). Bioleaching of copper from waste printed circuit boards by bacteria-free cultural supernatant of iron–sulfur-oxidizing bacteria. *Bioresources and Bioprocessing, 5*(1), 1–13.

Xiang, Y., Wu, P., Zhu, N., Zhang, T., Liu, W., Wu, J., & Li, P. (2010). Bioleaching of copper from waste printed circuit boards by bacterial consortium enriched from acid mine drainage. *Journal of Hazardous Materials, 184*(1–3), 812–818.

Yang, Y., Chen, S., Li, S., Chen, M., Chen, H., & Liu, B. (2014). Bioleaching waste printed circuit boards by *Acidithiobacillus ferrooxidans* and its kinetics aspect. *Journal of Biotechnology, 173*(1), 24–30.

Zhu, N., Xiang, Y., Zhang, T., Wu, P., Dang, Z., Li, P., & Wu, J. (2011). Bioleaching of metal concentrates of waste printed circuit boards by mixed culture of acidophilic bacteria. *Journal of Hazardous Materials, 192*(2), 614–619.

CHAPTER 9

Occupational health hazards associated with E-waste handling, treatment, management, and case studies

Loganath Radhakrishnan[1,2], J. Senophiyah Mary[1,2], Kumari Sweta[3], Arya Anuj Jee[3], Nityanand Singh Maurya[3], Anudeep Nema[4] and Dayanand Sharma[3,5]

[1]Institute for Globally Distributed Open Research and Education, Coimbatore, Tamil Nadu, India
[2]Environmental Consultant, Indian Green Service, Delhi, Delhi, India
[3]Department of Civil Engineering, National Institute of Technology, Patna, Bihar, India
[4]Department of Civil Engineering, School of Engineering, Eklavya University, Damoh, Madhya Pradesh, India
[5]Department of Civil Engineering, Sharda University, Greater Noida, Uttar Pradesh, India

9.1 Introduction

Electronic waste (E-waste) is defined as discarded electronic and electrical items, which are not able to fulfill their intended purpose. It may also be known as waste electrical and electronic equipment (WEEE). E-waste is considered one of the rapidly growing waste streams in the current world. E-waste contributes 8% of the municipal solid waste fraction. The reasons behind this large fraction are the industrial revolution, economic growth, lifestyle changes, and urbanization (Senophiyah-Mary, Loganath, & Meenambal, 2018; Senophiyah-Mary, Loganath, & Shameer, 2018). The different reasons for discarding the WEEE are shown in Fig. 9.1.

The basal convention classified E-waste as hazardous waste and has given the guidelines for controlling this waste. The informal sectors developed because of stringent government laws and environmental regulations for managing hazardous E-waste. These informal sectors generally follow inappropriate methods to treat the E-waste, such as open burning, heating, and acid bath and emit high amount of toxic gases to our environment. These kinds of informal sectors are functional for a long period and treat large volumes of E-waste without using pollution control devices and thus, releasing various types of toxic pollutants into the land, air, and water (Ádám et al., 2021; Patil & Ramakrishna, 2020; Perkins et al., 2014). For example, the used acid with residual metals is being discharged into the land directly and absorbed by the soil and groundwater.

Global E-waste Management Strategies and Future Implications
DOI: https://doi.org/10.1016/B978-0-323-99919-9.00014-3

153

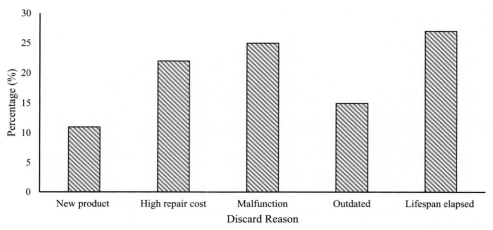

Figure 9.1 Schematic representation of different discard reasons for rapid WEEE generation.

E-waste comprises toxic and/or precious metals, chemicals, plastics, etc. whose practicing using inappropriate treatment methods lead to emitting toxic substance and thus dangerous to the environment as well as mankind. In India, <5% of E-waste was handling by authorized dealers/vendors (Balaji et al., 2021; Senophiyah-Mary, Loganath, & Shameer, 2018). Few inappropriate/unethical treatment methods are manual dismantling using readily available tools, removing different parts from the printed circuit board (PCB) by the charcoal grill, recovering gold and other valuable metals from PCB by dipping into the open-pit acid bath, melting and chopping plastics in poorly ventilated place, open burning of cables/wires to recover the copper, and dumping of remaining items on land or nearby riverbanks. These above unethical methods to recover valuable items from E-waste expose the laborers to toxic and hazardous chemicals such as polycyclic aromatic hydrocarbons, heavy metals, and inorganic acid, and create severe occupational health hazards (Gupta et al., 2014; Ilankoon et al., 2018; Needhidasan et al., 2014). The typical harmful pollutants and their occupational health hazards are presented in Table 9.1.

This study attempts to elaborate on various occupational health hazards in E-waste management, recycling, treatment, and handling along with few case studies.

9.2 Current options for E-waste treatment

E-waste treatment and recovery form E-waste includes collection, preprocessing, and end processing. For the recovery and recycling of valuable material (metals and other items), each of these processes is crucial. Effective government laws, efficient community outreach, and the installation of separate waste facilities in public areas all help expedite the collection of E-waste. Useful electronic components are returned to

Table 9.1 Harmful pollutants and their occupational health hazards.

S. No.	Name of the pollutant	Occupational health hazard
1.	Mercury (Hg)	Damages to respiratory, skin, and brain disorders due to their bioaccumulation.
2.	Brominated flame retardants (BFRs)	Infects liver, heart, and spleen disrupt endocrine systems.
3.	Lead (PB)	Affects children's brain development, blood system, kidney, and nerve damage.
4.	Hexavalent chromium (Cr) VI, barium (Ba)	DNA damage, asthmatic bronchitis, and muscle weakness.
5.	Cadmium (CD)	Adverse effects on the nervous system, liver, and kidney.
6.	Dioxins and furans	Immune system damage, interfere with regulatory hormones, and the reproductive system.

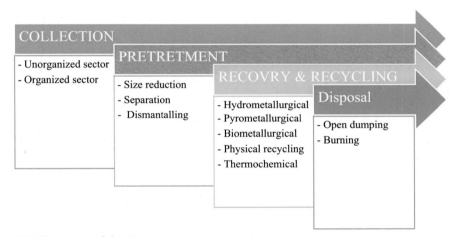

Figure 9.2 Flow chart of the E-waste treatment system.

consumer supply chains after sorting at the collection center (Khaliq et al., 2014). Fig. 9.2 shows a typical picture of E-waste treatment. In the collection center, recyclable and nonrecyclable components are safely separated from E-waste. E-waste follows a range of pathways after disposal, including formal and informal recycling, storage, and dumping, in both developed and less-developed country contexts. Developed countries are managing the formal collection sector with advanced manual and semi-automated dismantling and separation systems. On the other hand, developing countries are managing their E-waste by informal sectors. Some developing countries not even collecting their E-waste with proper precaution, although some backyard processes (such as open sky incineration, cyanide leaching, and simple smelters) are used directly to recover precious metals and discarding the residue with municipal solid waste at open dumps (Mmereki et al., 2016).

9.2.1 Informal treatment or recovery process

In developing countries, the informal sector treats a huge amount of E-waste. The recovery of metals in the informal process is done in nonconventional and nonscientific way. The informal way of recycling takes no safety measures and minimal control on the process to recover the material (Fu et al., 2013). Informal recycling involves fewer automatic procedures and health-protective measures with a much more decentralized unit. The most common processes are manual dismantling, sorting, open burning, and desoldering PCBs for informal recycling. These practices have a disproportionate impact on vulnerable populations because of the health consequences. The informal recovery process contaminates the surrounding environment and natural resources and also creates an adverse effect on animal health. In general, different valuable metals such as Cu, Al, Pb, Hg, Cd, and Pt are recovered by open burning of E-waste which releases huge amount of hydrocarbon and other toxic compounds (Song et al., 2015). Likewise, some metals such as Au, Ag, and Pd are recovered employing highly toxic acids such as HCl, HNO_3, H_2SO_4, and cyanide. The produced toxic gases/fumes are discharged into the environment without treatment, thus damaging the atmosphere and human health (Grant et al., 2013; Nachman et al., 2013).

Furthermore, the release of dioxins into the environment during the open combustion process degrades the air quality, and inhaling dioxin-contaminated air causes a variety of health problems, including birth abnormalities, brain, bone, renal, and liver damage (Fowler, 2017). The presence of Hg in E-waste harms the respiratory system and the digestive system (Luo et al., 2011). In addition, informal treatment sectors have little control over treatment residues generated in treatment and at the end of the process all residues directly discarded into the environment (land or water). These hazardous metals are also present in higher concentrations than what is allowed by environmental protection agencies and world health organization standards. All over the world, different researchers reported the presence of various toxic elements in natural land and water sources to cause by unethical E-waste treatment (Pradhan & Kumar, 2014; Song & Li, 2014). Heavy metal creates an adverse effect on human health such as Pb and Cd in humans irreversibly affecting the nervous system and leads to brain damage. Furthermore, Cd is one of the possible carcinogens just because it changes the human cell's deoxyribonucleic acid (Zhang & Zhang, 2018; Zheng et al., 2009). In addition, the fertility of soil and groundwater is affected by improper landfilling of E-waste by the informal treatment process. Heavy metal ions present in discarded E-waste may leak out and spread through the soil, eventually reaching water bodies or groundwater, as a result of such landfilling. Heavy metal ions can penetrate the soil—crop—food chain, presenting a significant hazard to public health. In addition, the extraction efficiency is also $<80\%$, whereas it can be achieved $>95\%$ informal processing of E-waste. Hence, the developments of appropriate strategies are the need of the hour to recover or recycle the E-waste to keep the ecosystem and its inhabitants safe.

9.2.2 Formal treatment or recovery process

Most developed countries (e.g., USA, Japan, Canada, and Sweden) and some developing countries (Colombia and China) have started formal E-waste. Dismantling and occasionally shredding of devices that need recycling is followed by sorting of recyclable elements using automated technology and manual process. Many formal recycling facilities are distributors of recovered materials such as plastics, glass, and metals to third parties. A typical formal E-waste recycling includes shorting, pretreatment, physical separation, and followed by the metal recovery operations such as pyrometallurgy, hydrometallurgy, and biohydrometallurgy. Previous studies in the literature on the recovery of metals include gold, silver, aluminum, copper, copper alloys, lead, tin, and zinc from E-waste. In a typical pyrometallurgical process, high-temperature furnaces are used to melt the E-waste. Then, the material properties are ejected as slag, which is handled to extract precious metals (Cayumil et al., 2014). The hydrometallurgical method uses certain lixiviants like thiourea, and sulfuric acid to leach out the metal. It produces a stable metal complex to leach out the metals simultaneously (Ding et al., 2017; Ghosh et al., 2015). Adding to the list, biohydrometallurgy is a method that uses certain microorganisms, and the metals can be recovered either with or without the presence of organisms in the extract medium (Nithya et al., 2018). Even at low concentrations, this microbially driven method is efficient, environmentally friendly, cost-effective, and energy-efficient, and requires only minimal industrial infrastructure to carry out the metal recovery process (Anjum et al., 2012; Natarajan & Ting, 2014; Vakilchap et al., 2016).

9.2.2.1 Pyrometallurgy method

Pyrometallurgy is an energy-intensive technique for processing or extracting nonferrous metals from metallurgical materials at high temperatures. Pyrometallurgical processes are widely employed for metal recovery from E-waste all around the world. When metals are mixed with nonmetals such as ceramics, it may be difficult to extract them using conventional recycling methods. In this case, pyrometallurgical techniques can be used. To extract, valuable metals involve incinerating, smelting, melting, and roasting (Kaya, 2016). The selection of the several essential processes is strongly influenced by the type of E-waste, as well as the need for smelting activities. The pyrometallurgy process may be used to recover metals such as Cu, Ag, Au, Pd, Ni, Se, Zn, and Pb from E-waste (Nithya et al., 2021). Minerals are extracted from the matrix using sophisticated smelting and refining processes (Khaliq et al., 2014). For example, PCBs are considered as complex E-waste, initially, they are shredded or chopped into desirable pieces and thereafter subjected to pyrometallurgical processes followed by electrochemical refining. The metals are collected in a molten bath and the oxides are obtained from the slag phase (Khaliq et al., 2014). The main drawback of this method includes highly elevated temperature with the considerable balance between the heat and the material under to be treated. Pyrometallurgical process releases toxic gases such as dibenzo-*p*-dioxin, biphenyl, anthracene, poly-brominated-di-benzofurans, and poly-brominated-di-benzodioxins.

Moreover, the formation of slag reduces the recovery yield due to unavoidable high temperature, dust, and smoke, thus also affects human health and the surrounding environment. The combustion process uses either natural air or pure oxygen. Natural air contains 79% of N_2 which reduces the thermal efficiency. On the other hand, using pure oxygen may enhance effectiveness and simultaneously lower flue gas generation. Because the metallurgical processes reach equilibrium at a fast rate, it is difficult to anticipate the mechanical elements of the recovery process. More research is needed for further improvement of recovery process and reduction in emission of harmful gases.

9.2.2.2 Hydrometallurgy method

This method includes pretreatment namely mechanical shredding followed by sorting process thereafter leaching process takes place. A leaching process includes the reaction of solid material with chemical reagents such as extractant or lixiviant to extract metals from dispersion medium. This low solute investment and the high recovery rate are the prime advantages and made them a superior technique to the former operation (Yazici & Deveci, 2014). Four types of common leaching processes, known as cyanide leaching, halide leaching, thiourea leaching, and thiosulfate leaching (Debnath et al., 2018). The solute transfer takes place in various liquid mediums such as halide, thiourea, thiosulfate cyanide, HCl, H_2SO_4, and HNO_3 in hydrometallurgy to selectively extract the desired metal from E-waste (Nithya et al., 2021). The required metals are precipitated due to gravity forces or applying centrifugal forces. Design parameters such as particle size, operating temperature, nature of lixiviants, and its sample ratio with leaching time are considered in this intermediate process. In the downstream, the metals from this suspension are recovered using solvent extraction, electrodeposition, ion exchange, and adsorption (Khaliq et al., 2014). Final forms of metals are achieved through electrorefining or chemical reduction processes. The downstream process is dependent on the type of metal to be recovered and the level of required purity. The advantage of this method includes high recovery rate, moderate temperature requirement, less energy consumable, minimal toxic gas emission, and secondary waste generation. The recovery efficiency of various metals such as gold 12%—98% (He & Xu,

2015; Quinet et al., 2005), copper 92%—96% (Cui & Anderson, 2020; Kamberović et al., 2018), silver 4%—100% (Oh et al., 2003), nickel 95% (Cui & Anderson, 2020), and palladium 58%—98% (Ha et al., 2014) has been reported. Recent literature indicates more emphasis on antimony and tin recovery/recycling (Barragan et al., 2020). The pyrometallurgy process has its limitations such as time-consuming process, shredded particle requirement, required toxic and expensive lixiviant with high safety standard, and thus the process becomes complicated and expensive. Moreover, it required corrosion resistive equipment for the leaching process with the possibility of metal loss through large number of downstream processes.

9.2.2.3 Biohydrometallurgy method

Biohydrometallurgy processing of E-waste is a technique in which microorganism helps dissolve elements from solid objects into liquids before the recovery of target metal employing separation. The less energy-intensive and minimal use of chemicals made this process a new and promising field. Biometallurgy also offers a wealth of possibilities in terms of both research and business. Furthermore, the technology is also referred to as clean and green. This method is used since the pre-Roman period for the recovery of copper, silver, and aluminum from Rio Tinto mines in the southwest part of Spain; however, it was commercially adopted few decades ago. It is classified as biosorption and bioleaching. Biosorption is the process of adsorbing metals using adsorbents made from waste biomass or abundant biomass. In the biosorption process, it utilizes *Aspergillus niger* (fungi), *Chlorella vulgaris* (agley), bacteria (*Penicilium chrysogenum*), hen eggshell membrane, ovalbumin, alfalfa, etc. (Debnath et al., 2018). The factor affecting biosorption includes the type of biological ligands, type of biosorbent, chemical stereochemical, coordination characteristics metal to be recovered, and characteristics of the metal solution (Tsezos et al., 2006). On the other hand, bioleaching refers to the mobilization of metal cations from insoluble materials through biological oxidation and other complexation processes (Ilyas & Lee, 2014). Bioleaching included four typical steps named as (1) acidolysis, (2) complexolysis, (3) redoxolysis, and (4) bioaccumulation.

Bioleaching includes three groups of bacteria known as autotrophic bacteria, heterotrophic bacteria, and heterotrophic fungi (Debnath et al., 2018). The most known autotrophic bacteria in the leaching process are *Acidobacillus thiooxidans*, *Acidobacillus ferrooxidans*, and *Sulfolobus* sp. (Natarajan & Ting, 2015). These bacteria obtain energy through the oxidation of ferrous ions as well as the reduction of sulfur-containing compounds. The heterotrophic bacteria (*Pseudomonas* sp. and *Bacillus* sp.) and fungi (*Aspergillus* sp. and *Penicillium* sp.) can translocate the metals from waste. The mechanism used for translocation is facilitated through acidolysis, metal complexation, redox reaction, or using bioaccumulation. Using thermophilic microorganisms, more than 80% recovery of copper, nickel, and zinc was observed (Ilyas et al., 2010), whereas using *A. thiooxidans* and *A. ferroxidans* consortium, more than 90% recovery of copper, nickel, zinc, and lead was observed. Metals recover efficiency by this process for various metals such as gold recover 8%—85% (Sheel & Pant, 2018), copper recover 23%—99% (Jagannath et al., 2017; Yang et al., 2009), silver recover 12%—41% (McDonald et al., 2011; Ruan et al., 2014), nickel recover 38%—96% (Horeh et al., 2016), and zinc recover 64%—95% (Ilyas et al., 2010) has been reported in previous from E-waste.

9.2.2.4 Physical recycling methods

E-waste is often treated with physical recycling processes, which assist liberate the metals and nonmetals embedded. After exploring numerous E-waste recycling facilities worldwide, the study concluded that physical recycling is the most systematic approach. This is one of the most influential metal recovery methods; it is often called a pretreatment step before further processing. In this method, the various processes included dismantling, shredding, chopping, disassembling, and many more. To separate metal and nonmetals, various mechanical machines such as shredders, pregranulators, and granulators are used to accomplish these stages. Different methods such as magnetic separation, eddy current separation corona discharge method, density-based separation, milling, froth floatation, and density separation are used. Using a combination of electrostatic and magnetic separation, which separates the metal from the nonmetal components, it is feasible to recover metal fractions comprising more than 50% copper, 24% tin, and 8% lead (Veit et al., 2005).

9.2.2.5 Thermochemical methods

When it comes to thermochemical processes, pyrolysis is a necessary process. Pyrolysis is a thermochemical process that ensures thermal degradation of a targeted material in the absence of air. Different kinds of pyrolysis, such as vacuum pyrolysis, microwave-induced pyrolysis, catalytic pyrolysis, and copyrolysis, have been documented (Debnath et al., 2018). E-waste pyrolysis is now restricted to the laboratory, although Jectec (a Japanese business) has previously deployed pyrolysis at their facilities. The plasma method used for E-waste treatment and metal recovery has also gotten much attention lately. Plasma technology is a high-temperature, environmentally friendly technology. E-waste has been processed using a high-enthalpy plasma jet, a plasma reactor, and a plasma torch (Debnath et al., 2018). PyroGenesis Canada Inc. is already putting it to good use despite the lack of information on the subject.

9.3 Utilization of E-waste as a construction material in concrete

Recently, the electrical and electronics industry has been growing at a very fast speed. The discarded solid waste of electrical or electronic equipment is known as electronic wastes or E-wastes. There are various techniques generally used for disposal of E-wastes such as reuse of E-devices, recycling, incineration (set fire at an elevated temperature of 900°C—1000°C in the absence of oxygen), acid bath (to extract copper from E-wastes), and landfilling (Luhar & Luhar, 2019). In which the reuse of E-devices is the best possible disposal technique that reduces the amount of E-wastes produced per year. The recycling of E-wastes is a modern technique of disposal next to its reuse possibilities in which various electrical and electronic devices are collected and separated into various parts and recycled using various recycling methods depending on metallic and nonmetallic characteristics of materials, the presence of precious and dangerous elements, etc. (Kang and Schoenung, 2005). The recycling of E-wastes is not only beneficial in terms of conservation of natural resources but also eliminate the dangerous elements precisely, which could originate health problems to the living things on the earth if directly discharge into the atmosphere (Shaikh et al., 2020). In addition, the precious elements present in the E-waste are also recovered by recycling which can be used for the manufacturing of new products.

The incineration method of disposal is quite beneficial as the amount of waste is reduced, and the energy obtained can also be used independently. However, this method has some drawbacks, such as emissions of harmful gases along with cadmium and mercury into the atmosphere. The landfilling method of waste disposal is the world's oldest method used by burying it between the layers of the earth. Nonetheless, disposal of E-waste through landfilling is not an environmentally sound technique because of its poisonous contents (cadmium, mercury, lead, etc.), which polluting soil and groundwater. In addition, landfills increase the risk of fires, which can release toxic fumes into the environment. So, there is no sound method for the disposal of E-wastes except recycling and only 20% of E-waste is handled appropriately worldwide (Evram et al., 2020). So, we need other sound methods for the disposal of E-wastes to minimize future risk.

Nowadays, the world infrastructures are considerably growing. Because of this, the requirement for construction materials is also rapidly increasing day by day for the construction of infrastructures. Of all those materials, concrete is the most universally consumed construction material and it is considered the second most used resource on earth after water. Which is consistent with good mechanical and durability properties, significantly low price, accessibility of constituent materials, and capability to be formed into any shape or size (Shaikh, 2016). Concrete is a composite material made with Portland cement, aggregate, sand (fine aggregate), water, and admixture (Neville & Brooks, 1987). In concrete, a major proportion in terms of volume is occupied by coarse aggregate and fine aggregate. On the one hand, the limited natural resources of aggregates are decreasing day by day and on the other hand, the current production process of Ordinary Portland Cement (OPC) requires natural coal as well as natural limestone. In addition to this, the production of Portland cement emits roughly 1.5 billion tons of greenhouse gases each year, or a normal of 6% of the complete outflows from various sectors throughout the planet (Dhakal, 2009). All these reasons are enough to prompt the incorporation of E-wastes as an alternative material that may employ aggregates or cement or both as partial replacements for the production of concrete. For this purpose, an attempt has been made by many researchers to use E-waste through partial replacement of ingredients for the production of concrete also known as green concrete (Akram et al., 2015; Mathur et al., 2017). As observed from the previous study, waste materials mainly plastics and glass obtained from E-waste are utilizing for the production of E-waste green concrete. The use of E-waste for the production of concrete cannot be considered 100% financially profitable due to the high cost of production. But then again because of the toxic and polluted gases that emit during the disposal of E-waste, the cost is worth considering.

9.3.1 Use of plastic E-waste for production of E-waste green concrete

The plastics obtained from WEEE are mainly used to replace the aggregate for the production of E-waste green concrete (Donadkar & Solanke, 2016). WEEE contains different forms of plastic such as nonflame retarded plastic, flame retarded plastic, and PCBs. The composition of WEEE is presented in Fig. 9.3.

There are two methods of using these plastics and PCBs as a partial replacement of aggregates. One of them is nonmanufactured plastic aggregates which consist of sorting, cleaning, and shredding or grinding the plastic waste and use it into concrete as a partial replacement of aggregates (Santhanam & Anbuarasu, 2020) as shown in Fig. 9.4. The other method is manufacturing the plastic aggregates mainly coarse

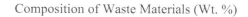

Composition of Waste Materials (Wt. %)

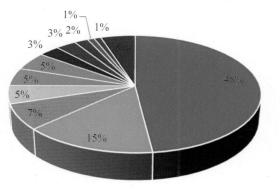

- 1 Iron and Steel
- 2 Nonflame Retarded Plastic
- 3 Copper
- 4 Glass
- 5 Flame Retarded Plastic
- 6 Aluminium
- 7 Printed Circuit Boards
- 8 Wood and Plywood
- 9 Concrete and Ceramic
- 10 Rubber
- 11 Other Nonferrous Materials
- 12 Others

Figure 9.3 Waste electrical and electronic equipment composition (Luhar & Luhar, 2019).

Figure 9.4 Nonmanufactured E-waste plastic aggregates as replacement of course aggregate in concrete.

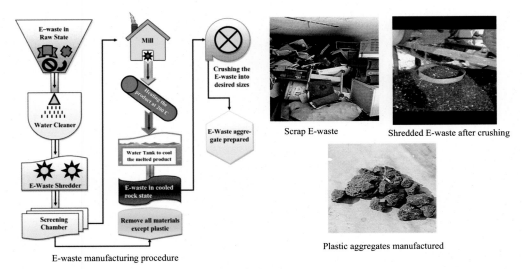

Figure 9.5 Manufactured E-waste plastic aggregates as replacement of course aggregate in concrete (Ullah et al., 2021).

aggregate of similar size and shape to the natural aggregates (Ullah et al., 2021) as shown in Fig. 9.5.

There are numerous benefits of using plastic E-waste aggregate for replacement of natural aggregate due to its lightweight characteristics such as transportation cost, production cost, provide sufficient thermal insulation, and reduce the impact of earthquakes (Akçaözoğlu et al., 2010; Colangelo et al., 2016). The mechanical properties such as compressive strength, tensile strength, and flexural strength are better or equal to that of conventional concrete due to the replacement of natural aggregate with plastic E-waste aggregate up to a certain limit mostly up to about 15%. On the other hand, with a replacement level of more than about 15%, the mechanical properties of plastic E-waste aggregate concrete were mostly found inferior as compared to conventional concrete (Alagusankareswari et al., 2016; Damal et al., 2015). However, the partial replacement of cement with industrial by-products such as ground granulated blast furnace slag, fly ash, and marble dust improves the strength properties of the concrete smashed with plastic E-waste aggregate (Nadhim et al., 2016). The reduction in strength is due to the less strength of plastic E-waste aggregate and due to poor bonding between the cement aggregate interfaces. The strength of plastic E-waste aggregate concrete may be low, but due to its lightweight characteristic, it can be used as lightweight concrete. Regarding the durability properties of plastic E-waste aggregate concrete, the resistance against the sulfate and chloride attacks, performance against alternate wetting—drying cycles, and abrasion resistance are higher and exhibited less permeability and sorptivity coefficient due to the replacement of natural aggregate with plastic E-waste aggregate, which makes it suitable for construction in the aggressive environment (Ferreira et al., 2012; Suchithra et al., 2015). On the contrary, plastic E-waste aggregate concrete exhibited less thermal stability, that is, the higher reduction in compressive strength of plastic E-waste aggregate concrete was noted when exposed to elevated temperature as compared to conventional concrete due to the flammable nature of the plastic E-waste aggregate. However, the reduction in compressive strength is mostly within the expectable limit (Lakshmi & Nagan, 2010; Mathew et al., 2013; Senthil Kumar & Baskar, 2018).

9.3.2 Use of glass E-waste for production of E-waste green concrete

Due to the nonflammable and nonbiodegradable nature, the removal of glass E-waste has become an enormous environmental problem. WEEE contains 5% of glass, which can be utilized for the production of E-waste green concrete. Glass obtained from WEEE is mainly consumed in powdered form to replace the cement and fine aggregate in concrete (Elaqra et al., 2019). There are several benefits of using E-waste glass powder in concrete, from the previous research, the replacement of up to 40% of fine aggregate with E-waste glass powder enhanced its internal microstructure and exhibited comparatively higher or equivalent mechanical and durability properties, and also showed higher resistance against crack as compared to conventional concrete (Chen et al., 2006; Gautam et al., 2012; Lakshmi & Nagan, 2011). The E-waste glass of cathode ray tube (CRT) is also a viable option as replacement material. Chemically, it is found rich in silica and showed pozzolanic properties which make it suitable for the replacement of fine aggregate and cement in concrete. The photographs of the generation process of E-waste glass of CRT from CRT are shown in Fig. 9.6. However, the E-waste glass of CRT contains high amount of lead (Pb) depending on its type which can leach from concrete during its service life. From the reported literature, the amount of Pb leaching is within the limit of the accepted emissions when used within the tolerable limits (Ling & Poon, 2012a, 2012b, 2014; Moncea et al., 2013).

The replacement of fine aggregate with E-waste glass of CRT untreated as well as treated improve the workability of mortar and concrete due to impervious smooth surfaces and low water absorption (Hui et al., 2013; Ling & Poon, 2011; T. Liu, Song, et al., 2018). The enhancement of workability could reduce the use of chemical admixtures like superplasticizer or water reducer for achieving the corresponding workability. Also, due to its properties of enhanced workability, the E-waste glass of CRT can be used for the production of self-compacting concrete. On the contrary, the use of E-waste glass of CRT boosts the alkali–silica reaction (ASR) expansion in concrete, but ASR expansion can be reduced by using supplementary cementitious materials such as fly ash, slag, silica fume, and metakaolin (Shi & Zheng, 2007). The mechanical properties of E-waste glass of CRT concrete improved when used very fine particles of it for

Figure 9.6 The generation process of CRT glass (Yao et al., 2018).

replacement due to its pozzolanic properties. However, the mechanical properties of E-waste glass of CRT concrete were found mostly within the expectable limit.

9.4 Category of E-waste recycling (collection, refurbishing, etc.) with case studies

Nowadays, the disposal of E-wastes has become a major challenging task in front of all official agencies. Fig. 9.7 shows the recycling process of electrical and electronic equipment (EEE) adapted from the UN Environmental Programme.

9.4.1 Collection of E-waste from different countries

In many countries, E-waste from the households is generally collected through the existing municipal schemes. As per quantities of E-waste generated in the European countries like Austria, Sweden, Belgium, France, and Germany, it was more than 20 kg/cap/year in the year 2012 while the collection was around 4—17 kg/cap/year only. It is further found that Sweden has shown the most successful schemes where about 17.5 kg/cap/year out of 22.2 kg/cap/year were collected (Salhofer, 2017). On the other hand, Balde et al. (2015) reported that European member states showed the average collection rate of 36% in the year 2014. Furthermore, China has reported the generation rate of E-waste in the year 2014 was about 6.0 mt/year while the collection rate was 1.3 mt/year only (Balde et al., 2015). In China, government are following "Old to New" schemes to the collection simpler and more attractive. Salholfer (2017) reported that the collection rate in the "Old to New" schemes was around 0.4—0.1 kg/cap/year. However, the problem for the formal schemes remained the collection cost which mainly originates due to the competition with the informal sectors. Further, Vietnam with a population of 93 million reported the generation rate

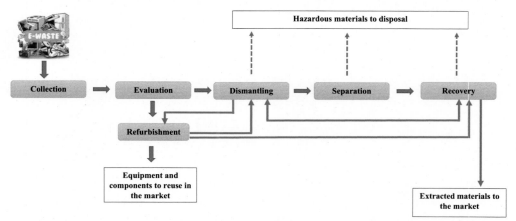

Figure 9.7 Flow diagram for environmental sound management of used electrical and electronic equipment within a recycling facility. *Adapted from UN Environmental Programme.*

of E-waste was around 2.6 kg/cap/year; however, the collection data are not available. It is to be mentioned that there are no formal collection schemes that are followed in Vietnam; however, the collection of E-waste was done through the informal collectors. The collection of E-waste in Vietnam is done by the peddlers who collect the disposed appliances from the end-users and sell them to the new buyers. The collection of E-waste in Vietnam is a major challenge as there is a lack of formal collection schemes and labor conditions in the informal sectors.

9.4.2 E-waste treatment

The treatment process of E-waste involves dismantling, separation, and recovery of useful materials. In dismantling, the hazardous and valuable components are separated from the E-waste. On the other hand, the materials are separated to release the materials from compounds and split in the separation process. Furthermore, the separated components were melted and the hazardous and the valuable components recovered from the separated components of the E-waste were sent to the hazardous waste management treatment and the market, respectively. In the bigger appliances, the hazardous components (like capacitor, Hg) are removed followed by mechanical separation using magnetic separation, eddy current separator, etc. Likewise, small appliances undergo manual dismantling followed by mechanical separation. Lightning equipment undergoes crushing and is then put into encapsulated vapor machine to avoid the emission of Hg vapor. In European countries, mechanical separation process is being developed to remove manual dismantling as the labor cost is very high.

Moreover, technologies have been developed to break the appliances in a rotating drum which is also known as "smasher" followed by separation of hazardous and valuable components. Hazardous materials generated from the E-waste are sent to dispose off in landfill or by incineration. Furthermore, PCBs obtained from the various appliances are sent to the metallurgic treatment plant. In Europe, there are three plants (namely Aurubis, Boliden, and Umicore) were established, where the smelting of PCBs is done and metals were recovered. According to Salholfer (2017), Umicore claims the recovery of metals from the E-waste is more than 95%. From 1980 onwards, a greater amount of E-waste was imported in China from USA and other European countries for recycling. Geeraerts et al. (2015) reported that China has imported the highest quantity of E-waste around 8 mt/year. Sepulveda et al. (2010) illustrated the E-waste treatment techniques and processes adopted by China and India. This process involves open burning of PCB to separate the recovered solder followed by leaching and then mixing of PCB for the recovery of precious metals.

The open burning of PCBs emits Pb, polybrominated diphenyl ether (PBDEs), dioxins, and furans in the form of fumes and ashes and from hydrometallurgical processes (like leaching, amalgamation, etc.). For the treatment of CRT TV sets, initially the metal frames, PCBs, and cables are separated then the glass body of the screen is divided into the front and cone glass. Furthermore, fluorescent powder is extracted using hot wire and laser cutting technology. In the case of the refrigerator and air conditioner, the coolant is initially extracted and then the body of the refrigerator is split in a closed system. It was found that treatment of E-waste in China is a major challenge due to the lack of supply of input materials in the recycling facilities. On the other hand, Vietnam is using "reuse and repair" idea for the E-waste management in their country. After the collection of the appliances, the appliances under repair or refurbishment process are then sold to the second-hand market. The appliances are initially dissembled in the repair shop and if it is impossible to repair then the appliances are sent to the dismantling workshop. After the dismantling process, the useful parts of the appliances are divided for further treatment or sale. Metals were recovered by the open burning of the materials and the residues released from the E-waste are disposed to the land, river, pond, etc. The present situation in Vietnam led to many serious risks associated to human health and the environment.

9.5 Health hazards

It is well known that there were various metals that are released from the E-waste which can risk the life of the human. However, if it is recycled correctly, rare metals such as gold, copper, or palladium can be extracted efficiently and can be used to maintain the economy of a country. On the other hand, the recycling process of E-waste can cause a serious risk to the environment and human health through the release of toxicants to air, water, and soil. There were varieties of toxicants released by direct and indirect exposure of the WEEE. Direct exposure is related to the inhalation of the fines, skin contact with toxic materials, etc. On the other hand, indirect exposure pathways include the consumption of toxicants through air, water, and soil. The unauthorized and unsafe recycling techniques aim to regain valuable materials from WEEE but therewith increase the risk for toxicant exposures (Perkins et al., 2014). Once these toxicants are consumed by the human body, they will rupture the fatty tissues of the human; hence, they cause severe health issues to people residing near the informal E-waste sectors (Y. Liu, Huo, et al., 2018; Zeng et al., 2017; Zhang et al., 2017). Furthermore, the people working in the E-waste recycling region are more at risk of life as they breathe contaminated fumes and dust. Therefore, if it is not disposed off in a suitable manner, it has the tendency to generate the contaminations through food-to-environment chain, simultaneously causing several hazardous impacts on the health of the general population (Li & Achal, 2020).

9.5.1 Effect on environment

It is well known that improper E-waste disposal and informal recycling of E-waste can lead to environment contamination. The release of the toxicants from the E-waste can be easily occupied into the environmental processes through bioaccumulation, food contamination hence endangering the life of the human. For instance, as per Wong et al. (2007), water contamination due to the release of toxic waste has been reported in the rivers in and around the recycling town in China. Robinson (2009) reported that the toxic particles in the form of fumes and dust emitted from the burning process settled over the river (Pearl River Delta Region) lead to effect the health of around 45 million people. Furthermore, the released toxic particles also follow soil—crop—food pathway to affect the health of the local residents. Hence, it is important to study the impact of various toxic substances released from the E-waste on the soil, air, animals, and hence on human health.

9.5.1.1 Soil contamination

It is already known that the release of hazardous substances from E-waste has been polluting the farmland soil (Meng et al., 2014; Zhang et al., 2015). The release of heavy metals from the E-waste transferred to the soil—vegetable and soil—grain system hence affecting human health. The heavy metals released from the E-waste got absorbed by the plant roots from the soil and further transferred to the stem and leaf of the plants. Ha et al. (2009) reported that soil near the E-waste recycling site in Bangalore Slum, India having various heavy metals up to 2850 mg/kg of Pb, 39 mg/kg of Cd, 957 mg/kg of Sn, 49 mg/kg of Hg, and 2.7 mg/kg of Bi. Ngoc et al. (2009) reported the heavy metals in the region of Guiyu, China up to 1.6 mg/kg of Cd, 140 mg/kg of Cu, and 93 mg/kg of Pb. Tóth et al. (2016) evaluated the concentration of heavy metals in the soil of the European Union up to 3.17 mg/kg of Cd, 252.53 mg/kg of As, 91.89 mg/kg of Co, 273.53 mg/kg of Cr, 9.62 mg/kg of Mn, and 1.63 mg/kg of Pb. Similarly, the concentration of heavy metals in the soil of Japan was found to be 64 mg/kg of Cr, 9 mg/kg of Co, 39 mg/kg of Ni, 99 mg/kg of Zn, 0.45 mg/kg of Cd, 0.32 mg/kg of Hg, and 29 mg/kg of Pb (Yamasaki et al., 2001).

9.5.1.2 Air contamination

The diffusion of heavy toxic metals into the air through dust and fumes generated due to uncontrolled burning of E-waste hence causes air pollution. The polluted air was further consumed by the human through inhalation, ingestion, and absorption in the skin hence affects their health. This contaminated air is mainly composed of hazardous components like polychlorinated dibenzo-p-dioxins (PCDDs), PBDEs, particulate matter, and persistent organic pollutants. Li et al. (2011) reported that the value of PCDDs in the air at the region of Guiyu, China to be $65-2765$ pg/m^3. Moreover, the uncontrolled incineration of the E-waste also leads to contribute in the production of a large amount of PBDEs which was measured to be about 16,575 pg/m^3 in the region of Guiya, China. Thacker et al. (2013) reported the ranges of the concentrations of PCDDs/DFs in India to be from 0.0070 to 26.8140 ng toxicity equivalent (TEQ)/Nm^3. On the other hand, the value of PCBs in India ranged from 0.0001×10^{-1} to 0.0295 ng TEQ/Nm^3. Tue et al. (2013) reported that the values of PCBs and PBDEs in the air at the region near the recycling sites in Vietnam were around 1000–1800 and 620–720 pg/m^3, respectively. It was found that the levels of these heavy toxicants are much higher in the air of the houses near the recycling sites than in the non-E-waste houses.

9.5.2 Impact on human health

It was found that the consumption of Pb in the people working in the informal recycling site is comparatively higher than the people working at formal sites (Phillips & Moya, 2014). The consumption of heavy metals in the human body is measured in terms of average daily dosage (ADD) which includes through ingestion, skin absorption, and inhalation. However, the average daily dose of heavy metals (Cd, Pb, Cu, and Cr) through ingestion is relatively higher (about 90%) than that through skin absorption and inhalation. Moreover, the people working in the mechanical workshop was found to be having higher ADD of heavy metals than those working in dismantling and operation sectors. Furthermore, people residing near the E-waste recycling sites are more prompt to expose to the toxicants like PCBs released from WEEE. Previous studies showed that people exposed to PCBs having a higher risk of cancer. Table 9.2 summarizes a list of hazardous substances that are released from the WEEE and its health impact on humans. Moreover, it was reported that release of the toxicants from the WEEE is increasing the possibility of premature births, still births, abortions, and reduced height and weight in the babies of pregnant women residing near the E-waste recycling areas (Song & Li, 2015). The birth affects due to the exposure of E-waste was increased from 1.03% to 4.72% (stillbirth), 1.57%–3.40% (lower birth height), and 3258–3168 g (mean birth weight). It was further seen that the exposure of the hazardous substances released from WEEE is affecting the thyroid-stimulating hormone and causing lung's problems in children (age ranging from 8 to 13 years). The children (age ranging from 8 to 9 years) having lung function with reduced forced vital capacity from 2121 to 1859 mL. In addition, the exposure of heavy metals is also rupturing the DNA of the human (J. Zheng, Chen, et al., 2013).

9.6 Occupational health hazards in formal and informal recycling of E-waste

It is a unique variable in itself in dealing with E-waste. The exposure implicated is always to be considered as a whole. There are three sources of exposure to occupational health hazards of E-waste recycling. They are formal recycling, informal recycling, and exposure to hazardous components of E-waste (Herat & Agamuthu, 2012;

Table 9.2 List of heavy metals released from the electronic waste and its health impact on human.

Heavy metals	Health impact	References
Lead (Pb)	Diminishes neurobehavioral development of children. Anemia. Kidney damage	Kumar and Singh (2014), Wu et al. (2019)
Mercury (Hg)	Diminishes neurobehavioral development of children. Anemia. Kidney damage	Shamim et al. (2015)
Chromium (Cr)	Carcinogenicity. Reproductive functions. Endocrine function	Banu et al. (2008), Quinteros et al. (2008)
Cadmium (Cd)	Damages lung, kidneys, and reproductive organs. Causes bone diseases like osteomalacia and osteoporosis	Kumar and Singh (2014)
Zinc (Zn)	Cu deficiency diseases like anemia and neurological abnormalities	Li and Achal (2020)
Nickel (Ni)	Causes respiratory diseases like asthma and chronic bronchitis, birth defects	Li and Achal (2020)
Aluminum (Al)	Skeletal development and diminishes metabolism rate, causes fetal toxicity	European Food Safety Authority (2004)
Arsenic (As)	Causes skin irritations, diabetes, cancer-like diseases, and decreased nerve conduction	Gamble et al. (2018)
Silver (Ag)	Skin irritations, respiratory diseases	Li and Achal (2020)
Lithium (Li)	Skin irritations, redness in eye, breathing shortness	Li and Achal (2020)
Iron (Fe)	Damages liver	European Food Safety Authority (2004)

Ilankoon et al., 2018; Senophiyah-Mary, Loganath, & Meenambal, 2019; Senophiyah-Mary, Thomas, et al., 2019). In informal recycling, the E-waste is dismantled after which the metallic parts are leached out using acid which is done with knowledge or without technical knowledge and they are exposed to dangerous toxic fumes (Daum et al., 2017; Grant et al., 2013). The most basic hazard that occurs in any industry is noise. Noise pollution is an important problem to be dealt with carefully as it may create a path to the generation of other health hazards. In a typical E-waste treatment, site drilling, blasting, cutting, materials handling, ventilation, crushing, and conveying are considered to be the major processes. It is important to control noise pollution as it may result in the loss of hearing. Vibration is considered to be an important criterion as it could damage the entire body and to be specific hands and spinal cord (Ilechukwu et al., 2021). Table 9.3 presents the process involved and two types of recycling sector along with their occupational health hazards.

Table 9.3 The process involved and two types of recycling sector along with their occupational health hazards.

Sl. No.	Process involved	Informal recycling Process involved	Occupational health hazard	Formal recycling Process involved	Occupational health hazard
1.	Dismantling	Manual process in an open environment. Workers use chisel-like tools to dismantle the electronic components.	Bromine emission from the BFRs of thermoset polymers and printed circuit boards (PCBs). During dismantling with tools, they will not be using any safety measures like heavy duty rubber gloves.	Semimanual process in a closed environment where electronic dismantling devices will be used and the emitted gases are trapped and sucked through pipes lines which will be treated with scrubbers and emitted under standard environmental conditions. Workers are asked to follow a standard procedure to dismantle the electronic appliance with which they are protected with heavy duty rubber gloves, ear muffs, etc.	People work under these environmental conditions do not inhale toxic fumes. They will be facing back, neck wrist pains which could be overcome by following proper ergonomics.
2.	Sorting	Manual process, where people will put the heap of dismantled E-waste in center and ladies, children will be assigned to segregate different pieces such as plastic, metals, capacitors, batteries, and glass.	There will be consistent contact between eye, ear, nose, mouth to hand with E-waste without safety measures.	Once the dismantled E-waste falls into the conveyor belt of the sorting machine, it turns to be completely automatic process which segregates all the different particles by magnetic, eddy current, and density separation processes.	Worker who lifts the dismantled waste might face back pain which can be reduced by using a trolley to mobilize the trolley to the conveyor belt. As there is no hand contact with the different particles, there will be very low or no exposure of toxicity to the workers.

| 3. | Treatment of PCBs | Acid leaching process is being followed in most of the informal sector in which the acid used is highly concentrated. Workers use stick or pipe to mix the acid with PCB to extract the metals. They will not be aware of extracting all the metals, rather they concentrate only on metals with high value where they leave the other in the lixiviant pool and drop them to the land, or pour it in the water body. | When worker induce the PCB into the acid solution huge fumes of toxic gases emerges which is inhaled by the people around the site. They will not be using any safety measures to protect themselves. The ill effects that come out of this process includes liver, kidney, brain damage, etc., while they also emit carcinogens which kills them in few years. It was also proved that the toxic fumes cause delivery of premature babies, still births, abortions among ladies, and lack of fertility among men. | Most of the formal recyclers do not treat the E-waste in their premise in India. They send them to Umicore technology, Belgium for smelting the waste. They get lump sum amount for the metals that are present in the PCB and other parts of electronic appliances. Although most of the send, few recycling companies use chemical methods to extract specific metals under controlled conditions. The fumes are trapped and connected to the chimneys where they are emitted through stacks. The chemicals are neutralized and disposed off in an environmentally sound manner. | Workers are safe from the emission of toxic fumes but they may face some problem in disposing the toxic chemicals. If they face any accidents, the apparatus used were ensured to be spill proof and they should not create problem to any mankind. The fumes have to be released at stand stack height as it may create problems to be nearby residents. They may cause eye irritation and blood contaminations with metals. |
| 4. | Tube lights and bulbs recycling | The tube lights and bulbs were broken down into pieces by kids and they | Mercuric dioxide and other metallic dioxide get emitted once the | It has a closed chamber in which the lights are carefully opened to | Only a small hole will be visible to drop the tube lights inside. There is a |

(Continued)

Table 9.3 (Continued)

Sl. No.	Process involved	Informal recycling Process involved	Occupational health hazard	Formal recycling Process involved	Occupational health hazard
		were thrown to nearby dust bins.	bulbs are broken down into pieces which may affect the brain immediately.	remove the trapped gasses and connected to a scrubber. The glass pieces are then crushed for the production of bangles.	chance of emission of few toxic gases outside the hole and thus it can cause health problems. But such visible hazards were not yet studied.
5.	Wires and cables	The wires and cables were stripped off using knife or blade to get the conducting metals. The wires were also completely soaked in acids solution to extract the metals out of wires. The wires were also burnt to get the metals out of the polymer.	Proper health and safety measures will not be taken by the workers and so they may result in cuts in hand. The acid-soaked wires emit fumes which are harmful to the environment. Burning of wires results in the emission of furans and dioxins.	Semiautomatic process in which the wires are inserted to a machine which strips off the plastic outer cover. The wires and metals of smaller size were separated by density separation.	During the stripping process, due to abrasion there is a mild emission of toxic fumes. The workers were given hard rubber gloves and masks, ear muffs as precautions. Noise pollution also occurs due to the running of machine.

Many factors affect the occupational health of the workers in the E-waste recycling industry in developing countries like India. The main problems associated with any industry are noisy environment, high-temperature conditions, and self-hygiene like toilet, bathrooms, and other facilities for workers (Shikdar & Sawaqed, 2003). Workers also experience body weakness, back pain, fatigue, eye irritation, breathing illness, itching, hair fall, hand and wrist pain, and headaches. Although these appear to be very casual, there are several serious health problems like decrease in sperm count which results in infertility, defective childbirth, cancer, loss of eyesight, lungs, liver, kidney, heart, brain damages, abortions, early puberty, immature birth, pre- and post-mature babies, thyroid dysfunction, and changes in cellular functioning (Dai et al., 2020; Igharo et al., 2018; Yu et al., 2020). However, these diseases arose for people who work for the prolonged period in the same environmental conditions. Although there are many investigations, these studies were not done preciously as most of them were predictions and assumptions. It is recommended that all the workers who work with heavy metals should be tested for the contamination with metals in body fluids and a full body check-up in regular intervals of time (Zhang et al., 2015, 2017; G. Zheng, Xu, et al., 2013; J. Zheng, Chen, et al., 2013). The specific source and route of exposure were not specific. Although few studies were conducted, there are some other factors like air pollution, smoking, and other habitual characteristics which were not considered in the measurement process.

9.7 Conclusion

It is important for an E-waste recycler to recover maximum resources out of E-waste with least or no emission of toxic gases by adopting cleaner production. They should also ensure health and safe working conditions among workers. There are few factors which have to be overcome by the investigators and researchers like data of epidemiology, associations which may result in biomagnification, cross findings, and understanding of the biological mechanism. Some precautionary actions have to be taken to reduce the adverse effects of the exposure of metals and metallic compounds among children. There is a large gap in the physical, mental, educational health, and behavioral outcomes due to the exposure of individual chemical compounds due to the recycling of E-waste. The inhibitor effects due to the mixture of chemicals, low dose, and long-term exposure to E-waste have to be studied. It is the duty of the formal E-waste recyclers to ensure whether they work in a safe environment. Although informal recyclers were known to work in very dangerous environments, the health risks were widespread to the entire area. If the formal recycling of E-waste is not improperly equipped and lacks health and safety measures, then it will damage the entire locality as most of the E-waste used here is bulk. It is high time to take necessary steps in action to prevent or reduce the pollution due to E-waste rather than creating policies.

References

Ádám, B., Göen, T., Scheepers, P. T., Adliene, D., Batinic, B., Budnik, L. T., Duca, R. C., Ghosh, M., Giurgiu, D. I., Godderis, L., & Goksel, O. (2021). From inequitable to sustainable e-waste processing for reduction of impact on human health and the environment. *Environmental Research*, *194*, 110728.

Akçaözoğlu, S., Atiş, C. D., & Akçaözoğlu, K. (2010). An investigation on the use of shredded waste PET bottles as aggregate in lightweight concrete. *Waste Management*, *30*(2), 285–290.

Akram, A., Sasidhar, C., & Pasha, K. M. (2015). E-waste management by utilization of E-plastics in concrete mixture as coarse aggregate replacement. *International Journal of Innovative Research in Science, Engineering and Technology*, *4*(7), 5087–5095.

Alagusankareswari, K., Kumar, S. S., Vignesh, K. B., & Niyas, K. A. H. (2016). An experimental study on e-waste concrete. *Indian Journal of Science and Technology*, *9*, 2.

Anjum, F., Shahid, M., & Akcil, A. (2012). Biohydrometallurgy techniques of low grade ores: A review on black shale. *Hydrometallurgy*, *117*, 1–12.

Balaji, R., Prabhakaran, D., & Thirumarimurugan, M. (2021). A novel approach to epoxy coating removal from waste printed circuit boards by solvent stripping using NaOH under autoclaving condition. *Cleaner Materials*, *1*, 100015.

Balde, C. P., Kuehr, R., Blumenthal, K., Fondeur Gill, S., Kern, M., Micheli, P., Magpantay, E., & Huisman, J. (2015). *E-waste statistics—Guidelines on classification, reporting and indicators* (p. 51 p.) Bonn, Germany: United Nations University, IAS - SCYCLE.

Banu, S. K., Samuel, J. B., Arosh, J. A., Burghardt, R. C., & Aruldhas, M. M. (2008). Lactational exposure to hexavalent chromium delays puberty by impairing ovarian development, steroidogenesis and pituitary hormone synthesis in developing Wistar rats. *Toxicology and Applied Pharmacology*, *232*(2), 180–189.

Barragan, J. A., Ponce de Leoón, C., Alemán Castro, J. R., Peregrina-Lucano, A., Gómez-Zamudio, F., & Larios-Durán, E. R. (2020). Copper and antimony recovery from electronic waste by hydrometallurgical and electrochemical techniques. *ACS Omega*, *5*(21), 12355–12363.

Cayumil, R., Khanna, R., Ikram-Ul-Haq, M., Rajarao, R., Hill, A., & Sahajwalla, V. (2014). Generation of copper rich metallic phases from waste printed circuit boards. *Waste Management*, *34*(10), 1783–1792.

Chen, C. H., Huang, R., Wu, J. K., & Yang, C. C. (2006). Waste E-glass particles used in cementitious mixtures. *Cement and Concrete Research*, *36*(3), 449–456.

Colangelo, F., Cioffi, R., Liguori, B., & Iucolano, F. (2016). Recycled polyolefins waste as aggregates for lightweight concrete. *Composites Part B: Engineering*, *106*, 234–241.

Cui, H., & Anderson, C. (2020). Hydrometallurgical treatment of waste printed circuit boards: Bromine leaching. *Metals*, *10*(4), 462.

Dai, Q., Xu, X., Eskenazi, B., Asante, K. A., Chen, A., Fobil, J., Bergman, Å., Brennan, L., Sly, P. D., Nnorom, I. C., & Pascale, A. (2020). Severe dioxin-like compound (DLC) contamination in e-waste recycling areas: An under-recognized threat to local health. *Environment International*, *139*, 105731.

Damal, V. S., Londhe, S. S., & Mane, A. B. (2015). Utilization of electronic waste plastic in concrete. *International Journal of Engineering Research and Applications*, *5*(4), 38.

Daum, K., Stoler, J., & Grant, R. J. (2017). Toward a more sustainable trajectory for e-waste policy: A review of a decade of e-waste research in Accra, Ghana. *International Journal of Environmental Research and Public Health*, *14*(2), 135.

Debnath, B., Chowdhury, R., & Ghosh, S. K. (2018). Sustainability of metal recovery from E-waste. *Frontiers of Environmental Science & Engineering*, *12*(6), 1–12.

Dhakal, S. (2009). Urban energy use and carbon emissions from cities in China and policy implications. *Energy Policy*, *37*(11), 4208–4219.

Ding, Y., Zhang, S., Liu, B., & Li, B. (2017). Integrated process for recycling copper anode slime from electronic waste smelting. *Journal of Cleaner Production*, *165*, 48–56.

Donadkar, M. U., & Solanke, S. S. (2016). Review of E-waste material used in making of concrete. *International Journal of Science Technology & Engineering*, *2*(07), 66–69.

Elaqra, H. A., Abou Haloub, M. A., & Rustom, R. N. (2019). Effect of new mixing method of glass powder as cement replacement on mechanical behavior of concrete. *Construction and Building Materials, 203*, 75−82.

European Food Safety Authority. (2004). Opinion of the scientific panel on dietetic products, nutrition and allergies [NDA] related to the tolerable upper intake level of iron. *EFSA Journal, 2*(11), 125. Available from https://doi.org/10.2903/j.efsa.2004.125.

Evram, A., Akçaoğlu, T., Ramyar, K., & Çubukçuoğlu, B. (2020). Effects of waste electronic plastic and marble dust on hardened properties of high strength concrete. *Construction and Building Materials, 263*, 120928.

Ferreira, L., de Brito, J., & Saikia, N. (2012). Influence of curing conditions on the mechanical performance of concrete containing recycled plastic aggregate. *Construction and Building Materials, 36*, 196−204.

Fowler, B. A. (2017). *Electronic waste: Toxicology and public health issues.* Academic Press.

Fu, J., Zhang, A., Wang, T., Qu, G., Shao, J., Yuan, B., Wang, Y., & Jiang, G. (2013). Influence of e-waste dismantling and its regulations: Temporal trend, spatial distribution of heavy metals in rice grains, and its potential health risk. *Environmental Science & Technology, 47*(13), 7437−7445.

Gamble, A. V., Givens, A. K., & Sparks, D. L. (2018). Arsenic speciation and availability in orchard soils historically contaminated with lead arsenate. *Journal of Environmental Quality, 47*(1), 121−128.

Gautam, S. P., Srivastava, V., & Agarwal, V. C. (2012). Use of glass wastes as fine aggregate in concrete. *Journal of Academia and Industrial Research, 1*(6), 320−322.

Geeraerts, K., Illes, A., & Schweizer, JP. (2015). *Illegal shipment of e-waste from the EU. A case study on illegal e-waste export from EU to China. A study compiled as part of the EFFACE project.* http://efface.eu/.

Ghosh, B., Ghosh, M. K., Parhi, P., Mukherjee, P. S., & Mishra, B. K. (2015). Waste printed circuit boards recycling: An extensive assessment of current status. *Journal of Cleaner Production, 94*, 5−19.

Grant, K., Goldizen, F. C., Sly, P. D., Brune, M. N., Neira, M., van den Berg, M., & Norman, R. E. (2013). Health consequences of exposure to e-waste: A systematic review. *The Lancet Global Health, 1*(6), e350−e361.

Gupta, S., Modi, G., Saini, R., & Agarwala, V. (2014). A review on various electronic waste recycling techniques and hazards due to its improper handling. *International Refereed Journal of Engineering and Science (IRJES), 3*(5), 05−17.

Ha, N. N., Agusa, T., Ramu, K., Tu, N. P. C., Murata, S., Bulbule, K. A., Parthasaraty, P., Takahashi, S., Subramanian, A., & Tanabe, S. (2009). Contamination by trace elements at e-waste recycling sites in Bangalore, India. *Chemosphere, 76*(1), 9−15.

Ha, V. H., Lee, J. C., Huynh, T. H., Jeong, J., & Pandey, B. D. (2014). Optimizing the thiosulfate leaching of gold from printed circuit boards of discarded mobile phone. *Hydrometallurgy, 149*, 118−126.

He, Y., & Xu, Z. (2015). Recycling gold and copper from waste printed circuit boards using chlorination process. *RSC Advances, 5*(12), 8957−8964.

Herat, S., & Agamuthu, P. (2012). E-waste: A problem or an opportunity? Review of issues, challenges and solutions in Asian countries. *Waste Management & Research, 30*(11), 1113−1129.

Horeh, N. B., Mousavi, S. M., & Shojaosadati, S. A. (2016). Bioleaching of valuable metals from spent lithium-ion mobile phone batteries using *Aspergillus niger. Journal of Power Sources, 320*, 257−266.

Hui, Z., Poon, C. S., & Ling, T. C. (2013). Properties of mortar prepared with recycled cathode ray tube funnel glass sand at different mineral admixture. *Construction and Building Materials, 40*, 951−960.

Igharo, O. G., Anetor, J. I., Osibanjo, O., Osadolor, H. B., Odazie, E. C., & Uche, Z. C. (2018). Endocrine disrupting metals lead to alteration in the gonadal hormone levels in Nigerian e-waste workers. *Universa Medicina, 37*(1), 65−74.

Ilankoon, I. M. S. K., Ghorbani, Y., Chong, M. N., Herath, G., Moyo, T., & Petersen, J. (2018). E-waste in the international context—A review of trade flows, regulations, hazards, waste management strategies and technologies for value recovery. *Waste Management, 82*, 258−275.

Ilechukwu, I., Osuji, L. C., Okoli, C. P., Onyema, M. O., & Ndukwe, G. I. (2021). Assessment of heavy metal pollution in soils and health risk consequences of human exposure within the vicinity of hot mix asphalt plants in Rivers State, Nigeria. *Environmental Monitoring and Assessment, 193*(8), 1−14.

Ilyas, S., & Lee, J. C. (2014). Biometallurgical recovery of metals from waste electrical and electronic equipment: A review. *ChemBioEng Reviews, 1*(4), 148−169.

Ilyas, S., Ruan, C., Bhatti, H. N., Ghauri, M. A., & Anwar, M. A. (2010). Column bioleaching of metals from electronic scrap. *Hydrometallurgy*, *101*(3−4), 135−140.

Jagannath, A., Shetty, V., & Saidutta, M. B. (2017). Bioleaching of copper from electronic waste using *Acinetobacter* sp. Cr B2 in a pulsed plate column operated in batch and sequential batch mode. *Journal of Environmental Chemical Engineering*, *5*(2), 1599−1607.

Kamberović, Ž., Ranitović, M., Korać, M., Andjić, Z., Gajić, N., Djokić, J., & Jevtić, S. (2018). Hydrometallurgical process for selective metals recovery from waste-printed circuit boards. *Metals*, *8*(6), 441.

Kang, H. Y., & Schoenung, J. M. (2005). Electronic waste recycling: A review of US infrastructure and technology options. *Resources, Conservation and Recycling*, *45*(4), 368−400.

Kaya, M. (2016). Recovery of metals from electronic waste by physical and chemical recycling processes. *International Journal of Chemical and Molecular Engineering*, *10*(2), 259−270.

Khaliq, A., Rhamdhani, M. A., Brooks, G., & Masood, S. (2014). Metal extraction processes for electronic waste and existing industrial routes: A review and Australian perspective. *Resources*, *3*(1), 152−179.

Kumar, U., & Singh, D. N. (2014). Electronic waste: Concerns & hazardous threats. *International Journal of Current Engineering and Technology*, *4*, 802−811.

Lakshmi, R., & Nagan, S. (2010). Studies on concrete containing E plastic waste. *International Journal of Environmental Sciences*, *1*(3), 270.

Lakshmi, R., & Nagan, S. (2011). Investigations on durability characteristics of E-plastic waste incorporated concrete. *Asian Journal of Civil Engineering (Building and Housing)*, *12*(6), 773−787.

Li, J., Duan, H., & Shi, P. (2011). Heavy metal contamination of surface soil in electronic waste dismantling area: Site investigation and source-apportionment analysis. *Waste Management & Research*, *29*(7), 727−738.

Li, W., & Achal, V. (2020). Environmental and health impacts due to e-waste disposal in China—A review. *Science of the Total Environment*, *737*, 139745.

Ling, T. C., & Poon, C. S. (2011). Utilization of recycled glass derived from cathode ray tube glass as fine aggregate in cement mortar. *Journal of Hazardous Materials*, *192*(2), 451−456.

Ling, T. C., & Poon, C. S. (2012a). Development of a method for recycling of CRT funnel glass. *Environmental Technology*, *33*(22), 2531−2537.

Ling, T. C., & Poon, C. S. (2012b). Feasible use of recycled CRT funnel glass as heavyweight fine aggregate in barite concrete. *Journal of Cleaner Production*, *33*, 42−49.

Ling, T. C., & Poon, C. S. (2014). Use of recycled CRT funnel glass as fine aggregate in dry-mixed concrete paving blocks. *Journal of Cleaner Production*, *68*, 209−215.

Liu, T., Song, W., Zou, D., & Li, L. (2018). Dynamic mechanical analysis of cement mortar prepared with recycled cathode ray tube (CRT) glass as fine aggregate. *Journal of Cleaner Production*, *174*, 1436−1443.

Liu, Y., Huo, X., Xu, L., Wei, X., Wu, W., Wu, X., & Xu, X. (2018). Hearing loss in children with e-waste lead and cadmium exposure. *Science of the Total Environment*, *624*, 621−627.

Luhar, S., & Luhar, I. (2019). Potential application of E-wastes in construction industry: A review. *Construction and Building Materials*, *203*, 222−240.

Luo, C., Liu, C., Wang, Y., Liu, X., Li, F., Zhang, G., & Li, X. (2011). Heavy metal contamination in soils and vegetables near an e-waste processing site, south China. *Journal of Hazardous Materials*, *186*(1), 481−490.

Mathew, P., Varghese, S., Paul, T., & Varghese, E. (2013). Recycled plastics as coarse aggregate for structural concrete. *International Journal of Innovative Research in Science, Engineering and Technology*, *2*(3), 687−690.

Mathur, A., Choudhary, A., Yadav, P. S., Murari, K., & Student, U. G. (2017). Experimental study of concrete using E-waste as coarse aggregate. *International Journal of Engineering Science*, *7*(5), 11244.

McDonald, R. I., Douglas, I., Revenga, C., Hale, R., Grimm, N., Grönwall, J., & Fekete, B. (2011). Global urban growth and the geography of water availability, quality, and delivery. *Ambio*, *40*(5), 437−446.

Meng, M., Li, B., Shao, J. J., Wang, T., He, B., Shi, J. B., Ye, Z. H., & Jiang, G. B. (2014). Accumulation of total mercury and methylmercury in rice plants collected from different mining areas in China. *Environmental Pollution, 184,* 179−186.

Mmereki, D., Li, B., Baldwin, A., & Hong, L. (2016). *The generation, composition, collection, treatment and disposal system, and impact of E-waste.* (Ed.),*E-waste in Transition: From Pollution to Resource*London: IntechOpen.

Moncea, A. M., Badanoiu, A., Georgescu, M., & Stoleriu, S. (2013). Cementitious composites with glass waste from recycling of cathode ray tubes. *Materials and Structures, 46*(12), 2135−2144.

Nachman, K. E., Baron, P. A., Raber, G., Francesconi, K. A., Navas-Acien, A., & Love, D. C. (2013). Roxarsone, inorganic arsenic, and other arsenic species in chicken: A US-based market basket sample. *Environmental Health Perspectives, 121*(7), 818−824.

Nadhim, S., Shree, P. N., & Kumar, G. P. (2016). A comparative study on concrete containing E-plastic waste and fly ash concrete with conventional concrete. *International Journal of Engineering Research, 4.*

Natarajan, G., & Ting, Y. P. (2014). Pretreatment of e-waste and mutation of alkali-tolerant cyanogenic bacteria promote gold biorecovery. *Bioresource Technology, 152,* 80−85.

Natarajan, G., & Ting, Y. P. (2015). Gold biorecovery from e-waste: An improved strategy through spent medium leaching with pH modification. *Chemosphere, 136,* 232−238.

Needhidasan, S., Samuel, M., & Chidambaram, R. (2014). Electronic waste—An emerging threat to the environment of urban India. *Journal of Environmental Health Science and Engineering, 12*(1), 1−9.

Neville, A. M., & Brooks, J. J. (1987). *Concrete technology* (438). Longman Scientific & Technical.

Ngoc, N., Agusa, T., Ramu, K., Phuc, N., Tu, C., & Murata, S. (2009). Chemosphere contamination by trace elements at e-waste recycling sites in Bangalore, India. *Chemosphere, 76*(1), 9−15. Available from https://doi.org/10.1016/j.chemosphere.2009.02.056.

Nithya, R., Sivasankari, C., & Thirunavukkarasu, A. (2021). Electronic waste generation, regulation and metal recovery: A review. *Environmental Chemistry Letters, 19*(2), 1347−1368.

Nithya, R., Sivasankari, C., Thirunavukkarasu, A., & Selvasembian, R. (2018). Novel adsorbent prepared from bio-hydrometallurgical leachate from waste printed circuit board used for the removal of methylene blue from aqueous solution. *Microchemical Journal, 142,* 321−328.

Oh, C. J., Lee, S. O., Yang, H. S., Ha, T. J., & Kim, M. J. (2003). Selective leaching of valuable metals from waste printed circuit boards. *Journal of the Air & Waste Management Association, 53*(7), 897−902.

Patil, R. A., & Ramakrishna, S. (2020). A comprehensive analysis of e-waste legislation worldwide. *Environmental Science and Pollution Research, 27*(13), 14412−14431.

Perkins, D. N., Drisse, M.-N. B., Nxele, T., & Sly, P. D. (2014). E-waste: A global hazard. *Annals Global Health, 80,* 286−295.

Phillips, L. J., & Moya, J. (2014). Exposure factors resources: Contrasting EPA's exposure factors handbook with international sources. *Journal of Exposure Science & Environmental Epidemiology, 24*(3), 233−243.

Pradhan, J. K., & Kumar, S. (2014). Informal e-waste recycling: Environmental risk assessment of heavy metal contamination in Mandoli industrial area, Delhi, India. *Environmental Science and Pollution Research, 21*(13), 7913−7928.

Quinet, P., Proost, J., & Van Lierde, A. (2005). Recovery of precious metals from electronic scrap by hydrometallurgical processing routes. *Mining, Metallurgy & Exploration, 22*(1), 17−22.

Quinteros, F. A., Machiavelli, L. I., Miler, E. A., Cabilla, J. P., & Duvilanski, B. H. (2008). Mechanisms of chromium (VI)-induced apoptosis in anterior pituitary cells. *Toxicology, 249*(2−3), 109−115.

Robinson, B. H. (2009). E-waste: An assessment of global production and environmental impacts. *Science of the Total Environment, 408*(2), 183−191.

Ruan, J., Zhu, X., Qian, Y., & Hu, J. (2014). A new strain for recovering precious metals from waste printed circuit boards. *Waste Management, 34*(5), 901−907.

Salhofer, S. (2017). E-waste collection and treatment options: A comparison of approaches in Europe, China and Vietnam. In R. Maletz, C. Dornack,, & L. Ziyang (Eds.), *Source Separation and Recycling: The Handbook of Environmental Chemistry, vol 63.* Cham: Springer.

Santhanam, N., & Anbuarasu, G. (2020). Experimental study on high strength concrete (M60) with reused E-waste plastics. *Materials Today: Proceedings, 22,* 919−925.

Senophiyah-Mary, J., Loganath, R., & Meenambal, T. (2018). A novel method for the removal of epoxy coating from waste printed circuit board. *Waste Management & Research, 36*(7), 645–652.

Senophiyah-Mary, J., Loganath, R., & Meenambal, T. (2019). *A performance study on the bioleaching process by the production of organic acids using HPLC/UV.* (Ed.),*Waste Valorisation and Recycling*Springer.

Senophiyah-Mary, J., Loganath, R., & Shameer, P. M. (2018). Deterioration of cross linked polymers of thermoset plastics of e-waste as a side part of bioleaching process. *Journal of Environmental Chemical Engineering, 6*(2), 3185–3191.

Senophiyah-Mary, J., Thomas, T., Loganath, R., & Meenambal, T. (2019). *Removal of copper from bioleachate of e-waste using orange activated carbon (OAC) and comparison with commercial activated carbon (CAC).* *Waste valorisation and recycling* (pp. 373–383). Springer.

Senthil Kumar, K., & Baskar, K. (2018). Effect of temperature and thermal shock on concrete containing hazardous electronic waste. *Journal of Hazardous, Toxic, and Radioactive Waste, 22*(2), p. 04017028.

Sepulveda, A., Schluep, M., Renaud, F., Streicher, M., Kuehr, R., Hagelueken, C., & Gerecke, AC. (2010). A review of the environmental fate and effects of hazardous substances released from electrical and electronic equipment during recycling: examples from China and India. *Environmental Impact Assessment Review, 30*, 28–41.

Shaikh, F. U. A. (2016). Mechanical and durability properties of fly ash geopolymer concrete containing recycled coarse aggregates. *International Journal of Sustainable Built Environment, 5*(2), 277–287.

Shaikh, S., Thomas, K., Zuhair, S., & Magalini, F. (2020). A cost-benefit analysis of the downstream impacts of e-waste recycling in Pakistan. *Waste Management, 118*, 302–312.

Shamim, A., Mursheda, A. K., & Rafiq, I. (2015). E-waste trading impact on public health and ecosystem services in developing countries. *Journal of Waste Resources, 5*(4), 1–18.

Sheel, A., & Pant, D. (2018). Recovery of gold from electronic waste using chemical assisted microbial biosorption (hybrid) technique. *Bioresource Technology, 247*, 1189–1192.

Shi, C., & Zheng, K. (2007). A review on the use of waste glasses in the production of cement and concrete. *Resources, Conservation and Recycling, 52*(2), 234–247.

Shikdar, A. A., & Sawaqed, N. M. (2003). Worker productivity, and occupational health and safety issues in selected industries. *Computers & Industrial Engineering, 45*(4), 563–572.

Song, Q., & Li, J. (2014). Environmental effects of heavy metals derived from the e-waste recycling activities in China: A systematic review. *Waste Management, 34*(12), 2587–2594.

Song, Q., & Li, J. (2015). A review on human health consequences of metals exposure to e-waste in China. *Environmental Pollution, 196*, 450–461.

Song, Q., Zeng, X., Li, J., Duan, H., & Yuan, W. (2015). Environmental risk assessment of CRT and PCB workshops in a mobile e-waste recycling plant. *Environmental Science and Pollution Research, 22*(16), 12366–12373.

Suchithra, S., Kumar, M., & Indu, V. S. (2015). Study on replacement of coarse aggregate by E-waste in concrete. *International Journal of Technical Research and Application, 3*(4), 266–270.

Thacker, N., Sheikh, J., Tamane, S. M., Bhanarkar, A., Majumdar, D., Singh, K., Chavhan, C., & Trivedi, J. (2013). Emissions of polychlorinated dibenzo-p-dioxins (PCDDs), dibenzofurans (PCDFs), and dioxin-like polychlorinated biphenyls (PCBs) to air from waste incinerators and high thermal processes in India. *Environmental Monitoring and Assessment, 185*(1), 425–429.

Tóth, G., Hermann, T., Da Silva, M. R., & Montanarella, L. J. E. I. (2016). Heavy metals in agricultural soils of the European Union with implications for food safety. *Environment International, 88*, 299–309.

Tsezos, M., Remoundaki, E., & Hatzikioseyian, A. (2006). Biosorption-principles and applications for metal immobilization from waste-water streams, October *Proceedings of EU-Asia workshop on clean production and nanotechnologies*, 23–33.

Tue, N. M., Takahashi, S., Suzuki, G., Isobe, T., Viet, P. H., Kobara, Y., Seike, N., Zhang, G., Sudaryanto, A., & Tanabe, S. (2013). Contamination of indoor dust and air by polychlorinated biphenyls and brominated flame retardants and relevance of non-dietary exposure in Vietnamese informal e-waste recycling sites. *Environment International, 51*, 160–167.

Ullah, Z., Qureshi, M. I., Ahmad, A., Khan, S. U., & Javaid, M. F. (2021). An experimental study on the mechanical and durability properties assessment of E-waste concrete. *Journal of Building Engineering, 38*, 102177.

Vakilchap, F., Mousavi, S. M., & Shojaosadati, S. A. (2016). Role of *Aspergillus niger* in recovery enhancement of valuable metals from produced red mud in Bayer process. *Bioresource Technology*, *218*, 991–998.

Veit, H. M., Diehl, T. R., Salami, A. P., Rodrigues, J. D. S., Bernardes, A. M., & Tenório, J. A. S. (2005). Utilization of magnetic and electrostatic separation in the recycling of printed circuit boards scrap. *Waste Management*, *25*(1), 67–74.

Wong, C. S., Duzgoren-Aydin, N. S., Aydin, A., & Wong, M. H. (2007). Evidence of excessive releases of metals from primitive e-waste processing in Guiyu, China. *Environmental Pollution*, *148*(1), 62–72.

Wu, Q., Leung, J. Y., Du, Y., Kong, D., Shi, Y., Wang, Y., & Xiao, T. (2019). Trace metals in e-waste lead to serious health risk through consumption of rice growing near an abandoned e-waste recycling site: Comparisons with PBDEs and AHFRs. *Environmental Pollution*, *247*, 46–54.

Yamasaki, S. I., Takeda, A., Nanzyo, M., Taniyama, I., & Nakai, M. (2001). Background levels of trace and ultra-trace elements in soils of Japan. *Soil Science and Plant Nutrition*, *47*(4), 755–765.

Yang, T., Xu, Z., Wen, J., & Yang, L. (2009). Factors influencing bioleaching copper from waste printed circuit boards by *Acidithiobacillus ferrooxidans*. *Hydrometallurgy*, *97*(1–2), 29–32.

Yao, Z., Ling, T. C., Sarker, P. K., Su, W., Liu, J., Wu, W., & Tang, J. (2018). Recycling difficult-to-treat e-waste cathode-ray-tube glass as construction and building materials: A critical review. *Renewable and Sustainable Energy Reviews*, *81*, 595–604.

Yazici, E. Y., & Deveci, H. A. C. I. (2014). Ferric sulphate leaching of metals from waste printed circuit boards. *International Journal of Mineral Processing*, *133*, 39–45.

Yu, Y. J., Lin, B. G., Qiao, J., Chen, X. C., Li, L. Z., Chen, X. Y., Yang, L. Y., Yang, P., Zhang, G. Z., Zhou, X. Q., & Chen, C. R. (2020). Levels and congener profiles of halogenated persistent organic pollutants in human serum and semen at an e-waste area in South China. *Environment International*, *138*, 105666.

Zeng, X., Duan, H., Wang, F., & Li, J. (2017). Examining environmental management of e-waste: China's experience and lessons. *Renewable and Sustainable Energy Reviews*, *72*, 1076–1082.

Zhang, B., Huo, X., Xu, L., Cheng, Z., Cong, X., Lu, X., & Xu, X. (2017). Elevated lead levels from e-waste exposure are linked to decreased olfactory memory in children. *Environmental Pollution*, *231*, 1112–1121.

Zhang, C. C., & Zhang, F. S. (2018). High copper recovery from scrap printed circuit boards using poly (ethylene glycol)/sodium hydroxide treatment. *Environmental Chemistry Letters*, *16*(1), 311–317.

Zhang, X., Zhong, T., Liu, L., & Ouyang, X. (2015). Impact of soil heavy metal pollution on food safety in China. *PLoS One*, *10*(8), p.e0135182.

Zheng, G., Xu, X., Li, B., Wu, K., Yekeen, T. A., & Huo, X. (2013). Association between lung function in school children and exposure to three transition metals from an e-waste recycling area. *Journal of Exposure Science & Environmental Epidemiology*, *23*(1), 67–72.

Zheng, J., Chen, K. H., Yan, X., Chen, S. J., Hu, G. C., Peng, X. W., Yuan, J. G., Mai, B. X., & Yang, Z. Y. (2013). Heavy metals in food, house dust, and water from an e-waste recycling area in South China and the potential risk to human health. *Ecotoxicology and Environmental Safety*, *96*, 205–212.

Zheng, J. C., Feng, H. M., Lam, M. H. W., Lam, P. K. S., Ding, Y. W., & Yu, H. Q. (2009). Removal of Cu (II) in aqueous media by biosorption using water hyacinth roots as a biosorbent material. *Journal of Hazardous Materials*, *171*(1–3), 780–785.

CHAPTER 10

Associated environmental threats due to incongruous E-waste management and a case study of southeast Asia

Mamta Tembhare, Deval Singh, Shashi Arya and Shilpa Vishwakarma
Waste Re-processing Division, CSIR-National Environmental Engineering Research Institute (CSIR-NEERI), Nagpur, Maharashtra, India

10.1 Introduction

The rapid generation of Waste Electronic and Electrical Equipment (WEEE) has contributed to global warming issues (Leung, 2019). Studies have stated that electronic waste (E-waste) "is the old, outdated, obsolete Electronics and Electrical (EE) products which have been discarded and not reused by owners" (UNU (United Nations University)/Step Initiative, 2014). The modern area of technology and urbanization have enhanced the consumer living standards, influencing the application of EE products to balance this growing demand. As per the study conducted by the United Nations Environment Program (UNEP) in 2019, the bulk quantity of WEEE produced in developed countries is sold and transported (unauthorized trading networks) to developing and less developed countries. Products such as used computers are sold at a scrap value from the United States to South Africa and Nigeria (Andeobu et al., 2021). Further, these products are processed by informal sectors (rag pickers and local vendors) to recover value-added metals. Later, the leftover waste residues are dumped at different low-lying areas, which may lead to various environmental hazards. The informal sector comprising adults and children may also face health hazards due to manual handling of toxic compounds.

In past few decades, the application of EE products has gained maximum demand among common residence. Therefore it is essential to develop management strategies to collect, processes, recover, and dispose E-waste. According to Forti et al. (2020), most of the developed countries have shown substantial increase in E-waste collection compared to the past few decades. However, the rate of production and quantity of E-waste have increased around the world as represented in Fig. 10.1.

It is predicted that 53 million metric tons (Mt) of E-waste generation in 2019 would upsurge 74.1 Mt by year 2030. Informal recycling process has inefficient waste

Global E-waste Management Strategies and Future Implications
DOI: https://doi.org/10.1016/B978-0-323-99919-9.00016-7

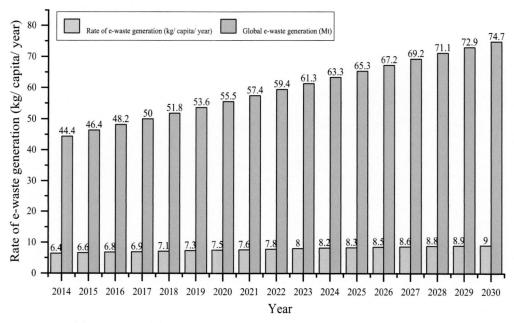

Figure 10.1 Global E-waste generation.

management practices compared to the formal sector around the world. Very small amounts of waste are treated with scientific method. The remaining waste is recycled using the informal method, and the leftover are incinerated or directly dumped at dumpsites. The global market for EEE has grown dramatically during the last two decades, but the life span has declined due to the usage of low-quality components. Globally, the number of EE gadgets will continue to rise, and microprocessors will be used in higher quantity. Manufacturing of EE product is one of the fastest expanding industries around the world (Needhidasan et al., 2014). Based on the report by Forti et al. (2020), WEEE can be majorly categorized as temperature exchange equipment, display units, lights, and information and telecommunication equipment (Forti et al., 2020). A large portion of E-waste also comprises mobiles, washing machines, printers, PCs, CPUs, dryers, and many other household appliances. Studies have also revealed that nonscientific treatment and disposal practices of WEEE have caused the threat to human health and the environment (Awasthi et al., 2016a, 2016b; Awasthi et al., 2017; Breivik et al., 2014; Chakraborty et al., 2017; Chakraborty, Khuman, et al., 2016; Chakraborty, Selvaraj, et al., 2016; Chakraborty, Zhang, et al., 2016; Li et al., 2012; Yu et al., 2014).

The majority of the pollutants and toxic compounds found in E-waste have serious environmental and health effects. Some contaminants are spread through the air, groundwater, or soil and can also be present in the surrounding air in regions near industrial facilities. In other situations, by-products are thrown directly using nonscientific methods into the soil or rivers, where future pollutant leaching might pollute the ecosystem and impact food chain sources. The nonscientific method of treatment can cause penetration of toxic compounds into the blood cells of E-waste handlers. Immediate human exposure to these pollutants (lead, barium, arsenic, nickel, cadmium, etc.) can have long-term health consequences such as cancer and skin disease. Such toxins can also have similar implications for the flora and fauna exposed to the E-waste recycling sites (Cayumil et al., 2016).

Furthermore, we should explore innovative solutions to "green" all stages of electrical device life. Awareness of the harmful effects of E-waste on biodiversity should be included in school and higher education curricula as a part of one training. Funds must also be made to develop new and improved E-waste management techniques (Moletsane & Venter, 2018). There is a need to study and implement proper formal recycling methods to mitigate environmental threats due to E-waste management. This chapter focuses on the characterization of different toxic components that are present in E-waste and also presents a brief overview of its associated environmental impact. A comprehensive analysis of the composition of several forms of E-waste, including big and small home appliances, IT and telecommunications equipment, and light equipment, among others, is provided with the goal of determining the dangerous chemicals contained in electronic equipment.

10.2 Hazardous contaminants in E-waste

The environmental hazards associated with E-waste recycling has caused significant concern to developed and developing countries. However, most developing countries (India, Pakistan, etc.) have transformed E-waste recycling facilities into a new business project (Leung, 2019). The discarded E-waste comprises precious metals (gold, silver, platinum, barium, copper, aluminum, steel, nickel, cadmium, etc.) and nonmetals (plastic, glass, silica). The informal sector engaged in these processing facilities has successfully gained economic viability by applying the extraction process. The leftover secondary raw material from these E-waste components is further processed at recycling facilities.

Moreover, these processing facilities practicing nonscientific E-waste handling and processing have caused the direct discharge of harmful contaminants in the air, water, and soil. However, among those, maximum substances are hazardous when they directly expose to the surrounding environment (UNEP, 2018). Table 10.1 presents list of the hazardous substances and their health effects.

Table 10.1 The hazardous substances and their health effects (Kumar et al., 2017).

S. No.	Name of substance	Applications	Health effects
1.	Copper	For conductivity of printed circuit boards, connecting the electrical circuit	Stomach ache, liver and kidney damage, anemia
2.	Mercury	Alkaline batteries, liquid crystal display	Increases blood pressure and pulse rate, disturbance in brain function as well as memory, minor brain, and kidney damage
3.	Chromium	To protect corrosion from galvanized steel plates, data plates	Damages DNA, kidney, liver
4.	Zinc	To protect corrosion coating on cathode-ray tubes (CRTs)	Damages skin
5.	Cadmium	Phosphor emitter in CRT screen, rechargeable Ni–Cd batteries	Damages kidney and lungs; makes bones fragile
6.	Lead	Used for soldering purpose in printed circuit board	Damages nervous system and circulatory system, adverse effects on brain growth as well as on endocrine system
7.	Rare earth metals	Used in CRT screen as a fluorescent layer	Causes damage to lungs, liver, and brain
8.	Polybrominated diphenyl ethers	Flame retardants for plastics	Damages endocrine system
9.	Polychlorinated dibenzo-p-dioxins and dibenzofurans	Type of combustion product	It affects the endocrine as well as reproduction functioning
10.	Polycyclic aromatic hydrocarbons	Type of combustion product	Carcinogenic for body

10.3 Environmental exposure to hazardous contaminants from E-waste processing sites

In most developing nations, E-waste processing sites are located at the outskirts of the cities and managed by the informal sector. These informal sectors perform the collection, processing, refurbishment, reprocessing, reuse, and disposal of E-waste. Apart from this, local scrap vendors and low-wage workers or technicians collaborate to repair and refurbish the used equipment for local resale. In some cases, scrap dealers collect old WEEE products from different households to resell these goods to local recyclers (Oteng-Ababio, 2012). However, the nonscientific method (manual disassembling, open burning, and acid leaching) of E-waste handling in open space may cause direct penetration of toxic compounds [polychlorinated biphenyls (PCBs), brominated flame retardants, dioxins, furans, and metals] into the environment (Steinbacher et al., 2009).

The open discarding of E-waste at the surface or banks of the river could result in dispersion and diffusion of toxic elements into the farmland and ruler arena. The study carried out in Ghana city, China, proved the presence of organic and inorganic contaminants in surface water bodies due to the E-waste recycling facility located nearby (Hosoda et al., 2014; Huang et al., 2014). Additionally, open dumping of E-waste at different low-lying areas causes the leaching of toxic metals into the soil and groundwater. The contaminants (toxic metals and organic compounds) released from E-waste recycling facilities are dispersed into the natural biota through transportation pathways such as soil, dust, food, surface, and groundwater run-off. Later, it gets accumulated in various biotic (plants and animals) and abiotic (soil, air, and water) constituents of environment. Simultaneously, these transportation pathways allow different contaminants to get exposed to various terrestrial and aquatic ecosystems, resulting in environmental hazards (Kiddee et al., 2013). Therefore it is essential to study different transportation pathways influencing contaminant flow from E-waste recycling facilities and their impact on natural biota such as air, water, and soil.

10.3.1 Dust/air

Dust is an environmental medium that allows a rapid flow and accumulation of contaminants into the different ecosystems. Samples collected from various locations help determine the concentration and dispersal of pollutants into a particular ecosystem (Leung et al., 2008). The flame released from processing sites due to the combustion process emits more than $16,575 \text{ pg/m}^3$ of polybrominated diphenyl ethers (PBDEs) into the natural atmosphere, approximately 300 times higher than normal atmospheric conditions (Deng et al., 2007). This flow of contaminants is dependent on microenvironmental factors such as wind direction, climate, topology, rainfall, and run-off (Wong et al., 2006). The dust of metal-laden released from recycling facilities is influenced by air current (wind speed and direction), resulting in contaminant depositing in natural topologies, including building walls, roads, topsoil, plants leaves, and roots (Wong et al., 2006). Later, these contaminants are dispersed into the air and water media. The higher concentration of metals in dust particulates can directly influence public health and safety through dust inhalation, ingestion, and dermal contact (Abrahams, 2002). These metal-laden dust may get mixed into the household dust, resulting in indirect exposure to metals by residents. The study carried out by Akangbe Yekeen et al. (2016) has proved a significantly high concentration of Pb near the E-waste recycling site in Guiyang, China.

Similarly, the air sample collected from the same town reported the highest ever concentration of polychlorodibenzo-p-dioxins in the atmosphere (Liu et al., 2008). Besides this, 11 sampling points from the residential area, school, garden, etc., were predefined to collect dust samples to identify the presence of metals in different areas. The results proved the presence of various metals in varying concentrations at other sites. In many cases, metals such as Cu, Cd, Hg, Ni, Pb, and Zn exceeded the permissible limit in dust concentration at different exposure points. Therefore it is essential to develop a cost-effective technology for countries practicing the informal method of E-waste processing; the idea is to limit the discharge of aerial particulate (smoke and dust) into the atmosphere from E-waste processing sites.

10.3.2 Surface water bodies

In most cases, E-waste processing sites near water bodies are the source of contaminant transmission. The processing site utilizes necessary water for metal extraction and E-waste dismantling, which leads to rivers, ponds, and lakes contamination. The open dumping of waste residue from these processing sites near the banks of the water bodies results in the leaching of contaminants. Later, these contaminants get induced into the aquatic system due to the natural run-off (Robinson, 2009). Besides this, acid disposal and hydrometallurgical recycling process also result in contaminant transfusion into the marine biota. Studies have reported that the diffusion of these contaminants into the water bodies may further contaminate groundwater, sediments, flora, and fauna (Wang & Guo, 2006). The sedimentation samples collected from the Odaw river, China, determined higher Pb, Cd, Cu, and Ni concentrations. It was observed that areas closer to E-waste burning sites have a higher concentration of heavy metals (Chama et al., 2014). While entering a water body, these metals tend to bind with sediment particles suspended into the water bodies, resulting in acidic conditions (Wong et al., 2007). The study carried out by Luo et al. (2007) reported the concentration of PBDEs in sediments (16,000 ng/g) and aquatic animals such as crab (766 ng/g) in river Nanyang, near Guiyu, China. Similarly, Wu et al. (2008) reported the presence of PCBs (up to 16,512 ng/g) and PBDEs (up to 1091 ng/g) in aquatic predators such as crucian carp, mud carp, prawns, and water snake.

Similarly, most developing countries such as India have traced the presence of harmful metals in the areas residing near the processing sites. Table 10.2 presents the reported concentration of toxic metals found in soil, water, and plants near the

Table 10.2 Concentration of toxic metals found in soil, water, and plants near recycling sites in Indian cities.

Sample	Cities	Residential/industry area	Ag	Al	As	Cd	Co	Cu	Cr	Fe	Hg	Mn	Mo	Ni	Pb	Sb	Se	Sn	Ti	V	Zn	References
Water	New Delhi	Mandoli	-	3.67	-	0.05	-	0.70	0.60	0.46	-	-	-	-	0.04	-	-	-	-	-	1.89	Pradhan and Kumar (2014)
	New Delhi	-	-	-	-	0.01	-	-	0.01	-	0.01	-	-	1.36	46.9	-	-	-	-	-	870	Brigden (2005)
	Kolkata	-	-	-	-	-	-	-	-	0.9	-	-	-	-	0.06	-	-	-	-	-	0.07	Kanmani and Gandhimathi (2013)
	Eastern side	-	-	-	-	-	-	-	-	0.02	-	-	-	-	0.03	-	-	-	-	-	-	
	North side	-	-	-	-	-	-	-	-	0.87	-	-	-	-	0.05	-	-	-	-	-	0.04	
	Tiruchirappalli	-	-	-	-	1.03	-	0.55	-	-	-	-	-	-	5.15	-	-	-	-	-	-	
Soil	Bangalore	-	-	1315	0.5	1	-	185	-	-	0.5	-	-	9	4	-	-	-	-	-	17	
	Bangalore	Slum area	14	-	-	2.33	11	592	73	-	1.8	449	1.8	-	297	14	-	86	0.4	30	326	
	Bangalore	Recycling facility	2.8	-	-	0.47	14	429	54	-	0.05	619	1.8	-	126	24	-	-	-	53	129	
	New Delhi	Ibrahimpur	12.38	-	-	66.6	-	500	293	-	0.3	-	-	-	3560	-	-	-	-	-	700	Brigden, 2005
	New Delhi	Mandoli	-	8.82	13	1.14	13	8734	83.6	4037	0.07	-	-	146	2133	-	12.3	-	-	-	416	Pradhan and Kumar (2014)
	Kolkata	Azad Metalwork	-	-	-	0.01	-	-	-	27	-	-	-	-	84	-	-	-	-	-	1.7	
	Hyderabad	Balanagar	-	-	-	-	-	6.86	4.41	-	-	-	-	-	-	-	-	-	-	-	12.2	Machender et al. (2011)
	Tiruchirappalli	-	-	-	-	30.58	-	39.3	-	-	-	-	-	-	291.3	-	-	-	-	-	-	Kanmani, and Gandhimathi (2013)
Plants	New Delhi	Dumping site	-	-	-	0.05	-	23	-	106.37	-	-	-	-	0.76	-	-	-	-	-	78.2	Pradhan and Kumar, 2014
	New Delhi	Arable land (50 m from processing site)	-	-	-	0.023	-	11	-	89.49	-	-	-	-	0.005	-	-	-	-	-	68	Pradhan and Kumar (2014)
	New Delhi	Arable land (100 m from processing site)	-	-	-	0.004	-	11	-	90.32	-	-	-	-	0.007	-	-	-	-	-	68	Pradhan and Kumar (2014)
	New Delhi	Residential area	-	-	-	0.003	-	11	-	88.47	-	-	-	-	0.006	-	-	-	-	-	68	Pradhan and Kumar (2014)

processing sites in Indian cities. From Table 10.2, it is conclusive that the E-waste recycling site in mandolin, New Delhi has a maximum concentration of Cu compared to other cities. Monika (2010) suggested that soil substrate with metal contamination can cause a decline in plant growth rate, followed by sudden death and drying of leaves. Generally, E-waste processing sites are more likely to cause metal contamination into the soil substrate. Later, vegetative crops and plants species translocate these metals into the tissue, vacuoles, and cell walls. John et al. (2009) suggested that plant species in contaminated soil strata are more likely to produce food with similar contamination. The soil sample collected from these processing sites had a higher concentration of Cu (3.15 mg/kg) and Cd (663.08 mg/kg) than its standard limits.

10.3.2.1 Effects of heavy metal on aquatic life

Studies have revealed that industrial discharge into the water bodies has caused heavy metals to accumulate into the aquatic species. These metals are highly toxic, persistent, and capable enough to induce high oxidative stress on marine species. There is direct and indirect uptake of these metals by the aquatic organism (Nammalwar, 1983). The direct uptake of these metals results in behavior, metabolism, migration, physiology, and reproduction changes. The indirect effects are through the food chain pattern, resulting in ecological stress. It may also cause water contamination and narrow the environmental diversity of different plants and animal species. Studies have reported that continuous exposure to Cu, Hg, Pb, and Ni can cause cytotoxicity to the plasma membrane, DNA, lipids, binding properties of protein and phospholipids; expression of the enzyme, transmembrane amino acid transport, Na and K-dependent ATPases; generation of reactive oxygen species can cause depletion of antioxidant enzymes (Leonard et al., 2004).

10.3.2.1.1 Impact of Pb on aquatic organism

Even at low concentrations, Pb tends to bioaccumulate in plants, animals, and microbes, resulting in chronic health hazards (Sindiku et al., 2015). The inorganic Pb is carcinogenic and causes disturbance in behavior, growth, metabolism, learning ability, and life span. Besides this, bioaccumulation of Pb into internal organs of aquatic species such as fishes causes scoliosis. Studies have also reported Pb bio-magnification into the marine ecosystem due to the continuous discharge of industrial effluent. In the case of algae, the dissolved Pb (>500 ppb) can cause a reduction in enzyme release, resulting in reduced photosynthesis (Rioboo et al., 2009). This process may hamper the growth of algae in water bodies, resulting in the decline of food availability for aquatic animals. Apart from this, fish's exposure to aqueous Pb^{2+} causes failure of internal organs such as male gonads, liver, heart, and kidney (Ebrahimi & Taherianfard, 2010). Continuous exposure to Pb^{2+} (>100 ppb) may also hamper the functioning of the fish gill. The high concentration of Pb exposure to aquatic fishes causes an increase in mortality rate, paralysis, damage to reproductive organs, neurological disorders, and muscular destruction. The average concentration of 100 to 100,000 µg/L of Pb has been reported in the marine food chain (Mager et al., 2010).

10.3.2.1.2 Impact of Cd on aquatic organism

Cd is found to have a small proportion in nature but has the most toxic effect on aquatic life. It can be a toxic element to the marine organism at each biological level, ranging from predators to small cellular microbes (Rashed, 2001). Even a small concentration can cause an imbalance in fish metabolism, resulting in abnormal behavior and locomotor anomalies (Bryan et al., 2011; Cicik & Engin, 2005). The long-term exposure (> 20 days) of Cd to aquatic animals such as juvenile fishes and adult rainbow trout (*Oncorhynchus mykiss*) has resulted in a decline in the growth rate of such species (Hayat et al., 2007). In the case of most aquatic fishes, Cd tends to accumulate in the liver, stomach, and gills, causing sudden death (Abu Hilal & Ismail, 2008). Besides, Cd causes a decline in plant growth rate and adverse effects on the entire marine (Solomon, 2008).

10.3.2.2 *Adverse effects of organic pollutants on aquatic lifecycle*
10.3.2.2.1 Impact of polybrominated diphenyl ethers on water bodies

The poisonous, bio-accumulative, and insistent nature of PBDEs allows significant hazards to the aquatic ecosystem. As per OECD (2003) report, PBDEs compounds are not naturally biodegradable concerning environmental chain. These compounds get bioaccumulated in fatty tissues of marine animals and result in bio-magnification via the food chain (Law & Herzke, 2010). Even though studies have suggested the minimum toxic effect of PBDEs on the aquatic ecosystem (Wollenberger et al., 2005), the quantity of research carried out is quite precise. Nevertheless, few studies have reported that PBDEs can develop neurological and endocrine disruption in aquatic animals (Branchi et al., 2005). PBDEs also tend to cause deformities and morphological abnormalities during embryogenesis (Lema et al., 2007). Similarly, Mhadhbi et al. (2010) reported pericardial edema and abnormal skeletal formations in turbot. Studies have also claimed that these compounds are teratogenic for the embryo–larval stages, resulting in partial improvement and mortality of embryo cells (Mhadhbi et al., 2010).

10.3.2.2.2 Impact of polychlorinated dibenzo-p-dioxins and dibenzofurans on aquatic organism

Polychlorinated dibenzo-p-dioxins and dibenzofurans (PCDD/Fs) are toxic and persistent organic pollutants (POPs) with a higher potential to bioaccumulate in aquatic species (Amirah et al., 2013). They tend to bio-magnify marine species at the higher trophic level of the food web. PCDD/Fs are hydrophobic, and it tends to bind affinity with soil sediments and organic particulates suspended in water bodies. Besides this, it also tends to bind affinity with lipid-rich tissues of aquatic animals. The continuous release of PCDD/Fs in marine biota may result in the uptake of these contaminants by aquatic species via soil sediments, water, and consuming contaminated prey (Strong & Eng, 2006). The most adverse effect of PCDD/Fs was found to be on fishes as it causes an imbalance in the reproductive cycle, thickening of eggshells, and damage to embryos (Andeobu et al., 2021). Therefore it is conclusive that long-term exposure to PCDD/Fs reduces aquatic species' life span and reproduction cycle (Strong & Eng, 2006).

10.3.3 Soil substrate

Various harmful by-products (heavy metal, organic and inorganic pollutants, etc.) gets settled into the soil strata via multiple environmental pathways. This also include natural (volcanic eruption), manmade (mining operation and smelting industries), and anthropogenic activities (incineration and burning of fossil fuels) (Caussy, 2003). The heavy metals settled in soil substrates are persistent and nonbiodegradable in the environment; they can pass through the human body via inhalation, soil ingestion, and food consumption (Li et al., 2014). The process of metal uptake via vegetative and nonvegetative plants through soil substrates in different terrain conditions results in the bioaccumulation of metals in plant tissues. However, it depends on plant and soil type (Liu et al., 2005). Studies have proved that sandy soil has more potential to bioaccumulate metals than clay-rich soil, increasing soil pH (Liu et al., 2005). Different species of plants, trees, vegetable crops, grasses, and weeds tend to accumulate wide varieties of metals (Kabata-Pendias, 1992). Wang et al. (2015) conducted a study using a buffer zone to understand the pollutant transportation from E-waste processing sites. The study proved that metal concentration was twice in plants species residing within the buffer zone, compared with the same plant species outside the area (Wang et al., 2015).

Similarly, Liang et al. (2007) collected samples from 18 plant species from the area near E-waste processing sites, and the aim was to determine the metal uptake process in these plant species. The study has suggested that plant root is the primary supporting parameter for metal uptakes responsible for bioaccumulation and contamination of food synthesizing tissues. Constant exposure to these contaminants through water and soil media results in chemical grazing land (Frazzoli et al., 2010); grazing animals exposed to these metal exposure lands may undergo health complications and sudden death (Kierkegaard et al., 2007).

Countries such as India and China practicing informal ways to manage E-waste processing have subsequently reported higher metal concentrations in soil substrates. The soil samples collected from government-authorized E-waste recycling facilities and backyard recycling sites reported the higher concentration of Mn, Cr, Cu, Co, Ag, Cd, In, Sn Sb, Pb, TI, and Bi in these countries (Kumar & Kumar, 2013). Therefore informal E-waste processing sites had majorly contributed to the bioaccumulation of metal contamination into the soil substrate. Similarly, in Guiyu, China, studies have reported 11 different categories of metals concentration (Be, Cr, Mn, Ni, Cu, Zn, As, Sb, Cd, Hg, and Pb) at E-waste processing sites, especially during the preliminary process of dismantling (Liu et al., 2005). The study conducted by Song et al. (2019) to determine the impact of E-waste contaminants on soil depth profile proved that metal concentration increases and total organic matter decreases in the soil depth profile of areas near the E-waste processing site. This impact on soil depth profile may cause groundwater contamination, metal uptake, and infertility in agricultural farmlands, as shown in Fig. 10.2.

10.3.4 Biota

Various plant species (trees, crops, grasses, and weeds) near E-waste processing sites are bound to accumulate a wide range of heavy metals. It may result in public health

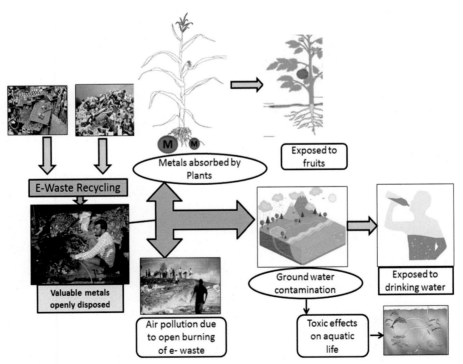

Figure 10.2 Effects on biodiversity due to heavy metal absorption.

implications due to the bioaccumulation of these metals in different environmental food chains. In some cases, dispersal and fall of plants leaf may also cause metal accumulation in natural biota such as soil. However, these metal species are highly influenced by different soil properties (pH, organic matter, and clay content plays). Similarly, metal distribution across different soil types and profiles, such as rooting depth, plays a vital role. Studies have proved that plant biota closer to E-waste processing sites has twice the rate of metal concentration when compared with farmland located outside the city (Wang et al., 2015). The uptake of these metals via plant roots from contaminated soils, water bodies, and contaminants deposition from the atmosphere onto plant surfaces can lead to plant contamination (Zhuang et al., 2009). Natural biota such as grassland and farms exposed to these E-waste processing sites via water and soil contamination can lead to chemical loading; intake of vegetative plants from these pastured land by diary ruminants can result in bioaccumulation of contaminants in tissues and animal-based by-products such as eggs and milk (Kierkegaard et al., 2007). Therefore it is essential to establish E-waste processing sites at a distance of >1 km from natural biota (Agyarko et al., 2010).

10.4 Case study on China's E-waste management

10.4.1 E-waste generation pattern

In 2016 more than 7.2 Mt of E-waste was produced in China, and it is estimated to surpass 15.5 Mt of E-waste by the year 2020. Based on this prediction, WEEE generation in China might reach 27 and 51 Mt by 2030 and 2050, respectively. Based on the five-year data collection, China Household Electric Appliance Research Institute (CHEARI) has identified an increase in five major EE appliances, such as washing machines, television, refrigerator, air conditioner, and computers. These appliances have contributed 4.06 and 5.38 million tons of WEEE in 2015 and 2017, respectively. Apart from this, nine different categories of household appliances have been identified based on their market demand and future availability. The overall E-waste generation was 62.4% from these five products and 36.4% from the remaining nine products. In 2017 mobile phones contributed 46% to overall E-waste generation, with more than 232 million units discarded per year. Recent technological advancements in monitors, TV screens, desktops, and portable devices such as fitness trackers, chargers, smart thermostats, power banks, and drones have also influenced the pattern of E-waste generation. Studies have suggested that the recycling and processing cost of these WEEE products require an investment of 42 billion dollars (Zeng et al., 2015). Therefore, to balance this change in E-waste generation, the Chinese government has implemented stringent laws and regulations regarding the nonacceptance of E-waste via import from foreign nations since January 1, 2018 (Fu et al., 2018).

10.4.2 Role of informal and formal sectors

In China formal and informal sectors are the primary E-waste handlers. However, the cost of processing was found to be higher in the case of formal sectors compared with its revenue collection from recycled EEE products. Therefore the lack of participation of private sectors in E-waste recycling has caused the open dumping of E-waste, which further led to the involvement of informal sectors. Before 2005, the decline in the formal sector in E-waste recycling has influenced informal sectors to participate in dismantling, precious metal extraction, and many other operations (Streicher-Porte & Geering, 2010). Small informal recyclers have transformed into formal recyclers due to external funding and technological advancement in the past few decades. In southeastern China Guangdong province has a small town of Guiya which has efficiently transformed itself into a formal E-waste recycler at the cost of human health and safety. The initial recycling activities in Guiya began in the late 1980s, and since the 1990s, many farmers and laborers have started to participate in strengthening their economy (He & Xu, 2014; Wang et al., 2020). However, the primary cause of environmental hazards in Guiya involved nonscientific E-waste recycling methods such as open burning and incineration, uncontrolled shredding and dismantling of E-waste components, acid washing, and other forms of hydraulic leaching. This has led to environmental contamination that has caused a threat to human and animal health (Puckett et al., 2002). Besides this, Taizhou, a small town located in Zhejiang Province of eastern China, has become a dominant town for E-waste recycling. More than 10% of residents have been involved in recycling practices since the 1970s (Chan & Wong, 2013). In the last few decades, children aged 10—20 have started to participate in these informal processing facilities in scrap yards and cottage industries. Therefore these sectors have given rise to an environmental and social problem, such as enlarging the gap between rich and poor, which might exaggerate social contradictions (Wilson et al., 2006). Most studies have reported that informal recyclers belong to poor groups and families engaged in E-waste recycling with a mindset to earn daily income. Besides, various start-ups with E-waste recycling facilities have emerged in the last few years, intending to generate a potential economy (Gu et al., 2010). As these recycling industries and start-ups have contributed to China's gross national product, the government has promoted such formal sectors to enhance their interest in the E-waste recycling industry to develop a full-time economic source (Schluep et al., 2009).

10.4.3 Treatment facilities

China has developed various E-waste processing units across 29 provinces within the central and eastern region (Song et al., 2019). This enhanced growth rate in formal treatment facilities has met the goal of preliminary recycling of E-waste. These newly established standard treatment units are mostly subsided and funded by government authorities focusing on enhancing mechanical treatment, resource recovery, and ultimate disposal of E-waste (Zeng et al., 2015). Besides this, the funded policy was implemented to enhance WEEE utilization for sustainable development (Wang et al., 2020). This initiative led by government authorities to prompt formal sectors to participate in comprehensive E-waste recycling through financial support will embrace a new boom. Broadly, China's E-waste treatment facilities have gone through four major stages of transformation: (1) Before 2005, formal and informal sectors were motivated to enhance on-site disassembling and disposal of E-waste components; (2) from 2005 to 2009, new enterprises and start-ups were funded to develop a formal treatment facility for household appliances; (3) from 2009 to 2011, based on the regulations formed by Household Appliances Old for New Rebate Program, more than 100 formal disassembly plants were established; (4) By 2015, more than 133 million units of WEEE were treated and processed by 109 formal enterprises established in China (Zeng et al., 2015). These four phases of transformation have made China efficient enough to manage its E-waste. It is essential for developing countries like India to motivate private enterprises and start-ups to participate in E-waste recycling through subsidies and funding schemes.

10.4.4 Impact on human health and safety

The E-waste component comprises hazardous organic and inorganic matters, which may cause a direct impact on human health residing near treatment facilities. In most cases, hazardous contaminants are exposed to the field workers and labors involved in informal E-waste recycling. Later, these contaminants get stored in fatty acids, resulting in health complications. This continuous cycle and transportation of contaminants may result in bioaccumulation, and this may cause contaminant exposure to remote areas. The working staff associated with this occupation are the primary victim of this contaminant exposure. In most cases, inhalation of toxic fumes and dust results in health hazardous. While, the lack of preventive measures such as barefooted working of staff and roaming of kids/animals was found to be predominant at E-waste processing sites (Ohajinwa et al., 2017). Besides this, people may get exposed to food-chain contamination through farms located near the E-waste treatment sites. The health hazards caused due to contaminant exposure can be categorized as inorganic and organic contaminants.

10.4.5 Impact of inorganic contaminants

Inorganic components such as carbon, iron, silicon, aluminum, tin, beryllium, and thermosetting plastics are the principal constituents of E-waste. Apart from this, also include a small proportion of heavy metals such as Hg, Pb, Cd, and thallium, and some amount of trace elements such as americium, antimony, arsenic, barium, bismuth, boron, cobalt, europium, gallium, germanium, gold, indium, lithium, manganese, nickel, niobium, palladium, platinum, rhodium, ruthenium, selenium, silver, tantalum, terbium, thorium, titanium, vanadium, and yttrium. The physio-chemical interaction of these listed inorganic components with natural biota may result in health hazards. During E-waste processing, workers and residents living near these processing sites have started exposure to these contaminants via biological pathways. The intake of this component is dependent on ingestion, inhalation, and absorption through the skin. The workers at mechanical processing sites have eight times higher heavy metals ingestion than those in manual E-waste dismantling sites. In different parts of China, studies have been conducted to understand the daily uptake of these components through the food supply chain. Studies carried at Taizhou suggested rice as a significant source of heavy metal intake (Fu et al., 2018). World Health Organization (WHO) and Food and Agriculture Organization (FAO) have defined the daily intake concentration for such metal. Fu et al. (2018) compared the results from these standards to evaluate their impact on human health. Pb and Cd intake from the rice was 3.7 and 0.7 mg/kg bw day, respectively, the lead uptake exceeded the FAO threshold standards of 3.6 mg/kg bw day, and cadmium uptake has reached 70% of standard threshold limit of 1 mg/kg bw day. This intake of metals can be from vegetables and other soil-food supply chains. A similar study was conducted for heavy metal intake from an E-waste site near Longtang town (Luo et al., 2007). It was found that Cd and Pb intake in most of the vegetables exceeded 4.7 and 2.6 times more compared to its threshold permissible limit. Generally, this intake of heavy metals is higher in the leafy vegetable plant than nonvegetative plants, and these green vegetable plants have rapid growth and transpiration rate. Therefore these plants favor maximum uptake of heavy metals from agricultural land, resulting in translocation in plant tissues. In addition, plants with broader leaves are more susceptible to physical contamination, such as dust accumulation. Studies have suggested that consuming vegetables with an intake of Pb and Cd might result in kidney damage, lung emphysema, damage to reproductive organs, osteomalacia, and osteoporosis disorders. Besides this, the study was carried out to determine the lead intake from dust and flume produced at printed circuit board recycling center located in Guiyu, approximately 175.7 mg/kg bw day (Leung et al., 2008). The oral average daily dose of Pb exceeded the required threshold limit. The impact of heavy metals was also predominant in children exposed to E-waste recycling areas. The study has reported concentrations of Cr, Mn, and Ni in 144 school students (age group of 8–13 years) residing near the E-waste recycling unit, resulting in inappropriate functioning of lungs (Zheng et al., 2013).

10.4.6 Impact of organic contaminants

In most cases, waste residue from E-waste processing sites is openly dumped and burned, which causes direct exposure of polyaromatic hydrocarbons (PAHs), PCBs, PBDEs, and dioxin into the ambient conditions. These contaminants are highly stable in the natural environment, resulting in bioaccumulation in living organisms through cyclic soil-food and water-food chains. People residing near E-waste recycling sites are exposed to PCBs via dietary (drinking water, fish, meat, etc.) and nondietary (inhalation and dermal exposure) intake (Meng et al., 2014). Studies have suggested that continuous exposure to PCBs can cause a higher cancer-causing index rate for areas closer to E-waste recycling centers (Meng et al., 2014). In the study conducted at Guiya open burning E-waste site, PBDEs concentration was highest (33,000—97,400 ng/g) compared to other organic pollutants. Similarly, concentration in soil and groundwater samples was 16,000 times higher compared to the control site. The soil samples collected from neighboring town Wenlin have the concentration of PCB at more than one hotspot area. It implies that PCBs in such regions are bound to travel faster to the nearby areas through the soil, water, and air media. In the last few decades, the concentration of PCBs has increased, which has caused hormonal disruption in children exposed to E-waste recycling units. Due to the inhalation of POPs, children aged 9—12 are highly prone to health hazards compared to adults. In comparison with controlled E-waste recycling sites, children working at E-waste recycling sites at Guiya are reported with higher serum concentrations of organ chlorine pesticide and PBDEs in blood vessels (Li et al., 2007). Similarly, the blood samples collected from Luqiao have the concentration of POPs in children residing near such recycling units. The concentration level of PCBs, dioxin, PBDE, and thyroid-stimulating hormone was found to be 484.00 ± 84.86 ng/g lipid weight, 26.00 ± 19.58 ng/g lipid weight, 664.28 ± 262.38 ng/g lipid weight, and 1.88 ± 0.42 µIU/mL, respectively (Li & Achal, 2020). While in areas with no E-waste recycling units, the pollutant index was found to be 255.38 ± 95 ng/g lipid weight, 39.64 ± 31.86 ng/g lipid weight, 375.81 ± 262.43 ng/g lipid weight, and 3.31 ± 1.04 µIU/mL, respectively (Li & Achal, 2020). It is conclusive that children with a higher intake of POPs might result in early-stage carcinogenicity in the liver, thyroid, and immune function; a neurological disorder such as lack of concentration and nerves break down can be identified.

10.4.7 Impact

Improper handling and uneven dumping of E-waste have caused environmental contamination. Contaminant exposure to soil, water, air, and dust has caused health issues and environmental hazards. Fig. 10.3 describes the impact of E-waste recycling activities on the ecological cycle. This process is driven by biogeochemical and physicochemical flux, later, resulting in leaching, atmosphere deposition, bioaccumulation of secondary products in plant uptake, biological and chemical degradation of natural resources.

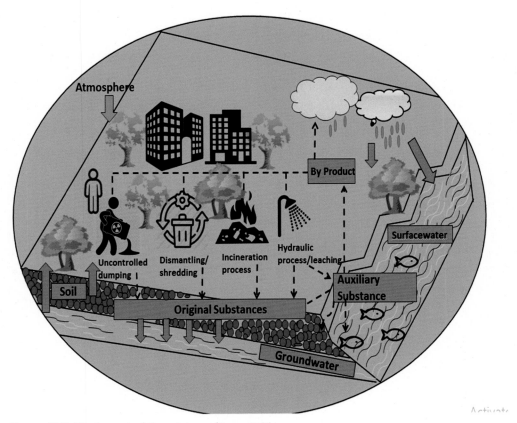

Figure 10.3 The impact of E-waste recycling activities.

10.4.8 Impact on soil substrate

Human activities have caused the discharge of harmful effluents and contaminants into the farmland from various processing industries. It has caused uptake of heavy metals and other toxic compounds from soil strata by vegetative plants (Kile et al., 2007). The plant roots can absorb and transfer these harmful contaminants (heavy metals) through stems and leaves, and soil is the intermedia source that allows contaminant transfer through the food chain. In most cases, the heavy metal concentration in soil substrate at control sites (no exposure to contaminant) was less when compared with areas located near E-waste recycling centers. In E-waste recycling, PCDD/Fs, PCBs, and PAHs are unintentionally released into the environment. In the soil samples collected from rice fields at Guiyu, the concentration of PCDD/Fs was found to be 2.73 ng/g and 11.7 pg·TEQ/g, respectively. However, after five years, the concentration of PCDD/Fs was found to be increased to 17.1 ng/g and 57 pg · TEQ/g. A similar study was initiated at Fengjiang town of Taizhou city in China, the concentration of PCBs and PCDD/Fs was found to be 167.7 ng/g and 271.0−932.5 pg/g of soil, respectively (Li et al., 2007). Besides this, the soil sample collected from an E-waste recycling center in Zhejiang Province, China, has the polybrominated biphenyl concentration of 27.18 ng/g. The range of concentration was higher than control sites (Zhao et al., 2008).

10.4.9 Impact on aquatic system

The leaching of harmful contaminants from E-waste recycling centers is a significant concern to aquatic life. The groundwater and surface water are the primary pathways to contaminant flow. Sediment samples collected from Nanyang River, near Guiyu town, had a PBDE concentration; this further led to bioaccumulation of PBDEs (766 ng/g) in carp present in river body (Luo et al., 2007). The continuous food chain dependence of top to bottom predators in an aquatic system allows bioaccumulation of harmful contaminants in living cells. Samples collected from top predators such as water snakes claimed higher concentrations of PBDEs (1091 ng/g) and PCBs (16,512 ng/g) (Wu et al., 2008). Similarly, the water sample collected from the downstream side of the river reported a higher concentration of PCB and PBDEs. Apart from this, heavy metal concentration in the downstream side of the water body near the E-waste recycling center at Guiyu town was 0.4 mg/L Pb, exceeding the threshold level of 0.05 mg/L Pb. Studies have reported the presence of heavy metals such as chromium, silver, molybdenum, lithium, tin, etc., in the streams situated near river Lianjiang (Wong et al., 2006). The Environmental Quality Standards for Surface Water reported that heavy metals concentration (Cd, Hg, Pb, and Cu) is higher in surface water near Guiyu town. In China drinking water has exceeded the heavy metal concentration of Ni beyond its threshold level. Sediment samples collected for different river streams exceeded the metal threshold level compared to its controlled sites.

10.4.10 Impact on air quality

The amount of dust and flumes released from E-waste recycling centers are the primary source of air pollution. The issue of open burning results in direct exposure of these contaminants to humans through inhalation, ingestion, and skin absorption. To validate the impact of these recycling centers on atmospheric exposure, air samples from Guiyu were collected to determine the presence of polychlorodibenzo-p-dioxins (65−2765 pg/m^3). Later, this was proved to be the highest dioxin level reported in China (Liu et al., 2008). While the open burning of E-waste in Guiyu has led to an increase in PBDEs concentration (16,575 pg/m^3), around 300 times higher concentration compared with Hong Kong city (situated 200 km away) (Deng et al., 2007). Besides this, the concentration of PBDEs varies in both day (11,000 pg/m^3) and night (5000 pg/m^3) (Chen et al., 2009). Since 2010, various E-waste recycling centers in China have increased the emission of POPs and particulate matter (PM) into the atmosphere. The air samples collected from multiple E-waste recycling sites with improper management and technological advancement have a high concentration of heavy metals and PM in atmospheric air. It has caused direct exposure of residents to these harmful contaminants residing near recycling centers (Gangwar et al., 2019). Some reports have suggested environmental pathways have caused an increase in PM, PCDD/Fs, PBDEs, and PCBs concentration in air ambient quality of areas residing near recycling centers (Zhang et al., 2017).

10.4.11 Impact on plants and foods

In the past few decades, the continuous release of POPs into the soil substrate have caused bioaccumulation, translocation, and transportation of toxic contaminant into the plant's cell (Liu et al., 2008; Shi et al., 2019). Compared with other toxic compounds, PBDEs tend to have a higher potential to translocate themselves into the plant cells through soils and sediments (Xu et al., 2013). The studies have proved that uptake of both organic and inorganic contaminants is significant in plants. The soil sample collected from these recycling sites has a reported concentration of 25,479 ng/g of PBDE. While plant species have also reported the concentration of PBDEs in leaves of bracken fern (*Pteridium aquilinum* L.) (144 ng/g), eastern daisy fleabane (*Erigeron annuus* L.) (326 ng/g), japanese dock (*Rumex japonicus* Houtt.) (278 ng/g), spider fern (*Pteris multifida* Poir.) (116 ng/g), and sorghum (*Sorghum bicolor* L.) (162 ng/g) (Yang et al., 2008). In China rice is a significant source of dietary constituents. The study was conducted on rice samples collected from farms located near the E-waste recycling center in Eastern China, Taizhou. The idea of the study was to determine the concentration of Pb and Cd in polished rice, which was 2−4 times higher than the threshold limit. The study conducted in the same town of China determined the presence of PBDE (18 ng/g) in poultry tissues and illustrated its impact on residents and the ecosystem (Fu et al., 2018). The rice field in Zhejiang provinces was also affected by contaminants because of nearby E-waste recycling facilities (Liang et al., 2008). Studies have also suggested the bioaccumulation of PCBs and halogenated flame retardants (HFRs) in forest area located near recycling facilities. The trace of HFRs and PCBs was found to be prevalent in Kingfisher bird of South China, while the concentration of PBDEs and PCBs ranged from 2.1×10^3 to 1.3×10^5 ng/g and 2.1×10^3 to 1.5×10^6 ng/g lipid mass, respectively (Liang et al., 2008). The lack of appropriate technology and management policies in China has caused a continuous hike in toxic contaminants in plants and animals. Therefore, in the past few years, the Chinese government has set certain buffer zones near E-waste recycling centers to prevent natural flora and fauna.

10.5 Conclusion

The general disposal of e-waste products has caused multiple environmental threats to human health and safety. Beside this, it has also caused an increase in deterioration of flora and fauna in the existing environment. The present chapter illustrates the fundamental penetration mechanism of toxic compounds and metals into the human body through inappropriate handling of e-waste. The study discusses the impact of these toxic compounds on different age groups residing near e-waste disposal sites. It was found that metals such as Pb and Cd are the major concern for aquatic biota, which results in bio-magnification. Finally, the discussed consequence of e-waste is interlinked to a case study, which highlights the factors influencing penetration of these toxic compounds into the environment. The case study covers essential areas of e-waste handling such as the collection process, treatment options, and government policies to mitigate future consequences.

References

Abrahams, P. W. (2002). Soils: Their implications to human health. *Science of The Total Environment, 291*, 1–32. Available from https://doi.org/10.1016/S0048-9697(01)01102-0.

Abu Hilal, A. H., & Ismail, N. S. (2008). Heavy metals in eleven common species of fish from the Gulf of Aqaba, Red Sea. *Jordan Journal of Biological Sciences, 1*.

Agyarko, K., Darteh, E., & Berlinger, B. (2010). Metal levels in some refuse dump soils and plants in Ghana. *Plant, Soil and Environment, 56*(2010), 244–251. Available from https://doi.org/10.17221/13/2010-PSE.

Akangbe Yekeen, T., Xu, X., Zhang, Y., Wu, Y., Kim, S., Reponen, T., Dietrich, K. N., Ho, S., Chen, A., & Huo, X. (2016). Assessment of health risk of trace metal pollution in surface soil and road dust from E-waste recycling area in China. *Environmental Science and Pollution Research International, 23*(17), 17511–17524. Available from https://doi.org/10.1007/s11356-016-6896-6.

Amirah, M. N., Afiza, A. S., Faizal, W. I. W., Nurliyana, M. H., & Laili, S. (2013). Human health risk assessment of metal contamination through consumption of fish. *Journal of Environment Pollution and Human Health, 1*(1), 1–5.

Andeobu, L., Wibowo, S., & Grandhi, S. (2021). An assessment of E-waste generation and environmental management of selected countries in Africa, Europe and North America: A systematic review. *Science of The Total Environment, 792*, 148078. Available from https://doi.org/10.1016/j.scitotenv.2021.148078.

Awasthi, A. K., Zeng, X., & Li, J. (2016a). Environmental pollution of electronic waste recycling in India: A critical review. *Environmental Pollution, 211*, 259–270.

Awasthi, A. K., Zeng, X., & Li, J. (2016b). Relationship between E-waste recycling and human health risk in India: A critical review. *Environmental Science and Pollution Research, 23*(12), 11509–11532.

Awasthi, A. K., Zlamparet, G. I., Zeng, X., & Li, J. (2017). Evaluating waste printed circuit boards recycling: Opportunities and challenges, a mini review. *Waste Management & Research, 35*(4), 346–356.

Branchi, I., Capone, F., Vitalone, A., Madia, F., Santucci, D., Alleva, E., & Costa, L. G. (2005). Early developmental exposure to BDE 99 or Aroclor 1254 affects neurobehavioural profile: Interference from the administration route. *Neurotoxicology, 26*, 183–192. Available from https://doi.org/10.1016/J.NEURO.2004.11.005.

Breivik, K., Armitage, J. M., Wania, F., & Jones, K. C. (2014). Tracking the global generation and exports of E-waste. Do existing estimates add up? *Environmental Science & Technology, 48*(15), 8735−8743.

Brigden, K. (2005). *Recycling of electronic wastes in China and India: Workplace and environmental contamination.* <http://www.greenpeace.org/raw/content/china/en/press/reports/recycling-of-electronic-wastes.pdf>.

Bryan, M. D., Atchison, G. J., & Sandheinrich, M. B. (2011). Effects of cadmium on the foraging behavior and growth of juvenile bluegill, *Lepomis* macrochirus. *Canadian Journal of Fisheries and Aquatic Sciences, 52*, 1630−1638. Available from https://doi.org/10.1139/F95-757.

Caussy, D. (2003). Case studies of the impact of understanding bioavailability: Arsenic. *Ecotoxicology and Environmental Safety, 56*(1), 164−173.

Cayumil, R., Khanna, R., Rajarao, R., Ikram-ul-Haq, M., Mukherjee, P. S., & Sahajwalla, V. (2016). Environmental impact of processing electronic waste − Key issues and challenges. *E-Waste in transition − From pollution to resource.* IntechOpen. Available from https://doi.org/10.5772/64139.

Chakraborty, P., Khuman, S. N., Selvaraj, S., Sampath, S., Devi, N. L., Bang, J. J., & Katsoyiannis, A. (2016). Polychlorinated biphenyls and organochlorine pesticides in River Brahmaputra from the outer Himalayan Range and River Hooghly emptying into the Bay of Bengal: Occurrence, sources and ecotoxicological risk assessment. *Environmental Pollution, 219*, 998−1006.

Chakraborty, P., Selvaraj, S., Nakamura, M., Prithiviraj, B., Ko, S., & Loganathan, B.G. (2016). E-waste and associated environmental contamination in the Asia/Pacific region (Part 1): An overview. In *Persistent organic chemicals in the environment: Status and trends in the Pacific Basin countries I Contamination Status* (pp. 127−138). American Chemical Society and Oxford University Press.

Chakraborty, P., Zhang, G., Cheng, H., Balasubramanian, P., Li, J., & Jones, K. C. (2017). Passive air sampling of polybrominated diphenyl ethers in New Delhi, Kolkata, Mumbai and Chennai: Levels, homologous profiling and source apportionment. *Environmental Pollution, 231*, 1181−1187.

Chakraborty, P., Zhang, G., Li, J., Selvaraj, S., Breivik, K., & Jones, K. C. (2016). Soil concentrations, occurrence, sources and estimation of air−soil exchange of polychlorinated biphenyls in Indian cities. *Science of the Total Environment, 562*, 928−934. Available from https://doi.org/10.1016/j.scitotenv.2016.03.009.

Chama, M. A., Amankwa, E. F., & Oteng-Ababio, M. (2014). Trace metal levels of the Odaw river sediments at the Agbogbloshie E-waste recycling site. *Journal of Science and Technology (Ghana), 34*, 1−8. Available from https://doi.org/10.4314/just.v34i1.1.

Chan, J. K. Y., & Wong, M. H. (2013). A review of environmental fate, body burdens, and human health risk assessment of PCDD/Fs at two typical electronic waste recycling sites in China. *Science of The Total Environment, 463−464*, 1111−1123. Available from https://doi.org/10.1016/J.SCITOTENV.2012.07.098.

Chen, D., Bi, X., Zhao, J., Chen, L., Tan, J., Mai, B., Sheng, G., Fu, J., & Wong, M. (2009). Pollution characterization and diurnal variation of PBDEs in the atmosphere of an E-waste dismantling region. *Environmental Pollution (Barking, Essex: 1987), 157*, 1051−1057. Available from https://doi.org/10.1016/J.ENVPOL.2008.06.005.

Cicik, B., & Engin, K. (2005). The effects of cadmium on levels of glucose in serum and glycogen reserves in the liver and muscle tissues of *Cyprinus carpio* (L., 1758). *Turkish Journal of Veterinary & Animal Sciences, 29*, 113−117.

Deng, W. J., Zheng, J. S., Bi, X. H., Fu, J. M., & Wong, M. H. (2007). Distribution of PBDEs in air particles from an electronic waste recycling site compared with Guangzhou and Hong Kong, South China. *Environment International, 33*, 1063−1069. Available from https://doi.org/10.1016/J.ENVINT.2007.06.007.

Ebrahimi, M., & Taherianfard, M. (2010). Concentration of four heavy metals (cadmium, lead, mercury, and arsenic) in organs of two cyprinid fish (*Cyprinus carpio* and *Capoeta* sp.) from the Kor River (Iran). *Environmental Monitoring and Assessment, 168*(1), 575−585.

Forti, V., Baldé, C. P., Kuehr, R., & Bel, G. (2020). *The global E-waste monitor 2020.* Bonn/Geneva/Rotterdam: United Nations University (UNU), International Telecommunication Union (ITU) & International Solid Waste Association (ISWA).

Frazzoli, C., Orisakwe, O. E., Dragone, R., & Mantovani, A. (2010). Diagnostic health risk assessment of electronic waste on the general population in developing countries' scenarios. *Environmental Impact Assessment Review, 30*(6), 388−399.

Fu, J., Zhang, H., Zhang, A., & Jiang, G. (2018). E-waste recycling in China: A challenging field. *Environmental Science & Technology, 52*, 6727–6728. Available from https://doi.org/10.1021/ACS.EST.8B02329.

Gangwar, C., Choudhari, R., Chauhan, A., Kumar, A., Singh, A., & Tripathi, A. (2019). Assessment of air pollution caused by illegal E-waste burning to evaluate the human health risk. *Environment International, 125*, 191–199. Available from https://doi.org/10.1016/J.ENVINT.2018.11.051.

Gu, Z., Feng, J., Han, W., Wu, M., Fu, J., & Sheng, G. (2010). Characteristics of organic matter in PM2.5 from an E-waste dismantling area in Taizhou, China. *Chemosphere, 80*, 800–806. Available from https://doi.org/10.1016/J.CHEMOSPHERE.2010.04.078.

Hayat, S., Javed, M., & Razzaq, S. (2007). Growth performance of metal stressed major carps viz. *Catla catla, Labeo rohita* and *Cirrhina mrigala* reared under semi-intensive culture system. *Pakistan Veterinary Journal, 27*, 8–12.

He, Y., & Xu, Z. (2014). The status and development of treatment techniques of typical waste electrical and electronic equipment in China: A review. *Waste Management & Research: The Journal of the International Solid Wastes and Public Cleansing Association, ISWA, 32*, 254–269. Available from https://doi.org/10.1177/0734242X14525824.

Hosoda, J., Ofosu-Anim, J., Sabi, E. B., Akita, L. G., Onwona-Agyeman, S., Yamashita, R., & Takada, H. (2014). Monitoring of organic micropollutants in Ghana by combination of pellet watch with sediment analysis: E-waste as a source of PCBs. *Marine Pollution Bulletin, 86*, 575–581. Available from https://doi.org/10.1016/J.MARPOLBUL.2014.06.008.

Huang, J., Nkrumah, P. N., Anim, D. O., & Mensah, E. (2014). E-waste disposal effects on the aquatic environment: Accra, Ghana. *Reviews of Environmental Contamination and Toxicology, 229*, 19–34. Available from https://doi.org/10.1007/978-3-319-03777-6_2.

John, R., Ahmad, P., Gadgil, K., & Sharma, S. (2009). Heavy metal toxicity: Effect on plant growth, biochemical parameters and metal accumulation by *Brassica juncea* L. *International Journal of Plant Production, 3*, 65–75.

Kabata-Pendias, A. (1992). Trace metals in soils in Poland—Occurrence and behaviour. *Soil Science, 140*, 53–70.

Kanmani, S., & Gandhimathi, R. (2013). Assessment of heavy metal contamination in soil due to leachate migration from an open dumping site. *Applied Water Science, 3*(1), 193–205.

Kiddee, P., Naidu, R., & Wong, M. H. (2013). Electronic waste management approaches: An overview. *Waste Management (New York, N.Y.), 33*, 1237–1250. Available from https://doi.org/10.1016/J.WASMAN.2013.01.006.

Kierkegaard, A., Asplund, L., de Wit, C. A., McLachlan, M. S., Thomas, G. O., Sweetman, A. J., & Jones, K. C. (2007). Fate of higher brominated PBDEs in lactating cows. *Environmental Science & Technology, 41*(2), 417–423.

Kile, M. L., Houseman, E. A., Breton, C. V., Smith, T., Quamruzzaman, Q., Rahman, M., Mahiuddin, G., & Christiani, D. C. (2007). Dietary arsenic exposure in Bangladesh. *Environmental Health Perspectives, 115*, 889–893. Available from https://doi.org/10.1289/EHP.9462.

Kumar, A., Holuszko, M., & Espinosa, D. C. R. (2017). E-waste: An overview on generation, collection, legislation and recycling practices. *Resources, Conservation and Recycling, 122*, 32–42.

Kumar, J., & Kumar, S. (2013). Environmental impact assessment and bioleaching of metals from electronic waste (E-waste).

Law, R.J., & Herzke, D. (2010). Current levels and trends of brominated flame retardants in the environment. In *Brominated flame retardants* (pp. 123–140). Springer. <https://doi.org/10.1007/698_2010_82>.

Lema, S. C., Schultz, I. R., Scholz, N. L., Incardona, J. P., & Swanson, P. (2007). Neural defects and cardiac arrhythmia in fish larvae following embryonic exposure to 2,2′,4,4′-tetrabromodiphenyl ether (PBDE 47). *Aquatic Toxicology (Amsterdam, Netherlands), 82*, 296–307. Available from https://doi.org/10.1016/J.AQUATOX.2007.03.002.

Leonard, S. S., Harris, G. K., & Shi, X. (2004). Metal-induced oxidative stress and signal transduction. *Free Radical Biology & Medicine, 37*, 1921–1942. Available from https://doi.org/10.1016/J.FREERADBIOMED.2004.09.010.

Leung, A. O. W. (2019). Environmental contamination and health effects due to E-waste recycling. *Electronic waste management and treatment technology* (pp. 335–362). Elsevier.

Leung, A. O. W., Duzgoren-Aydin, N. S., Cheung, K. C., & Wong, M. H. (2008). Heavy metals concentrations of surface dust from E-waste recycling and its human health implications in southeast China. *Environmental Science & Technology, 42*, 2674–2680. Available from https://doi.org/10.1021/ES071873X.

Li, H., Yu, L., Sheng, G., Fu, J., & Peng, P. (2007). Severe PCDD/F and PBDD/F pollution in air around an electronic waste dismantling area in China. *Environmental Science & Technology, 41*, 5641–5646. Available from https://doi.org/10.1021/ES0702925.

Li, J., Liu, L., Ren, J., Duan, H., & Zheng, L. (2012). Behavior of urban residents toward the discarding of waste electrical and electronic equipment: A case study in Baoding, China. *Waste Management & Research, 30*(11), 1187–1197.

Li, W., & Achal, V. (2020). Environmental and health impacts due to E-waste disposal in China — A review. *Science of the Total Environment, 737*, 139745.

Li, Z., Ma, Z., van der Kuijp, T. J., Yuan, Z., & Huang, L. (2014). A review of soil heavy metal pollution from mines in China: Pollution and health risk assessment. *Science of The Total Environment, 468*, 843–853.

Liang, S. X., Zhao, Q., Qin, Z. F., Zhao, X. R., Yang, Z. Z., & Xu, X. B. (2008). Levels and distribution of polybrominated diphenyl ethers in various tissues of foraging hens from an electronic waste recycling area in South China. *Environmental Toxicology and Chemistry/SETAC, 27*, 1279–1283. Available from https://doi.org/10.1897/07-518.1.

Liang, Y., Sun, W., Zhu, Y. G., & Christie, P. (2007). Mechanisms of silicon-mediated alleviation of abiotic stresses in higher plants: A review. *Environmental Pollution (Barking, Essex: 1987), 147*, 422–428. Available from https://doi.org/10.1016/J.ENVPOL.2006.06.008.

Liu, H., Probst, A., & Liao, B. (2005). Metal contamination of soils and crops affected by the Chenzhou lead/zinc mine spill (Hunan, China. *Science of the Total Environment, 339*, 153–166. Available from https://doi.org/10.1016/J.SCITOTENV.2004.07.030.

Liu, H., Zhou, Q., Wang, Y., Zhang, Q., Cai, Z., & Jiang, G. (2008). E-waste recycling induced polybrominated diphenyl ethers, polychlorinated biphenyls, polychlorinated dibenzo-p-dioxins and dibenzo-furans pollution in the ambient environment. *Environment International, 34*, 67–72. Available from https://doi.org/10.1016/J.ENVINT.2007.07.008.

Luo, Q., Wong, M., & Cai, Z. (2007). Determination of polybrominated diphenyl ethers in freshwater fishes from a river polluted by E-wastes. *Talanta, 72*, 1644–1649. Available from https://doi.org/10.1016/J.TALANTA.2007.03.012.

Machender, G., Dhakate, R., Prasanna, L., & Govil, P. K. (2011). Assessment of heavy metal contamination in soils around Balanagar industrial area, Hyderabad, India. *Environmental Earth Sciences, 63*(5), 945–953.

Mager, E. M., Brix, K. V., & Grosell, M. (2010). Influence of bicarbonate and humic acid on effects of chronic waterborne lead exposure to the fathead minnow (*Pimephales promelas*). *Aquatic Toxicology (Amsterdam, Netherlands), 96*, 135–144. Available from https://doi.org/10.1016/J.AQUATOX.2009.10.012.

Meng, M., Li, B., Shao, J. J., Wang, T., He, B., Shi, J. B., Ye, Z. H., & Jiang, G. B. (2014). Accumulation of total mercury and methylmercury in rice plants collected from different mining areas in China. *Environmental Pollution (Barking, Essex: 1987), 184*, 179–186. Available from https://doi.org/10.1016/J.ENVPOL.2013.08.030.

Mhadhbi, L., Fumega, J., Boumaiza, M., & Beiras, R. (2010). Lethal and sublethal effects of polybrominated diphenyl ethers (PBDEs) for turbot (*Psetta maxima*) early life stage (ELS). *Nature Precedings.* <https://doi.org/10.1038/npre.2010.4656.2>.

Moletsane, R. I., & Venter, C. (2018). Electronic waste and its negative impact on human health and the environment. In 2018 international conference on advances in big data, computing and data communication systems (icABCD)(pp. 1–7). <https://doi.org/10.1109/ICABCD.2018.8465473>.

Monika, J. K. (2010). E-waste management: As a challenge to public health in India. *Indian Journal of Community Medicine: Official Publication of Indian Association of Preventive & Social Medicine, 35*(3), 382.

Nammalwar, P. (1983). Heavy metals pollution in the marine environment. *Science Reporter, 20*(3) 158–160.

Needhidasan, S., Samuel, M., & Chidambaram, R. (2014). Electronic waste — An emerging threat to the environment of urban India. *Journal of Environmental Health Science and Engineering, 12*, 1–9. Available from https://doi.org/10.1186/2052-336X-12-36.

OECD. (2003). <https://www.oecd-ilibrary.org/economics/oecd-annual-report-2003_annrep-2003-en>.

Ohajinwa, C. M., Van Bodegom, P. M., Vijver, M. G., & Peijnenburg, W. J. (2017). Health risks awareness of electronic waste workers in the informal sector in Nigeria. *International Journal of Environmental Research and Public Health, 14*(8), 911.

Oteng-Ababio, M. (2012). Electronic waste management in Ghana — Issues and practices. In *Sustainable development — Authoritative and leading edge content for environmental management.* IntechOpen.

Pradhan, J. K., & Kumar, S. (2014). Informal E-waste recycling: environmental risk assessment of heavy metal contamination in Mandoli industrial area, Delhi, India. *Environmental Science and Pollution Research, 21*(13), 7913−7928. Available from https://doi.org/10.5772/45884.

Puckett, J., Leslie Byster, B., Sarah Westervelt, S., Richard Gutierrez, B., Sheila Davis, B., Asma Hussain, M., Madhumitta Dutta, S., & Link India, T. (2002). Exporting harm.

Rashed, M. N. (2001). Cadmium and lead levels in fish (*Tilapia nilotica*) tissues as biological indicator for lake water pollution. *Environmental Monitoring and Assessment, 68*, 75−89.

Rioboo, C., O'Connor, J. E., Prado, R., Herrero, C., & Cid, Á. (2009). Cell proliferation alterations in *Chlorella* cells under stress conditions. *Aquatic Toxicology (Amsterdam, Netherlands), 94*, 229−237. Available from https://doi.org/10.1016/J.AQUATOX.2009.07.009.

Robinson, B. H. (2009). E-waste: An assessment of global production and environmental impacts. *Science of The Total Environment, 408*, 183−191. Available from https://doi.org/10.1016/J.SCITOTENV.2009.09.044.

Schluep, M., Hagelüken, C., Meskers, C., Magalini, F., Wang, F., Müller, E., Kuehr, R., Maurer, C., & Sonnemann, G. (2009). Market potential of innovative E-waste recycling technologies in developing countries. In *R'09 world congress*, Davos, Switzerland.

Shi, J., Xiang, L., Luan, H., Wei, Y., Ren, H., & Chen, P. (2019). The health concern of polychlorinated biphenyls (PCBs) in a notorious E-waste recycling site. *Ecotoxicology and Environmental Safety, 186*, 109817. Available from https://doi.org/10.1016/J.ECOENV.2019.109817.

Sindiku, O., Babayemi, J. O., Tysklind, M., Osibanjo, O., Weber, R., Watson, A., Schlummer, M., & Lundstedt, S. (2015). Polybrominated dibenzo-p-dioxins and dibenzofurans (PBDD/Fs) in E-waste plastic in Nigeria. *Environmental Science and Pollution Research, 22*(19), 14515−14529.

Solomon, F. (2008). Impacts of metals on impacts of metals on impacts of metals on impacts of metals on aquatic ecosystems and.

Song, X., Lu, B., & Wu, W. (2019). Environmental management of E-waste in China. *Electronic waste management and treatment technology* (pp. 285−310). Butterworth-Heinemann. Available from https://doi.org/10.1016/B978-0-12-816190-6.00013-3.

Steinbacher, M., Henne, S., Emmenegger, L., Buchmann, B., & Hüglin, C. (2009). National Air Pollution Monitoring Network (NABEL). Empa, Swiss Federal Laboratories for Materials Science and Technology. <https://doi.org/10.5194/acp-3-2217-2003>.

Streicher-Porte, M., & Geering, A. C. (2010). Opportunities and threats of current E-waste collection system in China: A case study from Taizhou with a focus on refrigerators, washing machines, and televisions. *Environmental Engineering Science, 27*, 29−36. <https://doi.org/10.1089/EES.2009.0134>.

Strong, G., & Eng, P. (2006). Dillon consulting limited.

UNEP. (2018). *The United Nations and E-waste: System-wide action on addressing the full life-cycle of electrical and electronic equipment.*

UNU (United Nations University)/Step Initiative. (2014). *Solving the E-waste problem (step) white paper: One global definition of E-waste.* Bonn: UNU.

Wang, J., Liu, L., Wang, J., Pan, B., Fu, X., Zhang, G., Zhang, L., & Lin, K. (2015). Distribution of metals and brominated flame retardants (BFRs) in sediments, soils and plants from an informal E-waste dismantling site, South China. *Environmental Science and Pollution Research, 22*(2), 1020−1033.

Wang, K., Qian, J., & Liu, L. (2020). Understanding environmental pollutions of informal E-waste clustering in Global South via multi-scalar regulatory frameworks: A case study of Guiyu Town, China. *International Journal of Environmental Research and Public Health, 17*, 2802. Available from https://doi.org/10.3390/IJERPH17082802.

Wang, J. P., & Guo, X. K. (2006). Impact of electronic wastes recycling on environmental quality. *Biomedical and Environmental Sciences: BES, 19*, 137−142.

Wilson, D. C., Velis, C., & Cheeseman, C. (2006). Role of informal sector recycling in waste management in developing countries. *Habitat International, 30*, 797−808. Available from https://doi.org/10.1016/J.HABITATINT.2005.09.005.

Wollenberger, L., Dinan, L., & Breitholtz, M. (2005). Brominated flame retardants: Activities in a crustacean development test and in an ecdysteroid screening assay. *Environmental Toxicology and Chemistry/SETAC, 24*, 400−407. Available from https://doi.org/10.1897/03-629.1.

Wong, C. S. C., Li, X., & Thornton, I. (2006). Urban environmental geochemistry of trace metals. *Environmental Pollution (Barking, Essex: 1987), 142*, 1−16. Available from https://doi.org/10.1016/J.ENVPOL.2005.09.004.

Wong, M. H., Wu, S. C., Deng, W. J., Yu, X. Z., Luo, Q., Leung, A. O. W., Wong, C. S. C., Luksemburg, W. J., & Wong, A. S. (2007). Export of toxic chemicals − A review of the case of uncontrolled electronic-waste recycling. *Environmental Pollution, 149*(2), 131−140.

Wu, J. P., Luo, X. J., Zhang, Y., Luo, Y., Chen, S. J., Mai, B. X., & Yang, Z. Y. (2008). Bioaccumulation of polybrominated diphenyl ethers (PBDEs) and polychlorinated biphenyls (PCBs) in wild aquatic species from an electronic waste (E-waste) recycling site in South China. *Environment International, 34*, 1109−1113. Available from https://doi.org/10.1016/J.ENVINT.2008.04.001.

Xu, P., Tao, B., Li, N., Qi, L., Ren, Y., Zhou, Z., Zhang, L., Liu, A., & Huang, Y. (2013). Levels, profiles and source identification of PCDD/Fs in farmland soils of Guiyu, China. *Chemosphere, 91*, 824−831. Available from https://doi.org/10.1016/J.CHEMOSPHERE.2013.01.068.

Yang, Z. Z., Zhao, A. X. R., Zhao, A. Q., Qin, A. Z. F., Qin, A. X. F., Xu, A. X. B., Jin, A. Z. X., & Xu, A. C. X. (2008). Polybrominated diphenyl ethers in leaves and soil from typical electronic waste polluted area in South China. *Bulletin of Environmental Contamination and Toxicology, 80*, 340−344 <https://doi.org/10.1007/s00128-008-9385-x>.

Yu, L., He, W., Li, G., Huang, J., & Zhu, H. (2014). The development of WEEE management and effects of the fund policy for subsidizing WEEE treating in China. *Waste Management, 34*(9), 1705−1714.

Zeng, X., Gong, R., Chen, W.-Q., & Li, J. (2015). Uncovering the recycling potential of "New" WEEE in China. *Environmental Science and Technology*. <https://doi.org/10.1021/acs.est.5b05446>.

Zhang, B., Huo, X., Xu, L., Cheng, Z., Cong, X., Lu, X., & Xu, X. (2017). Elevated lead levels from E-waste exposure are linked to decreased olfactory memory in children. *Environmental Pollution (Barking, Essex: 1987), 231*, 1112−1121. Available from https://doi.org/10.1016/J.ENVPOL.2017.07.015.

Zhao, G., Wang, Z., Dong, M. H., Rao, K., Luo, J., Wang, D., Zha, J., Huang, S., Xu, Y., & Ma, M. (2008). PBBs, PBDEs, and PCBs levels in hair of residents around E-waste disassembly sites in Zhejiang Province, China, and their potential sources. *Science of the Total Environment, 397*, 46−57. Available from https://doi.org/10.1016/J.SCITOTENV.2008.03.010.

Zheng, G., Xu, X., Li, B., Wu, K., Yekeen, T. A., & Huo, X. (2013). Association between lung function in school children and exposure to three transition metals from an E-waste recycling area. *Journal of Exposure Science & Environmental Epidemiology, 23*, 67−72. Available from https://doi.org/10.1038/jes.2012.84.

Zhuang, J., Liu, Y., Wu, Z., Sun, Y., & Lin, L. (2009). Hydrolysis of wheat straw hemicellulose and detoxification of the hydrolysate for xylitol production. *BioResources, 4*(2), 674−686.

CHAPTER 11

E-waste policies, regulation and legislation in developed and developing countries

Rahul Rautela[1,2] and Bholu Ram Yadav[1,2]
[1]CSIR–National Environmental Engineering Research Institute (CSIR–NEERI), Nagpur, Maharashtra, India
[2]Academy of Scientific and Innovative Research (AcSIR), Ghaziabad, Uttar Pradesh, India

11.1 Introduction

During the COVID-19 pandemic, the world has experienced the biggest economic shutdown in decades. As a consequence, waste management has become a serious issue as people face difficulties to manage or dispose of their waste. According to the report of United Nation University Sustainable Cycles programme, Bonn, and the United Nations Institute for Training and Research, the electrical and electronic equipment (EEE) consumption in the first three quarters of 2020 was reduced due to the COVID-19 pandemic; as a result, around 4.9 million metric tons (Mt) of E-waste generation was reduced in 2020 (Baldé & Kuehr, 2021). Furthermore, the worldwide lockdowns due to the pandemic had imposed a radical shift in people's preferences for online platforms such as work from home, online shopping, home schooling, and online meetings. As a result, the consumption of small electronic gadgets like laptops, smartphones, electrical ovens, and game consoles, was increased by 0.3 Mt when compared to previous years.

The management of heterogenous waste like E-waste requires a well-organized management plan with stringent legislation. Also, proper infrastructure and an eco-friendly process are required to ensure sustainable management of E-waste (Dhir et al., 2021). Legislation may provide the framework to encourage cooperation and integration of informal sectors to formal sectors, to ensure the role of the informal sector (Vishwakarma et al., 2022). Legislation in E-waste management plays a crucial role in every stage of E-waste life cycle from manufacturing (EEE) to recycling. As of 2019 only 78 countries have enforced legislation, policy, and regulation for E-waste, which covers around 71% of the world population (Forti et al., 2020). The amount of global E-waste generation in 2019 was 53.6 Mt, which is expected to increase up to 74.7 Mt by 2030. Moreover, only 17.4% of global E-waste was treated formally and the remaining were dumped (Forti et al., 2020).

Global E-waste Management Strategies and Future Implications
DOI: https://doi.org/10.1016/B978-0-323-99919-9.00004-0
209

It is the need of the hour to implement and enforce the legislation for managing the E-waste and defining environment-friendly and suitable commercial strategies toward its recycling (Arya et al., 2021; Johri, 2008; Uddin, 2012). Most of the countries have drafted rules and regulations toward the management of E-waste systematically. However, waste management practices remain difficult due to ineffective implementation of the current legislative framework and social awareness initiatives, as well as workers' lack of understanding of occupational safety. (Arya & Kumar, 2020a). The conflicts with the E-waste management system in developing countries as compared to the European Union (EU) and Japan, which have already well-evolved consumer behavior and improvement in policies. Moreover, the illegal trading or transboundary movement of E-waste at the local and international level is a serious problem for E-waste management. As reported in the study by Global E-waste Monitor 2017, only 4 billion people are covered by the legislation related to E-waste management (Baldé et al., 2017). Therefore the need to implement effective strategies and promote the reuse, rehabilitation, or recycling of E-waste in special facilities are crucial to prevent contamination of the environment and the risks to human health (Fig. 11.1).

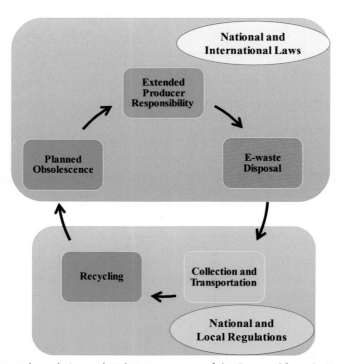

Figure 11.1 Rules and regulations related to every stage of the E-waste life cycle (Pont et al., 2019).

11.2 Policies and legislations

Globally, E-waste has become a serious concern as the amount of E-waste is increasing day by day. Therefore rules and regulations play a major role in managing E-waste across the world. Several policies and amendments were implemented across the world over the last few decades. Major laws and conventions related to E-waste are Waste Electrical and Electronic Equipment (WEEE) Directives, RoHS Directives, Battery Directives, Basel convention, extended producer responsibility (EPR), etc., which are designed to ensure the smoothening of E-waste management and lessen the associated risks of improper handling of E-waste. In the 1970s and 1980s, harmful waste like E-waste as used or second-hand products was transported from developed to developing countries. Following the implementation of the Basel Convention in 1992, transboundary movement of hazardous wastes was restricted due to the ban on the transportation or trading of e-waste from developed to developing countries (Shinkuma & Huong, 2009). Major policies and directives are discussed in the following sections and in Table 11.1:

11.2.1 WEEE Directives

The WEEE Directive came into force in 2003 by EU to ensure or regulate E-waste by environmentally sound techniques. The goal of the WEEE Directive is to reduce the amount of E-waste generation by promoting 3R's (reuse, reduce, and recycle) and ensure maximum resource recovery from E-waste (Shittu et al., 2021). It ensures the environmental performance of all stakeholders associated with the EEE life cycle, for example, producers, stakeholders, distributors, consumers, and recyclers which are particularly directly involved in the treatment of E-waste. The main principle of the WEEE Directive (Directive 2002/96/EC) was to implement EPR which mandates the producers to collect the E-waste after its end-of-life (EOL) from consumers (European Union, 2002).

11.2.2 RoHS Directives

RoHS stands for the restriction of hazardous substances, the first RoHS Directive or Directive 2002/95/EC came into force on January 27, 2003 in EU, to restrict the use of certain hazardous substances in the manufacturing of EEE products for the protection of human health and the environment. RoHS restricts the use of six substances (Hg, Cd, Cr (VI), Pb, PBDE, and PBB) in all EEE (Rautela et al., 2021). This directive is also focused on the substitution of harmful substances and the use of eco-friendly alternatives to ensure safety and minimize risks. RoHS 2 or Directive 2011/65/EU was introduced in July 2011, which covers all types of EEE and also a CE-marking directive for the EEE. In RoHS 3 or Directive 2015/863, bans on the use of four new substances (dibutyl phthalate, benzyl butyl phthalate, diisobutyl

Table 11.1 Existing E-waste rules and regulations.

S. No.	Rules and regulations	Remarks	Date of implementation	Status
1.	WEEE Directive '2012/19/EC (Amendment of WEEE Directive 2002/96/EC)	Establishes E-waste collection schemes, enhancement of 4Rs, and reduce the deleterious impact on human health and surroundings	August 13, 2012	Implement February 13, 2013
2.	RoHS 3 (EU 2015/863) (Amendment of RoHS 2002/95/EC)	Adds a ban on four new restricted substances: diisobutyl phthalate, dibutyl phthalate, benzyl butyl phthalate, and bis(2-ethylhexyl) phthalate, prohibiting the use of harmful substances in the manufacturing of different types of EEEs	March 31, 2015	This restriction has foreseen to come under effect until July 22, 2021
3.	Registration, Evaluation, Authorization, and Restriction of Chemicals (REACH) (EC 1907/2006)	Ensures a higher level of security for human health and the environment, while at the same time promoting competition, creativity, and the free exchange of substances on the internal market	April 2007	—
4.	Extended Producer Responsibility (EPR)	Extend the responsibilities of producers after the life cycle of EEE products to enhance the treatment or disposal of EEE products. Also, EPR is liable for financial and/or physical responsibility to producers to minimize the amount of waste and improve the design of products for the environment	Introduced in 1990 by the Swedish Ministry of the Environment	–

No.		Description		
5.	Polluter pays principle	Stipulates that any personnel or firm engaged directly or indirectly in polluting or contaminating the environment, posing threat to human health and safety is liable to bear the cost and expenses toward the damage	It was introduced in 1972 by the Organization for Economic Cooperation and Development (OECD)	It is mentioned in Rio Declaration on Environment and Development of 1992
6.	Basel Convention on the Control of Transboundary Movements of Hazardous Wastes and Their Disposal	Ensures minimize the wastes at source and transfer or trading of hazardous wastes between international level to reduce the impacts of hazardous wastes	March 22, 1989	This convention of 187 parties entered into force on May 5, 1992.
7.	Basel Ban Amendment	An agreement signed by Basel Convention parties to ban on exporting hazardous wastes from the members of the European Union (EU), OECD, and Liechtenstein to other countries	December 5, 2019	This amendment has been adopted by 99 parties
8.	Bamako Convention	This treaty prohibits the import of all hazardous and radioactive wastes into the African continent	Bamako Convention	
9.	Battery Directives 2006/66/EC	This directive sets the maximum limits of certain hazardous metals and chemicals in batteries. Also, it ensures the proper management of batteries and improvement of the performance of batteries	September 6, 2006	The directive was amended by Directive 2013/56/EU of November 20, 2013

EEE, electrical and electronic equipment; *WEEE*, waste electrical and electronic equipment.
Source: Adapted from Rautela, R., Arya, S., Vishwakarma, S., Lee, J., Kim, K. H., & Kumar, S. (2021). E-waste management and its effects on the environment and human health. *Science of The Total Environmental, 773*, 145623.

phthalate, bis(2-ethylhexyl) phthalate) in EEE (RoHS, 2011). RoHS regulation covers only EEE products, whereas regulations such as REACH (Registration, Evaluation, Authorization, Restriction of Chemicals) monitor the production and use of 197 substances that are substances of very high concern in different industries.

11.2.3 Extended producer responsibility

To streamline the E-waste management, the concept of EPR was introduced in 1972 by the OECD, which extends the responsibilities of producers to overlook the EEE after its EOL (E-waste). As per OECD, EPR is environmental legislation that is based on extending the responsibilities of the producers to look after the E-waste after the product's postconsumer stage (Méndez-Fajardo et al., 2020; Tran & Salhofer, 2018; Porwal & Chatterjee, 2019). EPR system widens the responsibilities of producers from product manufacturing to its postrecycling. The key points of EPR principles are:

1. To design eco-friendly products, to improve design and performance of EEE products.
2. Achieving a high product utilization rate.
3. Preserving materials through effective and environmentally sound collection, treatment, and reuse, and recycling.

However, the manufacturers or EPR mandates the product take-back system by three approaches: (1) product take-back mandate and recycling rate targets, (2) voluntary product take-back mandate and recycling rate targets, and (3) mandatory take-back and targets with a tradable recycling scheme (Gupt & Sahay, 2015). According to Lindhqvist (2000), for the implementation of a successful EPR model, five major responsibilities are essential, namely, liability, economic, physical, informative, and owner responsibilities. EPR-based regulations in different countries are discussed in Table 11.2.

11.2.4 Battery Directives

Battery Directives were introduced by EU in 2006 to protect the harmful impacts of waste batteries and accumulators into the environment. This directive implies all types of batteries and accumulators irrespective of their design, size, and nature. It prohibits the manufacturing, marketing, and use of waste batteries containing hazardous substances. Moreover, this directive aims to improve the environmental performance of batteries and accumulator's life cycle by connecting all the associated operators such as manufacturers, distributors, consumers, and recyclers. Producers of batteries and accumulators and other products that incorporate batteries are responsible for the management of the waste batteries and accumulators after EOL (Battery Directives, 2006).

Table 11.2 EPR-based regulations in different countries.

Countries	EPR-based regulations	Producer responsibility	Stakeholders responsibility
Netherlands	The Management of White and Brown Goods Decree, 1998	Define recycling targets and charge advance fees, i.e., charged from consumers	Consumer responsibilities to return the E-waste to collectors/retailers and PRO collect the E-waste from collectors/retailers
Japan	Law for the Promotion of Effective Utilization of Resources (LPUR) and Law for the Recycling of Specified Kinds of Home Appliances (LRHA)	Under LPUR, producers have voluntary participation in recycling with no recycling targets, whereas LRHA assigned mandatory participation with fixed recycling targets	Consumers have to return their E-waste directly to the producers physically or via post offices
Switzerland	Ordinance on 'The Return, the Taking Back, and the Disposal of Electrical and Electronic Appliances' (ORDEA)	Advanced recycling fee (ARF) collected is used to pay the fee to PRO	ARF was charged by retailers and PRO are responsible for recycling activities
South Korea	Producer Recycling system in 2003	Recycling fees are paid to PRO or commercial recyclers and recycling targets are up to 55%–70%	Either consumers sell to dealers directly or pay a fee to the municipalities to collect the waste
United States	The Electronic Waste Recycling Act, 2003	Producer responsibilities are not defined	Consumers have to pay recycling fees to retailers, which are paid to authorized recyclers
Taiwan	The Recycling Fund Management Committee (RFMC) introduced in Taiwan in 1998	Producers have no recycling targets and pay recycling fees to the government for recycling funds	Consumers are free to choose the disposal route and particularly designated recyclers are responsible for recycling
Colombia	Four specific resolutions focusing on WEEE streams	Producers defined the fixed targets and were responsible for financing the collection and treatment of E-waste	Consumers are responsible to separate E-waste from municipal solid waste (MSW) and return to collectors
India	E-waste Management Rules, 2016	PRO holds the liability of E-waste collection and also ensures its eco-friendly channelization	Consumers either sell to dealers or recyclers directly
China	Regulation on the Administration of the Recovery and Disposal of WEEE	No fixed recycling targets are mentioned by producers	Consumers return the waste directly or through informal collectors

PRO, producer responsible organization; WEEE, waste electrical and electronic equipment; MSW, municipal solid waste; ORDEA, Ordinance on "The Return, the Taking Back,old and the Disposal of Electrical and Electronic Appliances; RFMC, Recycling Fund Management Committee.
Source: Adapted from Gupt, Y., & Sahay, S. (2015). Review of extended producer responsibility: A case study approach. Waste Management & Research, 33(7), 595–611.

11.3 International processes and agreement conventions

Several international and regional agreements and conventions were discussed in Table 11.3 that are introduced to regulate E-waste management systematically in order to protect human health and the environment.

Table 11.3 E-waste associated with international bodies and conventions.

Rules and regulation	Remarks	Year
Basel Convention on the Control of Transboundary Movements of Hazardous Wastes and their Disposal	The goal of this convention is to protect human health and the environment from hazardous and other wastes. The main focus of this convention is to use eco-friendly management techniques, good waste disposal practices, and reduce transboundary movements. In Annex VIII of this convention, E-waste was added in the fourth meeting of the conference of the parties (Decision IV/9) in 1998. The ninth meeting of the conference of the parties to the Basel Convention in 2006 mandated to implement the work plan for the environmentally sound management of E-waste.	1989
Convention on Chemicals, concerning safety in the use of chemicals at work, International Labour Organization	This convention provides the right of the worker to know the nature of work and knowledge about the materials used at work and employers ensure proper standards training to workers and safety measures.	1990
Council Decision Waste Agreement, OECD	This agreement aims to prevent the illegal trading and transboundary movement of waste to recover valuable materials.	1992
United Nations Framework Convention on Climate Change (UNFCCC)	UNFCCC is not directly involved in the E-waste sector, despite their green initiatives such as teach or train waste pickers to collect, recycle, and dispose of different types of E-waste. This initiative provides green jobs and knowledge of safety measures from the risks of exposure to hazardous materials.	1994

(Continued)

Table 11.3 (Continued)

Rules and regulation	Remarks	Year
Rotterdam Convention	The convention promotes the shared responsibilities in the imports and exports of certain hazardous chemicals at the international level with the purpose of environmental and human health protection. Also, exporters should mention safe handling methods, proper labeling, and knowledge of hazardous chemicals.	1998
Stockholm Convention	The convention was designed for the pollutants which are widely distributed and persist for a longer period. These pollutants present in E-waste have perilous effects on human health and the environment. This convention measures and takes action to remove these pollutants from the wastes.	2001
Connect 2020 Agenda for Global Telecommunication/ICT Development	Connect 2020 Agenda was designed to share visions, targets, and goals of committed to transitioning to an information society, empowered by the interconnected world, where telecommunication/information and communication technology (ICT) enables and accelerates socially, economically, and environmentally sustainable growth and development for everyone. Within this specific Goal, target 3.2 addresses the issue of E-waste by reducing the volume of redundant E-waste by 50% by 2020.	2020
Waigani Convention, South Pacific	An agreement that prohibits the exports and imports of radioactive or hazardous waste to countries that are members of the Pacific Islands Forum.	1995
The Aarhus Convention, United Nations Economic Commission for Europe	The convention contains the Aarhus Protocol on Heavy Metals, which was one of the eight protocols which deal with the air quality problem within the European Union (EU).	2001
The Restricting of Use of Hazardous Substances in Electrical and Electronic Equipment (RoHS) Directive, EU	This directive bans or restricts the use of certain toxic substances in the production of different types of EEE.	2002

(Continued)

Table 11.3 (Continued)

Rules and regulation	Remarks	Year
The Waste Electrical and Electronic Equipment Directive, EU	This directive aims to upsurge the reuse of EEE and E-waste recycling and set up collection centers where consumers return their E-waste free of charge.	2003
The Durban Declaration, Africa	The declaration is known as the African regional platform/forum on E-waste alongside international bodies. The requirements of the declaration are as follows: amend existing legislation regarding E-waste management, improve their compliance with legislation, and countries must review existing legislation.	2008
The Libreville Declaration, Africa	This declaration came in the first interministerial conference in Africa to protect human health and the environment. Also, the declaration recognized the impacts of hazardous substances and associated risk factors on human health and the environment. These include risk factors for poor health which can arise from E-waste.	2008

EEE, electrical and electronic equipment.

11.4 Case studies

11.4.1 The United States

The United States is the second-highest generator of E-waste in the world, which was estimated at around 6.9 Mt in 2019. Initially, the United States set up regulations at a federal and state level to tackle waste problems. However, there is no authorized federal legislation for E-waste in United States, but some states have implemented legislation for E-waste. For federal legislation to begin the E-waste stewardship in United States, the National Strategy for Electronics Stewardship was introduced in 2011 by the General Services Administration, Council on Environmental Quality, and the Environmental Protection Agency (EPA). Only 25 out of 50 US states have adopted E-waste legislation and 23 of them implemented the concept of EPR in their regulations (Biedenkopf, 2020). National Strategy for Electronics Stewardship was launched in 2011 by EPA with four main objectives: (1) implementing federal laws, (2) collecting incentives for eco-friendly or green EEE products, (3) reducing impacts of exports of E-waste in United States, and (4) proper strategic and sustainable management of E-waste in United States (Hsain, 2019; NESE, 2011).

11.4.2 Japan

In Japan, two major fundamental laws for E-waste are the Law for the Promotion of Effective Utilization of Resources (LPUR) and the Law for the Recycling of Specified Kinds of Home Appliances (LRHA). LPUR law encourages manufacturers to reduce the amount of E-waste and promote recycling. In 2001 LPUR was revised to stipulate that producers have the obligation of resource recovery from home appliances of specific four types (air conditioners, washing machines, refrigerators, and televisions) after their EOL. Moreover, the cost of recycling and transportation are to be paid by the consumers after products EOL (Shinkuma & Huong, 2009). LRHA was introduced in April 2009 to increase the responsibilities of manufacturers and consumers for the recycling of home appliances. LPUR was designed based on the 3R's principle to establish a circular economy-based recycling system, whereas LHRA sets the recycling standards, and also consumers have to pay the necessary fee to retailers for collection and transfer of their appliances. As per LHRA, the minimum required recycling rate for washing machines is 65%, for refrigerators and freezers is 60%, for television is 55%, and for air conditioners is 70%.

11.4.3 China

China is the largest E-waste producing country in the world by generating more than 10 Mt of E-waste in 2019. To deal with this huge amount of waste, the Chinese government implemented some general laws such as Clean Production Law, the Solid Waste Pollution Control Law, and the General Environmental Law. In 2003 China's first regulation "Announcement of the State Environmental Protection Administration on Strengthening the Environmental Management of Electronic Waste" for E-waste management was issued by the State Administration of Environmental Protection (Wang et al., 2021). Under this regulation, consumers have to report about their E-waste generation, collection, disposal to the Chinese government authorities. Also, the use of discarded or outdated E-waste was banned and any organization associated with E-waste recycling activities needs permission from the environmental protection departments (Borthakur, 2020). In 2009 the Chinese government enacted the "Regulations on the Administration of the Recovery and Disposal of Waste Electrical and Electronic Products," which was based on the principles of EPR. "The Implementation Plan of the Extended Producer Responsibility System" was enacted by the State Council which was rooted in the principle of EPR. This regulation was implemented in July 2012 to promote more E-waste recycling and disposal with the concept of EPR (Wang et al., 2021).

The "Administrative Measures on the Collection and Use of WEEE Treatment Fund" was ratified in 2012 by the Ministry of Finance (Tong & Yan, 2013).

11.4.4 India

In 2011 the first policies and legislations for E-waste management came into effect as E-waste (Management and Handling) Rules, 2011. These rules cover all types of manufacturers or producers, consumers, purchase, sale, processing, collection, dismantler, and recycler. However, these rules were not applied to batteries, micro and small enterprises, and radioactive wastes (E-waste Rules, 2011). In 2016 the Government of India (GOI) introduced E-waste (Management) Rules, 2016, which makes the policies more constructive with the addition of a reverse chain system in the previous rules. The rules in 2016 ensured that all the stakeholders involved in the realm of E-waste management were included. At the same, the producer responsible organization was introduced. The organization is responsible for the collection of E-waste and ensures its systematic flow. The EPR was incorporated in the rules that producers are liable to buy the products back for safe recycling (E-waste Rules, 2016). Also, mercury lamps and compact fluorescent lamps (CFL) are included in the rules of E-waste. In 2018 E-waste Management (Amendments) Rules, 2018 expands the responsibilities of producers from the manufacturing of goods and its targets EPR in the financial year. The rules were focused on expediting and implementing the eco-friendly management of E-waste. All the associated stakeholders ensure the sustainable management and recycling of E-waste. The average life of goods for collection target was fixed by Central Pollution Control Board (CPCB) and CPCB is responsible for monitoring the product quality by conducting random sampling of EEE goods to ensure the RoHS provisions as per guidelines of GOI. As per Amendment, till 2023, 70% of E-waste generated shall be covered under the EPR plan. Moreover, from 2025 onward the collection target of E-waste by weight shall be 20% of the sales figure of the previous year (E-waste, 2018).

11.4.5 Vietnam

In Vietnam, the major Solid Waste Management regulation like Decree No. 59/2007/ND-CP which is known as Regulations of Hazardous Waste Management was amended as Environmental Protection Law, 2005 (Decision No. 155/1999/QD-TTg) which framework the country's legislation for waste management. For E-waste, Prime Minister's Decision No. 16/2015/QD-TTg was implemented to regulate the manufacturers and consumers of EEE (Tran & Salhofer, 2018). This decision was based on the EPR system for outdated EEE products and specifies all the duties to the producer on take-back and disposal of the outdated items (Nguyen et al., 2017). A Circular No. 36/2015/TTBTNMT issued by the Department of Natural Resources and Environment reported that all the E-waste generating industries have been submitted an annual report for their amount of

E-waste. Another regulation implemented by the Vietnamese government Decree No. 187/2013/ND-CP on the prohibition of imports of second-hand EEE. However, Vietnam still imports used EEE in illegal ways from other countries (Tran & Salhofer, 2018).

11.4.6 South Africa

South Africa has no specific regulations for E-waste management. However, a legislative framework was designed to manage the E-waste in the country with environmental sound techniques. The National Environmental Management Act came into force in 1998 to apply environmentally sound management techniques in waste management (Ecroignard, 2006). Other legislations which are associated with E-waste are the Hazardous Substances Act (1973), Occupational Health and Safety Act (1993), and Environment and Conservation Act (1989) (Ledwaba & Sosibo, 2016). These acts are implemented to manage and control all types of waste including E-waste by waste

Table 11.4 Global perspectives of existing policies for E-waste collection, import, and treatment of E-waste.

Country	Remarks	References
India	• E-waste Management (Amendments) Rules, 2018 • E-waste (Management) Rules, 2016 • E-waste (Management and Handling) Rules, 2011	Arya and Kumar (2020b)
Japan	• Law for the Recycling of Specified Kinds of Home Appliances (LRHA) • Law for the Promotion of Effective Utilization of Resources (LPUR) • Small Electrical and Electronic Equipment Recycling Act 2013 • Specified Home Appliance Recycling Law (SHARL), 1998	Patil and Ramakrishna (2020)
China	• Technical Policy for Pollution Prevention and Control of Waste Household Appliances and Electronic Products (2006) • Announcement of the State Environmental Protection Administration on Strengthening the Environmental Management of Electronic Waste • Administrative Measures for the Recovery of Renewable Resources (issued in 2007 and revised in 2019) • Measures for the Control of Pollution from Electronic Information Products • Administrative Measures for the Prevention and Control of Environmental Pollution by Electronic Waste (2008)	Wang et al. (2021), Wong (2018)

(Continued)

Table 11.4 (Continued)

Country	Remarks	References
	• Technical Specifications for Pollution Control in the Treatment of Waste Electrical and Electronic Products (HJ527-2010) • Regulation on the Administration of the Recovery and Disposal of Waste Electrical and Electronic Equipment (2009)	
South Korea	• Resource Recycling of Waste Electrical Electronic Equipment and Vehicles Act	Patil and Ramakrishna (2020)
Pakistan	• National Environmental Policy 2005 and Mid-Term Development Framework 2005–2010 are key policies that regulate the import and dumping of solid waste/hazardous waste	Imran et al. (2017)
Vietnam	• In 2015 Decision No. 16/2015/QD-TTg on the regulation of retrieval and disposal of discarded products • In 2013 Decision No. 50/2013 of the Prime Minister on prescribing retrieval and disposal of discarded products was approved and took effect	Tran and Salhofer (2018)
Singapore	• Singapore's National Environment Agency (NEA) implemented RoHS Directives on EEE products. Also, NEA will introduce a regulated E-waste management system by 2021 to ensure that E-waste is managed effectively and efficiently in Singapore through an EPR approach	Patil and Ramakrishna (2020)
Thailand	• Enhancement and Conservation of the National Environmental Quality Act (NEQA) 1992 • Economic Instruments for Environmental Management Act	Gupt and Sahay (2015)
Hong Kong	• No specific regulations for E-waste, However Promotion of Recycling and Proper Disposal (Electrical Equipment and Electronic Equipment) (Amendment) Ordinance 2016 was introduced for EEE	
Canada	• Electronics Product Stewardship Canada (EPSC) (2013)	Shittu et al. (2021)
United States	• Resource Conservation and Recovery Act (RCRA) • Responsible Electronics Recycling Act (HR 2284/S 1270)	Hsain (2019), Shittu et al. (2021)
Nigeria	• National Environmental (Electrical/Electronic Sector) Regulations, 2011	Shittu et al. (2021)

(*Continued*)

Table 11.4 (Continued)

Country	Remarks	References
Ghana	• Hazardous and Electronic Waste Control and Management Act, 2016 (Act 917)	Shittu et al. (2021)
United Kingdom	• Adopted EU Directives in 2007	Osibanjo and Nnorom (2007)
Sweden	• In 2005 the provisions WEEE Directive was transposed into its national laws	Osibanjo and Nnorom (2007)
France	• EU and RoHS directives were adopted in 2005 and 2002, respectively. In 2008 national environment policy Grenelle del' environment was implemented that focuses on the collection, eco-design efforts, and exchange rule, and national council on waste was created	–
Germany	• Electrical and Electronic Equipment Act and adopted EU directives in 2005	Shittu et al. (2021)
Belgium	• Adopted EU directives in 2007	Osibanjo and Nnorom (2007)
Balkans countries	• Albania, Bulgaria, Bosnia and Herzegovina, Montenegro, Macedonia, Serbia, and Slovenia have national legislation on E-waste in effect. Bulgaria and Slovenia adapted WEEE Directives	Shittu et al. (2021)
Australia	• National Television and Computer Recycling Scheme was implemented in Australia under the Australian Government's Product Stewardship Act 2011	Morris and Metternicht (2016)
New Zealand	• Waste Minimization Act 2008	Patil and Ramakrishna (2020)

EEE, electrical and electronic equipment; *EU*, European Union; *WEEE*, waste electrical and electronic equipment.

minimization and 3R's principles. South African E-waste Association is a working group which formed the blueprint of E-waste management in the country. These blueprints focused on the take-back system, advanced recycling fee (ARF), waste minimization, and waste disposal (Ichikowitz & Hattingh, 2020) (Table 11.4).

11.5 Major challenges faced by countries to implement E-waste legislation

The massive amount of e-waste generated every day has become a worldwide problem. Developed and developing countries are struggling to manage huge amounts of e-waste in a proper and sustainable manner. The major challenges are discussed below:

- *Streamline the collection process:* Most of the E-waste is dumped with municipal waste due to a lack of knowledge and awareness on the responsible disposal of E-waste. Also, the lack of technical standards for E-waste collection, transportation, and storage results in disorganized management.
- *Complex design and composition:* The E-waste recycling sectors pose various challenges due to the complex design and composition of the waste appliances as it includes many varieties of materials that are combined, fixed, hooked, fused, or soldered. Toxic materials are bound to nontoxic materials, making it tough to isolate the materials for reclamation, thus requiring exclusive manpower and cost-effective technologies.
- *Lack of investment to support profitable E-scrap recycling improvements:* Most of the nations have significant planning and management for MSW, but still E-waste recycling has remained unnoticed in many cities and towns in the developing nations. Recycling and treatment facilities require a high initial investment and only a few countries have allocated budget or funds toward E-waste management. This negligence has led the informal sectors to dominate the E-waste recycling activities in an unauthorized manner, resulting in loss of valuable and precious resources, energy consumption, deteriorated health conditions, and environmental pollution (Hicks et al., 2005).
- *Lack of stringent implementation of regulations specifically addressing E-waste:* There are a lack and inadequate enforcement of existing legislation toward the management of the transboundary movement of hazardous wastes and recyclables (Osibanjo & Nnorom, 2007). A large population in most of the developing nations are unaware of the term E-waste, its policies, rules and legislations, and its management (Singh & Amin, 2018). The implementation of a statutory framework for EPR in various developing countries is lacking and neglected (Leclerc & Badami, 2020).
- *Lack of awareness:* In developing countries, people generally store the old scraps or partially functional gadgets in their storehouse due to lack of information and awareness due to either emotional attachments or privacy concerns. In such scenarios, there is inaccuracy in data inventorization, leading to failure of efficient technologies and management systems.
- *Limited formal E-waste recycling units:* Due to a dearth of administrative enactment and inefficient technologies, most of the developing countries struggle to establish formalized recycling units. Most of the recyclers do not have authorized facilities, that is, industry certification programs like responsible recycling program and E-Stewards certification programs, which set guidelines for secure recycling and disposal of E-waste (Ceballos & Dong, 2016) and responsibilities toward the legal

procedures and precautionary measures. Hence, most of the recyclers opt for informal trading and business opportunities for E-waste recycling.

11.6 Conclusion and recommendation

Different policies and regulations are implemented across the world to ensure the smooth handling of E-waste. Nonetheless, e-waste management is a severe concern because regulations in most developed and developing countries are poor. Only 78 of the 193 countries have policies for E-waste management; hence, most of the E-waste undergoes informal recycling or transboundary movement to various developing countries such as China, India, Nigeria, and Ghana. E-waste management policies based on EPR and ARF concepts give financial support and improve e-waste recycling. However, many developing nations required stringent regulations at every stage of the EEE product life cycle from its production to recycling and disposal. Concepts like EPR, ARF, and polluter pay principle incorporated in policies can enhance and ease E-waste management.

References

Arya, S., & Kumar, S. (2020a). Bioleaching: Urban mining option to curb the menace of E-waste challenge. *Bioengineered*. Available from https://doi.org/10.1080/21655979.2020.1775988.

Arya, S., & Kumar, S. (2020b). E-waste in India at a glance: current trends, regulations, challenges and waste management strategies. *Journal of Cleaner Production*, *271*. Available from https://doi.org/10.1016/j.jclepro.2020.122707.

Baldé, C.P., Forti V., Gray, V., Kuehr, R., & Stegmann, P. (2017). *The global E-waste monitor 2017*. Bonn/Geneva/Vienna: United Nations University (UNU), International Telecommunication Union (ITU) & International Solid Waste Association (ISWA).

Baldé, C.P., & Kuehr, R. (2021). *Impact of the COVID-19 pandemic on E-waste in the first three quarters of 2020*. Bonn: United Nations University (UNU)/United Nations Institute for Training and Research (UNITAR) — co-hosting the SCYCLE Programme.

Battery Directives. (2006). Directive 2006/66/Ec of The European Parliament and of The Council of 6 September 2006 on batteries and accumulators and waste batteries and accumulators and repealing Directive 91/157/EEC. *Official Journal of the European Union*.

Arya, S., Rautela, R., Chavan, D., & Kumar, S. (2021). Evaluation of soil contamination due to crude E-waste recycling activities in the capital city of India. Process Safety and Environmental Protection. https://doi.org/10.1016/j.psep.2021.07.001.

Biedenkopf, K. (2020). *E-waste policies in the United States: Minimalistic federal action and fragmented subnational activities*. Handbook of electronic waste management (pp. 577–588). Butterworth-Heinemann.

Borthakur, A. (2020). Policy approaches on E-waste in the emerging economies: A review of the existing governance with special reference to India and South Africa. *Journal of Cleaner Production*, *252*, 119885.

Ceballos, D. M., & Dong, Z. (2016). The formal electronic recycling industry: Challenges and opportunities in occupational and environmental health research. *Environment International*, *95*, 157–166. Available from https://doi.org/10.1016/j.envint.2016.07.010.

Dhir, A., Malodia, S., Awan, U., Sakashita, M., & Kaur, P. (2021). Extended valence theory perspective on consumers' E-waste recycling intentions in Japan. *Journal of Cleaner Production*, *312*, 127443.

Ecroignard, L. (2006). E-waste legislation in South Africa. *Electronics Technical*. South African E-waste Association (eWASA).

European Union. (2002). Directive 2002/96/EC of The European Parliament and of The Council of 27 January 2003 on waste electrical and electronic equipment (WEEE). *Official Journal of the European Union*.

E-waste Rules, 2011. *E-waste (management and handling) rules, 2011*. Ministry of Environment and Forests, Government of India, New Delhi.

E-waste Rules, 2016. *E-waste (Management) Rules, 2016*. Ministry of Environment and Forest and Climate Change, Government of India, New Delhi.

E-waste, 2018. *E-waste (Management) Amendment Rules, 2017*. Ministry of Environment, Forest and Climate Change, Government of India, New Delhi.

Forti, V., Baldé, C.P., Kuehr, R., & Bel, G. (2020). *The global E-waste monitor 2020: Quantities, flows and the circular economy potential*. Bonn/Geneva/Rotterdam: United Nations University (UNU)/United Nations Institute for Training and Research (UNITAR).

Gupt, Y., & Sahay, S. (2015). Review of extended producer responsibility: A case study approach. *Waste Management & Research*, *33*(7), 595−611.

Hicks, C., Dietmar, R., & Eugster, M. (2005). The recycling and disposal of electrical and E-waste in China—Legislative and market responses. *Environmental Impact Assessment Review*, *25*(5), 459−471. Available from https://doi.org/10.1016/j.eiar.2005.04.007.

Hsain, H. A. (2019). *Electronic waste and the special case of lead-free piezoelectrics: A call for legislative action*. *2019 IEEE international symposium on technology and society (ISTAS)* (pp. 1−5). IEEE.

Ichikowitz, R., & Hattingh, T. S. (2020). Consumer e-waste recycling in South Africa. *South African Journal of Industrial Engineering*, *31*(3), 44−57.

Imran, M., Haydar, S., Kim, J., Awan, M. R., & Bhatti, A. A. (2017). E-waste flows, resource recovery and improvement of legal framework in Pakistan. *Resources, Conservation and Recycling*, *125*, 131−138. Available from https://doi.org/10.1016/j.resconrec.2017.06.015.

Johri, R. (2008). *E-waste: Implications, regulations, and management in India and current global best practices*. TERI, India.

Leclerc, S. H., & Badami, M. G. (2020). Extended producer responsibility for E-waste management: Policy drivers and challenges. *Journal of Cleaner Production*, *251*, 119657. Available from https://doi.org/10.1016/j.jclepro.2019.119657.

Ledwaba, P., & Sosibo, N. (2016). *E-waste management in South Africa: Case Study: Cathode Ray Tubes Recycling Opportunities*. Preprints.

Lindhqvist, T. (2000). *Extended producer responsibility in cleaner production*. Lund: Lund University, The International Institute for Industrial Environmental Economics.

Méndez-Fajardo, S., Böni, H., Vanegas, P., & Sucozhañay, D. (2020). *Improving sustainability of E-waste management through the systemic design of solutions: The cases of Colombia and Ecuador. Handbook of electronic waste management* (pp. 443−478). Elsevier Inc.. Available from http://doi.org/10.1016/b978-0-12-817030-4.00012-7.

Morris, A., & Metternicht, G. (2016). Assessing effectiveness of WEEE management policy in Australia. *Journal of Environmental Management*, *181*, 218−230.

NESE. (2011). *Interagency task force on electronics stewardship*. National Strategy for Electronics Stewardship.

Nguyen, D. Q., Ha, V. H., Eiji, Y., & Huynh, T. H. (2017). Material flows from electronic waste: Understanding the shortages for extended producer responsibility implementation in Vietnam. *Procedia CIRP*, *61*, 651−656.

Osibanjo, O., & Nnorom, I. C. (2007). The challenge of electronic waste (e-waste) management in developing countries. *Waste Management & Research*, *25*(6), 489−501.

Patil, R. A., & Ramakrishna, S. (2020). A comprehensive analysis of E-waste legislation worldwide. *Environmental Science and Pollution Research*, *27*(13), 14412−14431.

Pont, A., Robles, A., & Gil, J. A. (2019). e-WASTE: Everything an ICT scientist and developer should know. *IEEE Access*, *7*, 169614−169635.

Porwal, P., & Chatterjee, S. (2019). Extended producer responsibility on E-waste management in India: Challenges and prospects. *International Journal of Science and Research*, *8*, ISSN: 2319-7064.

Rautela, R., Arya, S., Vishwakarma, S., Lee, J., Kim, K. H., & Kumar, S. (2021). E-waste management and its effects on the environment and human health. *Science of The Total Environmental*, *773*, 145623.

RoHS. (2011). Directive 2011/65/EU f The European Parliament and of The Council of 8 June 2011 on the restriction of the use of certain hazardous substances in electrical and electronic equipment. *Official Journal of the European Union*.

Shinkuma, T., & Huong, N. T. M. (2009). The flow of E-waste material in the Asian region and a reconsideration of international trade policies on E-waste. *Environmental Impact Assessment Review*, *29* (1), 25–31.

Shittu, O. S., Williams, I. D., & Shaw, P. J. (2021). Global E-waste management: Can WEEE make a difference? A review of E-waste trends, legislation, contemporary issues and future challenges. *Waste Management*, *120*, 549–563.

Singh, Y. P., & Amin, N. (2018). Assessing the challenges and issues of e-wastemanagementfor cities in developing countries. *International Research Journal of Engineering and Technology (IRJET)*, *5*.

Tong, X., & Yan, L. (2013). From legal transplants to sustainable transition: Extended producer responsibility in Chinese waste electrical and electronic equipment management. *Journal of Industrial Ecology*, *17*(2), 199–212.

Tran, C. D., & Salhofer, S. P. (2018). Analysis of recycling structures for e-waste in Vietnam. *Journal of Material Cycles and Waste Management*, *20*(1), 110–126.

Uddin, M. J. (2012). Journal and conference paper on (Environment) E-waste management. *Journal of Mechanical and Civil Engineering*, *2*, 25–45.

Vishwakarma, S., Kumar, V., Arya, S., Tembhare, M., Rahul., Dutta, D., & Kumar, S. (2022). E-waste in Information and Communication Technology Sector: Existing scenario, management schemes and initiatives. *Environmental Technology and Innovation*, *27*.

Wang, R., Deng, Y., Li, S., Yu, K., Liu, Y., Shang, M., & Liang, Q. (2021). Waste electrical and electronic equipment reutilization in China. *Sustainability*, *13*(20), 11433.

Wong, N. W. (2018). Electronic waste governance under "one country, two systems": Hong Kong and mainland China. *International Journal of Environment Research and Public Health*, *15*, 2347.

CHAPTER 12

E-waste management policies: India versus other countries

Somvir Arya, Ajay Gupta and Arvind Bhardwaj
Industrial and Production Engineering Department, Dr. B.R. Ambedkar National Institute of Technology, Jalandhar, Punjab, India

12.1 Introduction

Electronic products became an essential part of our daily routine life. These products make our life more convenient and we cannot assume our life without the use of these gadgets. Our dependency on these products is also increasing with technological advancement. On the other side, life span of electronic gadgets is becoming shorter due to invention in technology.

As per Arya, Patel, et al. (2021) and Arya, Rautela, et al. (2021), informal E-waste recycling has become a very serious issue to health of human beings and surroundings. In India, Delhi has become the hub of informal E-waste recycling as approximately 95% of discarded electronic equipments are managed here. Authors carried out a study at the informal recycling units of Delhi region to examine the soil contamination and found that these recycling divisions have greater opportunities for the extraction of valuable and precious elements. As per the authors, discarded electronic products have the great potential to match the requirement of parts and components in secondary market at much cheaper rates. Authors collected the soil samples from surroundings of various informal recycling units. They compared the observed values of hazardous metals produced from recycling of E-waste with standard/desired values and concluded the observed values are much higher than the acceptable values of soil contamination. This is due to the contamination of heavy metals such as lead, cadmium, nickel, and zinc from e-scrap. Authors suggested encouraging the switching of informal E-waste recycling units to more latest recycling techniques. Furthermore, authors emphasizes on the urgent need to take strong actions on policy and decision-making and also suggested to adopt the environmental friendly recycling techniques to control the ill effects of discarded electrical and electronics equipment (EEE). They, also, pointed out the illegal imports of E-waste from the developed countries. Likewise Arya, Patel, et al. (2021) and Arya, Rautela, et al. (2021) stated that financial benefits, dismantling of outdated PCs has become the best approach for management of recyclable metals and nonmetals. It is also a great opportunity for the entrepreneurs to

Global E-waste Management Strategies and Future Implications
DOI: https://doi.org/10.1016/B978-0-323-99919-9.00009-X
229

work in the field of e-scrap recycling. Residuals of E-waste which are dumped any-where in the surroundings can cause serious threats to human health and environment. Most of the work, in developing countries, such as transportation, handling, disman-tling, segregation, and recycling is carried out manually even sometimes with bare hands. Authors studied the toxicity characteristics of pollutants in the residuals of the discarded printed circuit boards (PCBs) by toxicity characteristics leaching proce-dure test. As per their study, authors highlighted that the waste residue of Cu and Pb is hazardous and other elements stand nonhazardous. This study also emphasized the residuals obtained, after the final processing of discarded electronics gadgets, can also be considered for recovering some valuable materials.

E-waste is a rich source of precious metals and nonmetals as highlighted by many authors in literature. E-waste has become a viable source for urban mining and busi-ness opportunities as well. But around the globe, E-waste has become the greatest threat to the mankind owing to the presence of toxic substances. The exponential production of E-waste has created multidimensional challenges to the human health and environment. Developed countries readily exported their headache of recycling E-waste to poor countries. Due to the unavailability of proper technical infrastructure, developing countries pollute the environment to a great extent. In India, the negli-gence seen from the governing bodies is not having accurate data for E-waste genera-tion and recycling. Authorities are still lacking in maintaining sufficient records of the production and sales of the e-items to quantify E-waste generation. Authors recom-mended the conversion of informal recycling system into formal one in a transparent manner (Arya & Kumar, 2020a). A systematic investigation of literature related to waste electrical and electronics equipment (WEEE) from the point of view of Circular Economy by Bressanelli et al. (2020) was focused wherein 115 research papers were analyzed and studied further according to four aspects:

- Objectives and methodology,
- Geography and approach,
- Actors and life cycle phases,
- Circular Economy 4Rs (Reduce, Reuse, Remanufacture, and Recycle)

Authors identified several research gaps in the literature which further needs to work upon. This review article provides a means for e-equipments producers, service channels, End of Life (EOL) actors, and policymakers to use the evidence of previous research to take their decisions for minimizing e-scrap. This literature review also sug-gested the designers of e-equipments to consider the cost of gadgets in order to pro-vide economic benefits to end users along with environmental benefits as well. Authors considered only those research papers which are published in highly reputed journals with high impact factor and excluded other research articles (conference papers and technical reports) that may also have considerable contribution on the issue of formal E-waste recycling.

Maphosa and Maphosa (2020) explained that developing countries have become the primary destination for WEEE exported by the developed nations; making E-waste management a critical issue in developing countries. The E-waste management practices in Sub-Saharan Africa by critically examining the literature on e-scrap recycling. Sufficient studies on e-scrap recycling are available for developed countries; however, very limited literature is available for developing countries inspite of continuously receiving huge quantities of used electronic gadgets from developed countries. Authors searched for research papers on Web of Science, EBSCO Host along with Sabinet databases, and reviewed them using the systematic literature review process. Based on the study, it was reported that more than 80% of the E-waste management articles in Sub-Saharan African region were published by only three countries of the region namely Ghana, South Africa, and Nigeria. This paper also highlighted the lack of legislations and availability of limited infrastructure was the key barriers in adopting formal E-waste management system. Arya and Kumar (2020b) highlighted that globally, E-waste management has become essential due to material recovery and to reduce its adverse effects on surroundings. The PCB is a rich source of metals and nonmetals. For materials recovery, hydrometallurgy and crude recycling practices are commonly used in these casual recycling units. However, these traditional processes of material recovery result into release of hazardous gases to the environment which greatly pollutes the environment and ultimately affects the wellness of human beings. Bioleaching is considered as an advanced process of materials recovery as compared to conventional ones. Utilizing bioleaching technology for resource recovery has a significant positive impact on environmental sustainability, public health, and economic prosperity. For eco-friendly disposal of e-scrap and to minimize the limitations of current recycling processes, authors focused on the use of advance technologies in e-scrap management. Also, the needs of cheap spare parts, working components, and extraction of valuable materials have contributed to a great extent in making eco-friendly disposal of E-waste more lucrative and economical.

Amer et al. (2019) explored that effective reverse supply chain and consumers awareness play a key role in managing E-waste through formal channel and are helpful for governments and policymakers to develop E-waste recycling channel in sustainable and environmentally safe manner. However, Otieno and Omwenga (2015) asserted that the main concern is the fact that accumulation, storage, exchange, and disposal of WEEE in developing nations have not been streamlined and overseen in a strong manner. This is also helpful in reutilization of electronic gadgets, preservation of the environment, and protection of the overall public. Agreeing with the study led by Ghadimzadeh et al. (2014), discovered that E-waste disposal is actually performed using some exceptionally challenging and uncontrolled means. These authors further reported that the absence of information on volume of E-waste being generated and the absence of conducive policies and rules at the state level are major factors for E-waste bungle. It was certainly claimed by Balde et al. (2015), Nigeria is the African nation to enforce the e-scrap legislation. Additional conflicts are connected with lack of understanding and ability for handling and managing E-waste. With this regard, Tyagi et al. (2015) mentioned that many companies in unorganized sector are recycling major part of e-scrap in India. Collection of E-waste has been carried out by area scrap dealers. Collection of E-waste is followed by the managing process entails disassembling and segregation of the e-items. Primitive methods are actually utilized in the specific procedure, which include (1) dismantling of electronic gadgets; (2) heating or disassembling of PCBs with bare hands; (3) recovering copper and aluminum from cables; (4) melting or breaking plastic parts; (5) toner sweeping; and (6) open acid leaching process for recovering the valuable metals from E-waste. The writer points out that the majority of the rag pickers are not educated and the individuals managing e-trash are also not skilled and untrained. Additionally, they perform the segregation and refurbishment of used items that will be sold in refurbished market. They simply work with conventional illegal strategies of burning the items for extracting the metals; they are unaware of the health risks involved. Over the past few decades, the world has revolutionized due to developments in electronic industry that has resulted in design and production of more and more useful electrical and electronics products that are dangerous for today's life around the universe (Dwivedy & Mittal, 2013). Recent addition to the dangerous waste stream is E-waste in the form of discarded electrical and electronic appliances. Mundada et al. (2004) explained that E-scrap comprises of both precious materials and perilous materials. Hence it always requires unique handling and special managing techniques to evade from environmental pollutants and harmful impact on public wellbeing (Robinson, 2009). Over the last 2 years, electrical and electronic appliance's manufacturing and utilization has increased due to developments in information and telecommunication technology. India is one of the fastest growing economies in the world, where the demand for domestic consumer e-items has been rising swiftly (Sinha-Khetriwal et al., 2005). India and China, which are emerging economies, have enormous e-scrap generators and have rapid development arcades for electrical and electronic gadgets (Sinha-Khetriwal et al., 2005; Widmer et al., 2005). The main stakeholders in the E-waste supply chain are traders, scrap dealers, smelters, consumers, retailers, exporters, recyclers, producers or manufacturers, and disassemblers or dismantlers. Although E-waste is an emerging problem, yet commercially it is as an opportunity of considerable significance, given the quantity of e-scrap being produced as it also contains precious materials. The precious materials in E-waste include gold, silver, copper, iron, aluminum, and other useful metals which are over 60% of the total weight, whereas pollutants comprise 2.70% (Widmer et al., 2005). Hence the process of recycling e-scrap is a key issue not only for the waste management perspective but also for recovering precious materials. The extended producer responsibility (EPR) enactment greatly influences E-waste management. Bahers and Kim (2018) explicated WEEE chain and movement by means of material flow analysis (MFA) and consider the EPR enactment. Bahers and Kim (2018) conducted a meticulous case study by implementing EPR for E-waste in the regions of Midi-Pyrenees and Toulouse's urban area. On the basis of MFA of WEEE chain, outcome depleted awareness in operational activities commence with exchange of waste material.

12.2 Comparisons of existing E-waste policies and important dimensions

Majority of E-waste in India is currently processed through informal channels. This is causing huge financial and environmental loss to the country. It may be due to the lack of inclusion of some important points in the policy of Government for E-waste or failure to effectively implement the existing policy. To identify such gaps, it is important to compare the Indian government policy on E-waste management with policies of other countries. This will help finding the limitations in our existing system and suggest the changes in the existing system to improve E-waste management strategy in India.

After studying the E-waste policies of a number of developing and developed countries, we identified six important dimensions for comparison purpose. These dimensions are:

- E-waste management system
- Responsibility
- Available mechanism and financial schemes
- Policies and regulations
- Challenges and drawbacks
- Technology

In India, most of the E-waste find their final destination through informal channels as shown in Table 12.1. This causes serious health and environmental hazards, which needs to be addressed. Government of India must take strict action against the informal recyclers. Formal collection, handling, and disposal of the electronic waste is necessary to minimize the negative effects of E-waste. Recycling fee should be added to the cost of the equipment and charged from the consumers in advance at the time of sale of the equipment. India is generating electronic waste at alarming pace. A ticket number can be allotted to every electronic gadgets at the time of sale of the product, which is helpful in tracking the products after the completion of its useful life. Also developed countries such as the United States and the United Kingdom are illegally dumping their electronic squander to India and other developing nations, this needs serious attention by the government. Manufacturers need to develop their own channel in collaboration with local authorities and recyclers to collect and manage the E-waste. Majority of E-waste are collected by scrap dealers and are managed by

Table 12.1 E-waste systems followed by different nations.

E-waste management system

United States	United Kingdom	Japan	Germany	China	Kenya	India	Remarks
1. Majority of E-waste outsource to developing countries. 2. Government developed electronic recycling program. End users are free to dump their discarded electronic products disregarding the manufacturer and location from where the products were bought. Consumers drop their e-products for a best buy gift card and new electronic products can be purchased from best buy (Namias, 2013). 3. Drop-off centers generally include retail	1. Before implementation of the WEEE directive in the United Kingdom, electronic waste was disposed-off with household waste. 2. Some distributors collect the old equipment at the time of home delivery of new electronic product. 3. Distributors, local authority offers the collection facility to end user without any cost (Waste Electrical and Electronics Equipment, 2014).	1. Home recycling ticket system was developed by the Association for Electric Home Appliances to ensure the formal recycling of home/ organizations electronic equipment. 2. Retailer takes responsibility for taking back any of the appliances, it sold and for delivering it to its manufacturer. 3. Manufacturer (importer) properly treats and recycles any of the appliances, it manufactured or imported (Bo &	1. Divided product responsibility was developed for EEE recycling operation. 2. Two pillars, (a) formal recyclers and (b) electronics products manufacturer, are identified which play a significant role in electronics waste disposal. Formal recyclers have to accept e-scrap at their recycling units. 3. Manufacturers are required to provide their own collection, handling, and recycling system. End users are forced by laws to dump their E-waste at	1. Major part of the E-waste in China is recycled by informal channel. They segregate different elements from electronic devices through manual dismantling and chemical processing. 2. This involves steps like heating circuit boards, chipping and melting plastics, and precious metal extraction using acid leaching and incineration 3. There is relatively little environmental impact associated with the collection of E-waste; therefore,	1. Initially government of Kenya introduced WEEE centers for school only. They named it as Computers for Schools Kenya (CFSK). 2. This project initially focused on the recycling of school computers only, but now this system has started to recycle E-waste from colleges, organizations, NGOs, etc. 3. After the success of this project, government is planning to import e-scrap from other African countries also. 4. Producers shall ensure that E-waste returned	1. No formal effective E-waste management system is present. 2. Majority of E-waste (from houses, organizations) is collected by informal channels (Scrap Dealers; Rajya Sabha Secretariat, 2011).	1. Effective formal collection system. 2. Advanced Recycling Fee (ARF) 3. A ticket may be issued at the time of sale of the e-product to track it. 4. Ban on illegal import of WEEE from developed countries. 5. Door step pick up service is required.

stores, government sites, and charitable drop-off centers. Collection efforts in the United States have been insufficient due to the lack of Federal legislation, and also due to poor awareness level of end users (Kahhat et al., 2008).	Yamamoto, 2010).	these facilities (Li et al., 2015).	central government has not explicitly banned it (Chi et al., 2011).	is not disposed at municipal sites (Environmental Management and Coordination Regulations, 2013).

informal channels in India. Another issue is that end users mix and dump the electronic waste with household wastes. This type of practices shall be avoided to minimize the adverse effects of E-waste on human livings and on surroundings. Government needs to strengthen and to develop an effective monitoring system to keep an eye on formal and informal channels. Also, like letter boxes, drop boxes can also be placed at different locations of the city especially in the authorized service centers of the various companies. The consumers can drop their e-scrap in these drop boxes at free of cost.

Extended producers responsibility (EPR) is the key to success for formal E-waste management. EPR makes the manufacturer responsible for collection, handling, and disposal of the discarded products manufactured by them. EPR is an effective step for formal disposal of E-waste and has been successfully implemented by developed countries such as the United States, the United Kingdom, and Japan as depicts in Table 12.2. Although EPR exists in India but due to lack of interest of government in E-waste recycling, this concept is not effectively implemented. There is a need to enforce it effectively. Similarly, on the lines of German government, Indian government can also introduce the concept of enhanced product life and motivate the manufacturers to modify the design of existing products to increase the useful life.

Table 12.3 depicts the available mechanism and financial schemes to manage e-scrap in different countries. Consumers need to be awarded with considerable amount if they return their old discarded products at the time of purchase of new equipment. Financial schemes play a significant role toward formal collection of E-waste and will be helpful in getting E-waste out of the houses. EPR shall be effectively implemented for successful formal collection, handling, and disposal. Manufacturers shall finance for collection and recycling of e-scrap.

As can be seen from Table 12.4, governments of some countries are directly offering financial incentive to the end users (China), while some governments (e.g., Japan) are offering incentives to communities to undertake these activities. A few countries are facilitating the consumers by providing free collection facilities but are not paying directly anything. Government of India needs to motivate the NGOs to actively participate in the collection, transportation, and recycling of E-waste. Also, government needs to give considerable subsidy on purchase of machines and equipment to establish the E-waste recycling unit. NGOs can also play a vital role in educating the end users about the negative effects of the electronic scrap. Government may share their responsibility with municipalities and instruct them to develop their channels for collection, handling, and transportation of e-scrap effectively.

To increase formal collection, manufacturers can introduce attractive incentive or exchange schemes as successfully implemented by other countries. This will be feasible only if they understand the financial benefits of such activity. Government can help by supplying the needed information to these manufacturers.

Table 12.2 Responsibility to recycle E-waste.

Responsibility	United States	Japan	Germany	China	Kenya	India	Remarks
Extended producer's responsibility (EPR) (Shumon & Ahmed, 2013)	1. EPR 2. Retailers and distributors are the part of formal recycling chain. They provide take-back scheme for consumers and then forward the collected e-scrap to formal recyclers (Parliamentary Office of Science and Technology of the UK Parliament POST Note-292, 2007).	1. Consumers are required to work together with retailers for the collection of their WEEE (Bo & Yamamoto, 2010). 2. Retailers are enforced by laws to accept the E-waste on customer's request. Retailers have to transfer the collected WEEE to the producer's combined collection points (Shumon & Ahmed, 2013).	1. Extended Producer's responsibility 2. Closed Substance Cycle and Waste Management Act were imposed on manufacturers in 1996. Producers have to design their electronic products in such a manner that their useful life can be increased. 3. Reduction of Hazardous Substance (RoHS) 4. Producers and retailers are also responsible for the recovery of electronic items after the completion of their useful life.	1. EPR was introduced. 2. Role of all the key players such as consumers, manufacturers, recyclers, and authorities is well defined for effective system.	1. Individuals take-back schemes was introducing. 2. Producer Responsibility Organization (PRO) was introduced to strengthen EPR. 3. Consumers have the choice to dispose e-scrap to the E-waste collection centers (National Environmental Management Act, 1998).	1. EPR was introduced to fix the responsibility of manufacturers. 2. PRO has been established to strengthen EPR. PRO would be financed and authorized by a single manufacturer or a number of manufacturers collectively for collection, handling, and disposal of e-scrap in environmentally friendly manner (Sharma & Hussain, 2018). 3. RoHS	1. Responsibility of every stakeholder is clearly defined. 2. Government needs to encourage the manufacturers to design the products for enhanced life.

Table 12.3 Available mechanism and financial schemes to manage E-scrap.

Available mechanism and financial scheme

United States	United Kingdom	Japan	Germany	China	Kenya	India	Remarks
1. EPR 2. Retailers offer discounts on exchange of old devices. 3. Best buy scheme offers gift cards or cash incentive for customers who bring their old devices to drop box (Namias, 2013).	1. Retailers collect the E-waste form houses for free. 2. Distributors should make arrangement with producer's compliance schemes (PCSs) to collect WEEE. 3. Consumers can drop their electronic scrap at designated collection facilities free of cost (Waste Electrical and Electronics Equipment, 2014).	1. Local authorities are totally responsible for collection of e-scrap at user's request. 2. Collection fee is imposed on consumers. (Lee & Na, 2010). 3. The Small Home Appliance Recycling Act in Japan supports an incentive-oriented system that enables interested players (consumers, business operators, municipalities, retailers, etc.) to develop their own waste collection, handling, and recycling units (Ministry of the Waste Management in Germany, 2018).	1. Producers shall finance the whole recycling facility. Collection, handling, and environmentally friendly disposal of E-waste from private households/ organizations. 2. Manufacturer shall finance the recycling of the E-waste of its own manufactured products sold in the market after August 13, 2005 (Deubzer, 2011) 3. Consumers are educated about the adverse effects of E-waste on human health and on environment (Environmental Audit Committee EAC of UK, 2019).	1. In 2009, The National Development and Reform Commission of China introduced the National Old-for-new Home Appliance Replacement Scheme (HARS). According to this scheme, consumers get subsidy worth 10% of the new electronics product's price (OECD, 2010). Only condition to get this subsidy is that end users need to sell their discarded electronic items to authorized recyclers (Tong & Yan, 2013). 2. EPR	1. Government provides incentives to consumers to dispose their electronic scrap to formal channel. 2. The government facilitates NGOs, local investors, and private organizations by providing tax exemptions on E-waste recycling equipment and lands. 3. The government provides incentives for all players involved in collection, handling, and environmentally friendly disposal (Ministry of Environment and Forestry-National E-waste Mgt. Strategy, 2019).	1. The E-waste management and handling rules, 2011 instruct the manufacturers of electronic products to achieve 100% collection of their discarded products. 2. Reduction of hazardous substances (Central Pollution Control Board of India, Ministry of Environment and Forest, 2011) 3. No formal incentive scheme is available. 4. Exchange offer is available at retailers end at very low exchange value.	1. Effective implementation of EPR 2. Advanced Recycling Fee (ARF) can be imposed. 3. Retailer may offer subsidy if old e-product sold to formal recyclers. 4. Government can introduce minimum price for formal disposal of various E-waste items. 5. Manufacturers/ retailers need to introduce attractive exchange offers to make the collection and processing viable, government can introduce incentive schemes for manufacturers and retailers also.

Table 12.4 Regulations and policies to manage E-waste.

Policies and regulations

United States	United Kingdom	Japan	Germany	China	Kenya	India	Remarks
1. Electronic products are designed, sold, and managed in a more sustainable manner. 2. Protect health of human beings and the surroundings in the United States and abroad. 3. Promote new and innovative technologies. 4. Encourage electronic producers to expand their collection channel (Northeast Supply Enhancement, 2017)	1. Retailers must accept WEEE for free from private household customers at their retail site. 2. RoHS directions 3. Establish better product administration 4. Sway the behavior of users toward more environmentally correct method 5. Popularize the best reuse, refurbishing, recovery of materials, handling, and disposal practices 6. Political and institutional support for management of e-scrap in environmentally friendly manner (Umesh, 2013)	1. Japanese Home Appliance Recycling Law target only four main types of equipment namely: washing machines, TV, air conditioners, and refrigerators. The law forces the producers including importers of these appliances to recycle them (Tomohiro et al., 2005). 2. Computer Recycling Law promotes the effective utilization of available resources to take-back computers from houses and establish a formal reverse supply chain system.	1. Landfilling Limiting Policies 2. Producers Responsibility Policies 3. Closed substance Cycle and Waste management Act—in order to sustain natural resources and to establish environmentally friendly recycling of e-scrap (German Government's Policy on waste, 2017).	1. Regulation on recycling and disposal of WEEE enforced implementation of EPR, special fund to support e-scrap management. 2. Laws on promoting the development of circular economy, suggested the use of 3Rs (Recover, Reuse, and Recycle) (Chenyu et al., 2015).	1. Ban on open burning 2. Not to abandon E-waste anywhere other than collection centere. 3. Ban on imports of WEEE having cathode tube excluding medical equipment (Environmental Management and Coordination Regulations, 2013)	1. EPR 2. RoHS guidelines (Singh, 2012) 3. Central Pollution Control Board of India, Ministry of Environment and Forest, 2011 4. E-waste management rule 2016	1. Landfilling should be banned. 2. Rules should be implemented effectively. 3. Closed system (for monitoring) must be adopted to get proper feedback

Although policies and regulations exist in India, these are not implemented effectively. The poor implementation of the policies results into recycling of e-scrap through informal channels. Government must consider the hazardous effects of E-waste seriously, also rules and regulations shall be imposed effectively. Proper monitoring system is also needs to be developed. Due to the lack of awareness of hazardous effects of E-waste on human livings and on surrounding, we dump them to open environment. Landfilling of e-scrap causes sand and water pollution too. For the same reason, landfilling should be avoided.

India is facing a number of challenges in implementing effective E-waste management system. Major challenges are listed in Table 12.5. Many of these challenges are faced by other developing countries such as China and Kenya. Even many developed countries are facing the issues like nonachievement of targets and gradual decrease in volume of E-waste collected through formal channels. This shows that developing and implementing an effective system will be challenge for our country. Government needs to improve the awareness level of citizens and awareness programs need to be organized to make the consumers aware about ill effects of improper E-waste disposal. Legislations must be imposed effectively to force the end users to choose formal recycling channels. Incentive schemes should be of such magnitude that it motivates the people to dispose their discarded products to formal recyclers.

Electronic products are the assembly of thousands of components. It has both valuable and hazardous materials. Valuable materials need to be recovered and hazardous materials need to be recycled in environmentally correct manner. Latest tools and techniques must be used to recycle the E-waste to minimize the negative effects of it. Also we need to adopt new methods and tools such as shredders, eddy current separator, and suspension magnet for ecofriendly E-waste disposal system as indicated in the remarks column of Table 12.6.

12.2.1 Concluding remarks—comparison of policies

Although formal channels exist in India but due to lack of effective policies, most of the E-waste do not find its way to it. Government needs to make serious efforts to educate individuals, organizations, and recyclers about hazardous effects of informal recycling of E-waste, enforce laws to curb informal recycling. Government needs to enforce the EPR very strictly for formal E-waste collection, handling, and disposal. Another choice with the government is to impose the recycling fee on the buyers at the time of purchase of the electronic products. After collecting the recycling fee, it becomes the responsibility of the manufacturer to collect and dispose-off the E-waste. Manufacturers need to develop their own channels for collection, handling, and environmentally friendly disposal. Manufacturers may be asked to either set up their own collection centers in every city or provide door to door collection facility for

Table 12.5 Challenges and drawbacks during E-waste disposal.

Challenges and drawbacks

United States	United Kingdom	Japan	Germany	China	Kenya	India	Remarks
1. E-waste in a sustainable approach includes broad diversity of challenges as well as opportunities for prime stakeholders like consumers and organizations. Awareness campaign and training programs help to achieve sustainability targets. 2. Political pledge is urgently required (Shamim et al., 2015).	1. Volume of E-waste generated 2. Lack of availability of adequate formal infrastructure 3. Public awareness 4. Collection targets not achieved (Environmental Audit Committee EAC of UK, 2019).	1. Major challenge is that the nation is observing a decrease in the quantity of E-waste collection (Yolin, 2015). 2. Collection efficiency 3. Roles and responsibilities of all stakeholders	1. Efficiency of available recourse. 2. Quality and quantity of collection (Waste Management in Germany, 2018).	1. Environmental pollution 2. The absence of management regulations fits for China and ineffective formal collection system. 3. How to regulate waste electronic products. 4. Difficulties in collection, storage, transportation, and treatment process 5. Unfair competition caused by weak implementation of laws and regulations due to regional disparity 6. Collection system of waste (Guomei, 2006)	1. Awareness level about negative effects of WEEE is very poor. 2. Most of the consumers are not capable to purchase new electronic products resulting in a considerable number of end users going for used or refurbished products which are cheaper but have a shorter life cycle. 3. The government recycling agencies dealing with electronic scrap have limited capacity to deal with E-waste. These agencies are also not working in an organized manner. 4. Lack of rules and regulations to deal with WEEE effectively (Ministry of Environment Report, Japan, 2010)	1. Volume of E-waste generated 2. Involvement of child labor 3. Ineffective legislation 4. Lack of available infrastructure 5. Lack of incentive schemes 6. Poor level of awareness 7. Illegal E-waste imports 8. Unwillingness of authorities involved 9. High cost involved in setting up a recycling unit (Kumar et al., 2014)	1. Run awareness programs 2. Effective legislation 3. Attractive incentive schemes 4. Strict monitoring system 5. Formal collection 6. Ban on illegal E-waste imports

effective collection of E-waste. Manufacturers need to make some arrangements for collection of the old electronic items at their service centers. There must be a strong tracking system of sold electronic items so that recovery becomes easier. A ticket number can be issued at the time of purchase to track the product and collect the product when its shelf life is over. Although comprehensive rules and regulations exist in India, they are not effectively implemented by the authorities. For effective E-waste management system, rules and regulations need to be effectively enforced.

Financial incentive is an important factor which can play a key role in facilitating formal collection and disposal of E-waste. Due to the lack of adequate financial incentive, most of the end users are discarding their electronic products to scrap dealers, although the scrap dealers are not offering suitable price. To encourage disposal through formal collection channel, recyclers need to pay adequate money to end users at the time of disposal. Financial incentives will motivate the end users to dispose-off their E-waste through proper channels. Government needs encourage development of formal collection system with the collaboration of producers, distributors, and retailers for effective E-waste collection.

From our earlier finding, it is clear that awareness level of consumers (individuals and organizations) about negative effects of E-waste on human health and on surroundings is very poor. The awareness level of the end users must be improved in order to increase the collection through formal channels. Awareness programs should be organized to enhance the level of awareness. Strict monitoring system is the demand of time.

After studying the E-waste management policies of the countries listed in Tables 12.1–12.6, a list of salient points for proper E-waste management was prepared. Table 12.7 indicates the presence/absence of each of the salient point in the policies of various countries. Last column indicates the action that needs to be taken by Indian Government.

It is clear from Table 12.7 that existing policy of E-waste management in India covers most of the important points, but the same has not been implemented effectively by the government. Due to the lack of effective monitoring system, majority of the organizations are not serious about submitting annual reports of discarded electronic equipment/waste. The middle man like retailers and authorized service centers are also receiving the E-waste in terms of discarded electronic gadgets under exchange schemes and they are selling this discarded e-scrap to the informal channel to achieve some financial gain.

Government needs to enforce the organizations to maintain a proper record of their discarded electronics equipment and develop a close and effective monitoring system. Another option for developing effective E-waste recycling system is to impose penalties to the end users for disposing their electronic gadgets to informal channels/landfills. Also, by providing considerable financial incentive in term of discounts, gift vouchers etc., the end users need to encouraged to dispose-off their electronic items through formal channel. The money need for this incentive may be charged from the users at the time of purchase

Table 12.6 Technology used to recycle E-scrap.

Technology							
United States	United Kingdom	Japan	Germany	China	Kenya	India	Remarks
1. Export to developing countries 2. Processing via pyrometallurgical processing methods 3. Acid leaching (Namias, 2013; Shamim et al., 2015)	1. Currently, treatment of E-waste is done by manual disassembly to segregate mercury from the other waste steam. 2. Formal electronics scrap recycling units use a wide variety of machinery/equipment such as crushing machine, grinding machine, conveyor belts, compacting, and palletizing machines to recover the valuable materials (Waste Electrical and Electronics Equipment, 2013).	The prevailing methods for recycling used home appliances are to first break them apart with a simple machine and then sort out the recyclable parts using different processes. Magnetic separator and wet gravity separator are used (World Intellectual Property, 2013).	1. Computers and TVs are first of all manually dismantled, and then trained persons segregate the pollutants containing components. 2. The discarded products are segregated on the basis of their size, shapes, etc. 3. Shredding process is used. 4. Automatic separation process (Environmental Audit Committee EAC of UK, 2019)	1. Refurbishing and repairing of Used Electrical and Electronics Equipment (UEEE) 2. Recyclable fraction sold to recyclers 3. Formal recyclers use mechanical dismantling of electronic scrap. 4. Informal recyclers use traditional techniques to recycle the discarded electronic equipment. They usually do acid leaching and open burning of E-waste to recover valuable metals (Ghosh et al., 2016)	Inadequate skills and techniques to deal with E-waste	1. Formal recyclers mainly use mechanical dismantling, segregation, and recovery of valuable materials. 2. Informal recyclers use traditional techniques to recycle the discarded electronic equipment and recover the valuable metals (Rajya Sabha Secretariat, 2011)	1. Latest machines (such as shredders, eddy current separator, and suspension magnet) should be used for recycling of E-waste. 2. Chemical leaching process may also be used to recover valuable materials instead of open burning process.

Table 12.7 Comparison of policies with respect to various factors.

Factors	United States	United Kingdom	Japan	Germany	China	Kenya	India	Remarks
Extended producers responsibility (EPR)	Yes	Yes	Yes	Yes	Yes	Yes	Yes	Although EPR policy exists in India, the same has not been implemented effectively.
Advanced Recycling Fee (ARF)	No	No	No	Yes	No	No	No	ARF can be enforced on consumers by the government. This money can be used for implementing EPR.
Restriction of Hazardous Substances (RoHS)	Yes	Yes	Yes	Yes	Yes	No	Yes	RoHS is effectively implemented in India by the government.
Collection methods	– Curbside collection – Drop boxes	At retailers end	Local government and producers	Formal sector	Formal (very less) and informal sector (majority)	Collection centers	Formal (very less quantity) and informal sector (majority)	Some sorts of facilities such as drop boxes and door step collection should be provided to end users for effective collection.
Existence of formal sector	Yes	Yes	Yes	Yes	Yes	Yes	Yes	Although formal collection channel exists in India, yet majority of E-waste is collected by informal channels. Steps need to be taken to enforce the use of formal channels.
Imports from other countries	No	No	No	No	Yes	Yes	Yes	Imports from the developed countries must be banned considering the negative impact on environment and health of the human beings.

Infrastructure available	Insufficient	Insufficient	Available	Available	Available	Available	Available	Infrastructure is available in India, but due to lack of collection, the entrepreneurs are not in a position to run their plants throughout the year.
Involvement of child labor in E-waste management	No	No	No	No	Yes	No	Yes	Although child labor is banned in India, even then child labor is currently involved in E-waste management practices in India.
Effective implementation of rules and regulations	Yes	No	Yes	Yes	No	No	No	Rules and regulations for E-waste management exist in India but most of them are not implemented effectively.
Technique used to dismantle E-waste	Chemical processing	Manual	Machine	Machine	Manual	Manual	Manual	Manual dismantling, currently being used in India, is very dangerous to the workers involved in E-waste management and all the valuable materials present in E-waste cannot be fully recovered by this method. So, it needs to be replaced with latest technology-based methods.
Effectiveness of formal collection/disposal system	Yes	No	Yes	No	No	No	No	Majority of E-waste in India is collected by informal channel (Kabadi walas). Formal collection is almost absent in India.

(Continued)

Table 12.7 (Continued)

Factors	United States	United Kingdom	Japan	Germany	China	Kenya	India	Remarks
Incentives	Discount on exchange	No incentive scheme	No incentive scheme	No incentive scheme	Flat 10% discount on purchase of new e-product	No	Discount on exchange	Exchange offers are given by selected brands/retailers during festival seasons only. This incentive needs to be extended throughout the year.
Consideration of life span of E-product	Yes	Yes	Yes	Yes	No	No	No	Life spans of the e-products are not taken into consideration. Some mechanism needs to be developed to ensure the disposal of E-waste items at the expiry of useful life.

in the form recycling fee. As the volume of E-waste being processed through formal channel will increase, establishing such units will become more attractive for the entrepreneurs and more such units will be established by the entrepreneurs.

12.3 Conclusion

India is a leading producer of e-scrap in the world. However, the process of managing E-waste has not been given the needed attention. The responsibility of the government is to frame and implement necessary policies and rules for effectively managing E-waste. On comparing state wise, in India, northern region states are more predominant for E-waste production. It has been studied that many processing units for managing E-waste are present at many places in Delhi. As the northern regions of India are performing the major E-waste management practices, the research explores the E-waste management process in northern states of India. As illegal import of E-waste is regularly taking place in many developing nations' including India, it is resulting in many-fold increase in E-waste volume. Informal practices are being carried out as the E-waste consists of some valuable elements such as gold, copper, and aluminum. Industries are performing the informal procedures for recycling and are ignoring the presence of the hazardous elements along with the valuable elements. These hazardous elements are causing harms to the people working in these informal recycling units and to the environment as well. Hence government needs to provide a proper regulation procedure for E-waste management and find out the ways to stop this illegal import of E-waste through strong law enforcements in order to protect people from various hazardous issues. There is also a need to use efficient practices for increasing the awareness about hazardous effects of E-waste and to educate people regarding proper disposal of electronic gadgets instead of just discarding them into garbage.

Indian E-waste management policies have been compared with the policies of other countries such as the United States, the United Kingdom, Japan, Germany, China, and Kenya on various key aspects of E-waste management. With the extensive study of their policies, we have found some limitations in our existing policies. Comparisons showed that although the existing E-waste management policy of India covers many important aspects of good E-waste management, they are not effectively implemented. Government needs to impose Advanced Recycling Fee on the consumers and develop a strong system to monitor the flow of electronic products from their initial manufacturing stage to their final disposal. Roles and responsibilities of each and every stakeholder (individuals/organizations/recyclers) need to be clearly defined and a close system is required to monitor their activities.

References

Amer, Y., Doan, L. T. T., Lee, S. H., & Phuc, P. H. K. (2019). Strategies for E-waste management: A literature review. *International Journal of Energy and Environmental Engineering, 13*(3), 157−162.

Arya, S., & Kumar, S. (2020a). E-waste in India at a glance: Current trends, regulations, challenges and waste management strategies. *Journal of Cleaner Production, 271*, 122207.

Arya, S., & Kumar, S. (2020b). Bioleaching: Urban mining option to curb the menace of E-waste challenge. *Bioengineered., 11*, 640−660.

Arya, S., Patel, A., Kumar, S., & Loke, S. P. (2021). Urban mining of obsolete computers by manual dismantling and waste printed circuit boards by chemical leaching and toxicity assessment of its waste residues. *Environmental Pollution, 283*, 117033.

Arya, S., Rautela, R., Chavan, D., & Kumar, S. (2021). Evaluation of soil contamination due to crude E-waste recycling activities in the capital city of India. *Process Safety and Environmental Protection, 15*, 641−653.

Balde, C., Wang, F., Kuehr, R., & Huisman, J. (2015). The global E-waste monitor 2014: Quantities, flows and resources. *United Nation University (UNU) Institute for the Advanced Study of Sustainability*, 01−80.

Bo, B., & Yamamoto, K. (2010). Characteristics of E-waste recycling systems in Japan and China. *International Journal of Environmental, Chemaical, Ecological, Geological and Geophysical Engineering, 4*, 89−95.

Bressanelli, G., Saccani, N., Pigosso, D. C. A., & Perona, M. (2020). Circular economy in the WEEE industry: A systematic literature review and a research agenda". *Sustainable Production and Consumption, 23*, 174−188.

Chenyu, L., Zhang, L., Zhong, Y., Wanxia, R., Tobias, M., Mu, Z., Ma, Z., Geng, Y., & Xue, B. (2015). An overview of E-waste management in China. *Journal of Material Cycles and Waste Management, 17*, 1−12.

Chi, X., Streicher-Porte, M., Wang, M., & Reuter, M. (2011). Informal electronic waste recycling: A sector review with special focus on China. *Waste Management, 31*, 731−742.

Deubzer, Otmar (2011). E-waste Management in Germany. United Nations University Institute for Sustainability and Peace (UNU-ISP).

Dwivedy, M., & Mittal, R. K. (2013). Willingness of residents to participate in E-waste recycling in India. *Environmental Development, 6*, 48−68.

CPCB (Central Pollution Control Board) of India, Ministry of Environment and Forest, Government of India, New Delhi Rules. (2011), pp. 01-33.

Environmental Audit Committee (EAC) of UK, UK Parliament, UK. (2019).

Environment Management and Coordination Regulations for Kenya (2013), Ministry of Environment, Water and Natural Resources, Nairobi. pp. 01−70.

German Government's Policy on Waste. (2017).

Ghadimzadeh, A., Askari, A., Gomes, C., & Ishak, M. D. (2014). E-waste management: Towards an appropriate policy. *European Journal of International Management, 6*(1), 37−46.

Ghosh, A., Pratt, A. T., Soma, S., Theriault, S. G., Griffin, A. T., & Trivedi, P. P. (2016). Mitochondrial disease genes COA6, COX6B and SCO2 have overlapping roles in COX2 biogenesis. *Human Molecular Genetics, 25*(4), 660−671.

Guomei, Z. (2006). Promoting 3R strategy: E-waste management in China. *State Environmental Protection Administration of China*, 01−23.

Kahhat, R., Kim, J., Xu, M., Allenby, B., & Williams, E. (2008). Proposal for an E-waste management system for the United States. *IEEE international symposium on electronics and the environment* (pp. 01−06).

Kumar, M. J., Krishna, S., & Raj. (2014). E-waste, a new challenge to the environmentalists. *Nature Environment and Pollution Technology, 13*, 333−338.

Lee, S. C., & Na, S. (2010). E-waste recycling systems and sound circulative economies in East Asia: A comparative analysis of systems in Japan, South Korea, China and Taiwan. *Sustainability, 2*, 1632−1644.

Li, J., Zeng, X., Chen, M., Ogunseitan, O. A., & Stevels, A. (2015). "Control-Alt-Delete": Rebooting solutions for the E-waste problem. *Environmental Science and Technology, 49*(12), 7095−7108.

Maphosa, V., & Maphosa, M. (2020). E-waste management in Sub-Saharan Africa: A systematic literature review. *Cogent Business & Management, 7*, 01−17.

Ministry of Environment and Forestry-National E-waste Mgt. Strategy (2019). Cabinet Secretary, Ministry of Environment and Forestry. pp. 01–43.

Ministry of Environment Report, Japan. (2010). *History and current state of waste management in Japan* (pp. 01–32).

Mundada, M., Kumar, S., & Shekdar, A. (2004). E-waste: A new challenge for waste management in India. *International Journal of Environmental Studies, 61*, 265–279.

Namias, J. (2013). *The future of electronic waste recycling in the United States: Obstacles and domestic solutions.* Columbia University.

National Environmental Management Act. (1998). *Amendments to environmental impact assessment regulations* (ACT NO. 107 OF 1998).

Northeast Supply Enhancement (NESE) Project Report (2017), Williams, pp. 01–04.

Otieno, I., & Omwenga, E. (2015). E-waste management in Kenya: Challenges and opportunities. *Journal of Emerging Trends in Computing and Information Sciences, 6*(12), 661–666.

Parliamentary Office of Science and Technology of the UK Parliament (POST) Note-292. (2007). *Radio spectrum management.*

Rajya Sabha Secretariat Research Unit Report. (2011). *E-waste in India.*

Robinson, B. H. (2009). E-waste: An assessment of global production and environmental impacts. *Science of the Total Environment, 408*(2), 183–191.

Shamim, A., Mursheda, A., & Rafiq, I. (2015). E-waste trading impact on public health and ecosystem services in developing countries. *International Journal of Waste Resources, 5*(4), 01–12.

Sharma, R., & Hussain, S. (2018). India: E-waste management in India. *Mandaq Sheshu Babu, Sabrang India article.*

Shumon, R.H., & Ahmed, S. (2013). Sustainable WEE management in Malaysia: Present scenarios and future perspectives. In *IOP conference series: Materials science and engineering* (pp. 01–10).

Sinha-Khetriwal, D., Kraeuchi, P., & Schwaninger, M. (2005). A comparison of electronic waste recycling in Switzerland and in India. *Environmental Impact Assessment Review, 25*, 492–504.

Tomohiro, T., Atsushi, T., & Yuichi, M. (2005). Effective assessment of Japanese recycling law for electrical home appliances: Four years after the full enforcement of the law. In *Proceedings of IEEE international symposium on electronics and the environment* (pp. 243–249).

Tong, X., & Yan, L. (2013). From legal transplants to sustainable transition – Extended producer responsibility in Chinese waste electrical and electronic equipment management. *Journal of Industrial Ecology, 17*(2), 199–212.

Tyagi, N., Baberwal, S. K., & Passi, N. (2015). E-waste: Challenges and its management. *Journal of Undergraduate Research and Innovation, 1*(3), 108–114.

Umesh, K. (2013). E-waste management through Regulations. *International Journal of Engineering Inventions, 3*(2), 06–14.

Waste Electrical and Electronics Equipment (WEEE) Regulations for United Kingdom (2013). Department of Business, Innovation and Skills. pp. 01–38.

Waste Electrical and Electronics Equipment (WEEE) Regulations Government Guidelines Notes (Government of UK) (2014). Department of Business, Innovation and Skills. pp. 01–38.

Waste Management in Germany. (2018). Federal Ministry for the Environment, Nature Conservation and Nuclear Safety, BMU.

Widmer, R., Krapf, O. H., Khetriwal, S. D., Schnellmann, M., & Boni, H. (2005). Global perspectives on E-waste. *Environmental Impact Assessment Review, 25*(5), 436–458.

World Intellectual Property Report (2013). WIPO Economics & Statistics Series. pp. 01–208.

CHAPTER 13

Development of success factors for managing E-waste in Indian organizations: an empirical investigation

Somvir Arya, Ajay Gupta and Arvind Bhardwaj
Industrial and Production Engineering Department, Dr. B.R. Ambedkar National Institute of Technology, Jalandhar, Punjab, India

13.1 Introduction

On concern with global e-scrap recycling, the first country to implement and manage E-waste is Switzerland. In Switzerland and other developing nations, the backbone for E-squander recycling system is the advance recycling fee and extended producer responsibility. Nations like Germany, the United Kingdom, the United States, and France are producing approximately 1.5—3 million tonnes of E-waste every year and are among major E-waste generators in the world. These nations follow homogeneous E-waste management processes. An appropriate E-waste recycling starting from an efficient collection to extraction and discarding of materials has guaranteed lucrative prospect for business (Khan, 2017).

A large number of workshops as well as research have been carried out by various organizations like Basel Convention to know about the various problems associated with adoption of environmentally sound management (ESM) of e-scrap in Asian nations. From the researches, it has been identified that there is a shortage of E-waste catalog, lesser number of experienced staff to execute ESM practices, insufficient infrastructure, and insufficient knowledge about the impact of E-waste on the environment (Saoji, 2012). With the developments in lifestyle as well as contemporary technology, the demand for electric home appliances and electronic products is growing rapidly. These gadgets have become an integral part of our lives. These gadgets have both positive and negative effect on our lives. These devices contain toxic metals which are dangerous for human life and if they mix with the environment, they are likely to cause serious harmful effects. Electronic business is the world's fastest and largest growing trade. Rapid development along with urbanization as well as increased interest in consumer goods; all these factors lead to rise in the manufacturing and usage of electrical and electronic appliances (Garlapati, 2016). This study is an attempt to assess the performance of E-waste management strategies by using structural equation modeling (SEM).

13.2 Research framework

For this study, survey questionnaire has been prepared. The questionnaire is pretested for content validity by experienced managers of industry of Punjab. Furthermore, construct, convergent, SEM, and discriminant validity has been checked by using AMOS software. From the exhaustive literature review, different constructs have been identified. Then the relationship between these constructs has been identified (with or without mediation effect). The development and comparison of different SEM models with and without mediation effect has been done and best-fitting model for Indian context has been identified. The measurement of factors has been done on 5-point like scale
- No Idea
- Little Knowledge
- Average Knowledge
- High Knowledge
- Highly Updated Knowledge
  Fig. 13.1 shows the research framework used for our study.

13.3 Exploratory factor analysis

This study focuses on exploring the awareness about the E-waste management by the organizations. This study uses exploratory factor analysis (EFA) method in order to explore the factors affecting the awareness level for E-waste management by the different organizations. This study uses *EFA method to identify the different latent components of awareness about the E-waste management by* the different organizations in order to manage the E-waste. In order to measure the different dimensions of the E-waste management used by the organizations, *twenty-one statements* related to E-waste management are included in the questionnaire. In order to identify the correlation structure between the statements and to identify the latent variables of the E-waste management, EFA method is applied on the 21 statements of E-waste management. The EFA methods facilitate in identifying the correlation relationship among the variables taken into consideration for the study. The EFA statistical technique examines the *correlation* among all the pairs of the variables considered in the analysis and tries and reduces the variables into few meaningful latent variables. *These latent variables are also known as factors.* These latent factors in the study represent a set of statements having significant correlation among them. EFA requires the fulfillment of few assumptions such as the availability of sampling

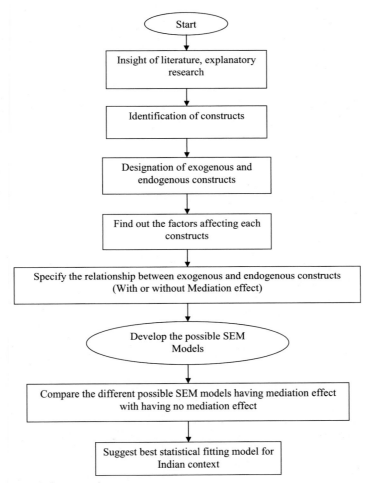

Figure 13.1 Research framework.

adequacy and the presence of significant coefficient of correlations among the pair of variables considered in the analysis. The Kaiser—Meyer—Olkin (KMO) measure and Bartlett's test of sphericity are used in the factor analysis procedure in order to check the presence of required sampling adequacy and the correlation structure between distinctive pair of variables. The statistical result of KMO measures of sampling adequacy and Bartlett check of sphericity are presented in Table 13.1.

The statistical result of KMO test indicates that the KMO statistic is 0.833 indicating the presence of *required sampling adequacy in the dataset*. The KMO statistic of 0.833 also represents the adequacy of sufficient variance in the responses toward the statements measuring the different components of E-waste management used by the organizations. This is the necessary condition for applying EFA to the statements. The Bartlett's analysis of

Table 13.1 Kaiser–Meyer–Olkin measure of sampling adequacy and Bartlett test of sphericity.

Kaiser–Meyer–Olkin measure of sampling adequacy		0.833
Bartlett's test of sphericity	Approx. Chi square	1343.416
	df	210
	Sig.	0.000

Table 13.2 Communalities value.

	Initial	Extraction
BT_1	1	0.682
BT_2	1	0.774
BT_3	1	0.799
BT_4	1	0.721
K_Pr_1	1	0.779
K_Pr_2	1	0.747
K_Pr_3	1	0.714
K_Pr_4	1	0.699
K_Bn_1	1	0.747
K_Bn_2	1	0.843
K_Bn_3	1	0.734
K_Br_1	1	0.781
K_Br_2	1	0.721
K_Br_3	1	0.756
K_Br_4	1	0.503
K_Br_5	1	0.853
K_Br_6	1	0.737
Aw_1	1	0.664
Aw_2	1	0.652
Aw_3	1	0.751
Aw_4	1	0.767

sphericity indicates the existence of significant correlation between the set of statements included in the analysis. *The null hypothesis of the Bartlett test assumes that the correlation matrix indicating the coefficient of correlation between all pair of variables is an identity matrix.* The results of Bartlett test indicate that probability value of Chi-square statistic is less than five percentage level of significance. Thus, with 95% confidence level, it could be concluded that the correlation matrix representing the coefficient of Pearson correlation is not an identity matrix. For this reason, it could be concluded that there exist significant relations between different pair of statements included in EFA analysis.

Table 13.2 represents the communalities or the variance explained of the included statements before and after the factor extraction. The initial communality (earlier than extraction) is always assumed to be 1. But, after the extraction, the communality will depend upon the amount of variance explained of the selected variable by the extracted factors. Initially 100% variance is available but the some portion of variance is lost in the process of extraction using EFA method. For this reason, it is required to investigate the remaining variance available for the evaluation. The communality of the variable is presented in Table 13.2 which indicates the percentage of variance defined by the variables before and after the extraction process in EFA analysis.

The result shows that the initial communalities of every variable are 1; however, the extracted communalities are much <1. The result of communalities indicates that the extracted variance of all the variables is more than 50%. The extracted communalities indicate the goodness-of-fit of the factor analysis. Higher the value of extracted communalities of variables better it is. As a result, all the variables considered in the factor analysis contribute in the factors extracted. The one of the most important key outputs of EFA analysis is the Eigen values of the components. The number of Eigen values is equal to the number of statements included in the analysis. However, only factors with Eigen value >1 is chosen for the further analysis. The results of factor analysis after applying principal component analysis are shown in Table 13.3.

The results of principal component analysis indicate that 21 statements considered for the EFA can be reduced to 5 factors having Eigen values >1. These five factors explain approximately 73% of the total variance of the included statements. Assuming that the explained variance of 73% is sufficient, the extracted factors can be used for further analysis. In the factor analysis, orthogonal rotation (Varimax method) is applied in order to improve the explanation of the variance of the factors. The rotated component matrix (RCM) represents the factor loading of each variable to the extracted factors.

The component loadings can be described as the correlation between the factors and the variables. The high-factor loadings of the statement with one single factor indicate that the variables taken into consideration for the analysis have significant correlation with that factor and insignificant component loadings to all different extracted factors. The result of the RCM is presented in Table 13.4.

The result of RCM indicates that 21 statements of the E-waste management can be reduced to 5 extracted components. It is also found that all the variables of E-waste management have significant factor loadings to only one single factor and insignificant factor loadings to other extracted factors. It is also observed from the results of EFA that the significant factor loadings for each factor are found to be >0.5. Thus it can be concluded from the results that extracted factors/components from the included statements fulfills the assumptions of convergent and discriminant validity. Observing the meanings of the statements used in the analysis, these factors can be named as:

- Basic Terminology
- Knowledge of Practices
- Knowledge of Benefits

Table 13.3 Eigen values of the components and total variance explained.

Component	Initial Eigen values			Extraction sums of squared loadings			Rotation sums of squared loadings		
	Total	% of variance	Cumulative %	Total	% of variance	Cumulative %	Total	% of variance	Cumulative %
1	6.736	32.076	32.076	6.736	32.076	32.076	4.266	20.312	20.312
2	3.028	14.418	46.493	3.028	14.418	46.493	3.044	14.493	34.805
3	2.482	11.82	58.313	2.482	11.82	58.313	2.97	14.143	48.949
4	1.909	9.088	67.401	1.909	9.088	67.401	2.651	12.624	61.572
5	1.272	6.055	73.456	1.272	6.055	73.456	2.496	11.884	73.456
6	0.788	3.754	77.21						
7	0.609	2.901	80.11						
8	0.516	2.458	82.568						
9	0.487	2.32	84.888						
10	0.444	2.116	87.004						
11	0.389	1.854	88.858						
12	0.381	1.812	90.671						
13	0.304	1.449	92.12						
14	0.284	1.352	93.472						
15	0.276	1.312	94.784						
16	0.243	1.155	95.94						
17	0.215	1.025	96.965						
18	0.202	0.961	97.926						
19	0.161	0.766	98.692						
20	0.147	0.702	99.394						
21	0.127	0.606	100						

Table 13.4 Rotated component matrix.

	Component				
	1	2	3	4	5
BT_1			0.81		
BT_2			0.865		
BT_3			0.865		
BT_4			0.789		
K_Pr_1		0.841			
K_Pr_2		0.837			
K_Pr_3		0.828			
K_Pr_4		0.772			
K_Bn_1					0.78
K_Bn_2					0.865
K_Bn_3					0.831
K_Br_1	0.871				
K_Br_2	0.821				
K_Br_3	0.862				
K_Br_4	0.58				
K_Br_5	0.905				
K_Br_6	0.804				
Aw_1				0.73	
Aw_2				0.602	
Aw_3				0.731	
Aw_4				0.823	

- Knowledge of Barriers
- Awareness

13.4 Validity of construct

Construct validation consists of two steps, that is, face validity and construct validity.

13.4.1 Face validity

Face validity is assessed by looking at the measure and assessing whether "on its face" the measure seems a good reflection of the construct (Trochim et al., 2015). In the opinion of experts from industry and academia, the questions were relevant for each construct and thus the face validity is established.

13.4.2 Construct validity

The construct validity of the different dimensions representing the different aspects of all the constructs from the organization's perspective is analyzed with the help of confirmatory factor analysis (CFA) method. The construct validity represents the validity of the scale developed in the study and confirms the presence of convergent validity and discriminant validity. The convergent validity indicates the presence of high correlation among different statements with the related construct. The convergent validity is measured with the help of construct loading of the different statements within the construct, composite reliability (CR) statistics of the different constructs in the scale, and average variance extracted (AVE) statistics for each construct in the scale. It is expected that in order to fulfill the requirement of convergent validity, the standardized slope coefficients of each statement should be >0.7 (Hair et al., 2010). The CR estimate of each factor explaining the different aspects of all the constructs about E-waste management from the organization's perspective is required to be >0.7 (Gupta & Singh, 2015; Hair et al., 2010). Also, AVE statistics of each extracted factor needs to be >0.5 (Gupta & Singh, 2015; Hair et al., 2010). The presence of convergent validity indicates that all the selected statements to represent the construct are significantly representing it. The presence of discriminant validity indicates that the different constructs, explaining the different aspects of all the constructs from the organization's perspective included in the scale, do not have significant correlation among them. The discriminant validity indicates that the responses received against the different statements explaining the various aspects of the constructs from the organization's perspective are not similar. The discriminant validity is measured with help of correlation between different pairs of constructs in the scale. It is expected that the correlation among the constructs will not be very high. In CFA analysis, correlation maxima between each construct and all other constructs is calculated and is termed as maximum shared variance (MSV). The values of AVE should be greater than the values of MSV for each construct (Dangi & Dewen, 2016; Hair et al., 2010). Correlation coefficients between different constructs should be positive, moderate, and <0.7 and square root estimates of AVE should be greater than the correlation estimate (Hair et al., 2010; O'Rourke & Hatcher, 2013).

To test the convergent and discriminant validity, measurement model is created using AMOS and is shown in Fig. 13.2. The construct loadings along with correlation coefficients between the different pairs of factors are also shown in Fig. 13.2.

The standardized regression weight should be >0.60 but ideally it should be ≥ 0.7 (Hair et al., 2010). For our analysis, we have considered 0.7 as critical value for each factor loading. As can be seen from Fig. 13.2, all the factor loadings between constructs and their measured items are above 0.7 except one value, the fourth measured item of construct namely *knowledge of barriers* in implementation of E-waste management which has 0.54 construct loading, which is <0.7. It indicates that the correlation, between the responses of various organizations to this question and the questions measuring barriers to E-waste management, is not strong. So the fourth item representing the statement, "To what extent poor attitude of the end users towards E-waste

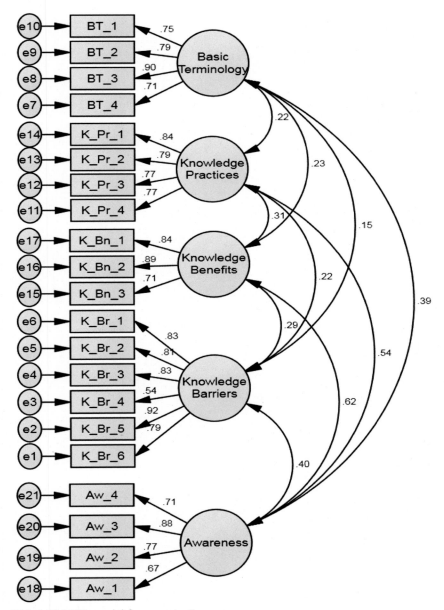

Figure 13.2 Initial CFA model for organizations.

management is a barrier in adopting formal E-waste management practices?" was removed. After removing this measurement item, various factor loadings and correlation values are calculated again by using AMOS and are shown in Fig. 13.3.

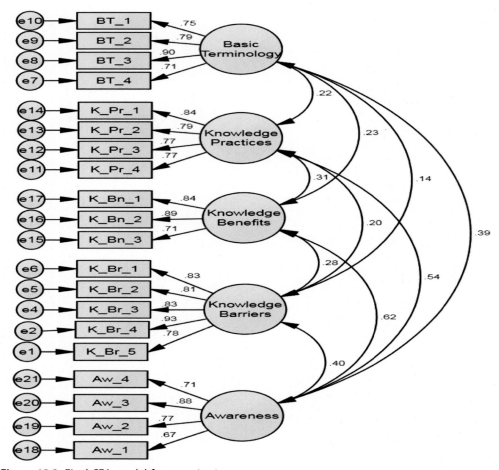

Figure 13.3 Final CFA model for organizations.

As can be seen from Fig. 13.3, all the factor loadings are above 0.7. Table 13.5 presents the statistical significance of all the measured items associated with each dimension of E-waste management.

13.4.2.1 Convergent validity
Table 13.5 presents the results of the calculated values of standardized construct loadings (standardized slope coefficient), unstandardized slope coefficients, standard error of the regression weight, critical ratio, and probability of occurrence of each regression coefficient due to sampling from a population in which there is no correlation between a measured item and the construct. As can be seen, P values are almost zero indicating that there is significant relationship between the constructs and the

Table 13.5 Regression weight for organizations (CFA).

			Standardized slope coefficient	Unstandardized regression weight	Standard error	Critical ratio	P
K_Br_5	<---	Knowledge of barriers	0.783	1.00		10.816	***
K_Br_4	<---		0.932	1.278	0.118	9.356	***
K_Br_3	<---		0.831	1.128	0.121	8.998	***
K_Br_2	<---		0.806	1.061	0.118	9.385	***
K_Br_1	<---		0.833	0.926	0.099		***
BT_4	<---	Basic terminology	0.714	1.00		8.16	***
BT_3	<---		0.895	1.186	0.145	7.503	***
BT_2	<---		0.794	0.949	0.126	7.08	***
BT_1	<---		0.746	0.976	0.138		***
K_Pr_4	<---	Knowledge of practices	0.767	1.00		7.802	***
K_Pr_3	<---		0.771	1.049	0.134	8.046	***
K_Pr_2	<---		0.794	1.185	0.147	8.468	***
K_Pr_1	<---		0.838	1.104	0.13		***
K_Bn_3	<---	Knowledge of benefits	0.713	1.00		7.992	***
K_Bn_2	<---		0.893	1.147	0.143	7.787	***
K_Bn_1	<---		0.837	0.87	0.112		***
Aw_1	<---	Awareness	0.671	1.00		6.763	***
Aw_2	<---		0.768	1.237	0.183	7.429	***
Aw_3	<---		0.880	1.539	0.207	6.343	***
Aw_4	<---		0.712	1.294	0.204		***

*** P-values are almost zero indicating that there is significant relationship between the constructs and the measured items.

Table 13.6 Convergent validity statistics.

	Composite reliability (CR)	Average variance extracted (AVE)
Awareness	0.888	0.666
Knowledge of barriers	0.915	0.685
Basic terminology	0.867	0.620
Knowledge of practices	0.863	0.613
Knowledge of benefits	0.843	0.643

measured items. These highly positive and significant values of standardized slope coefficients indicate that all the statements included in the study do reflect the respective dimension of E-waste management which they are supposed to measure. Similarly, the critical ratios for all the statements are found to be >1.96 and P values for all the statements are significantly $<.05$ indicating that the probability of getting these values of factor loadings as a matter of chance due to random sampling variation is negligible. Therefore convergent validity of each of the four constructs of E-waste management used in the present study is established.

Overall ability of a construct to explain the variability of measurement items is also measured with the help of CR and AVE. These values indicate the extent of variability of different measurement items explained by the construct to which they are hypothesized to be related. In the study, the CR statistic and AVE are estimated for all the constructs of E-waste management. These indicators of convergent validity are presented in Table 13.6.

The results indicate that the CR estimate of all the constructs of E-waste management calculated from the responses of individual users is >0.7 and AVE estimate of all the constructs is found to be >0.5. Hence, it can be concluded from the study that the all-measurement questions converge to their respective factors.

13.4.2.2 Discriminant validity

The objective of establishing discriminant validity is to ensure that different factors/dimensions are sufficiently different from one another, although it may not be possible to identify factors that are 100% different from one another. To establish discriminant validity, correlations between each pair of constructs/factors used in the current study for measuring the level of awareness of individuals about E-waste management are calculated. These calculated correlation values between the different constructs of E-waste management are presented in Table 13.7.

The results of the correlation analysis as shown in Table 13.7 indicate that Pearson's coefficient of correlation between different pairs of E-waste management dimensions namely basic terminology, knowledge of different practices for E-waste management, knowledge of benefits of waste management, knowledge of barriers in

Table 13.7 Correlation estimates between the constructs.

			Estimate
Knowledge of barriers	< – >	Basic terminology	0.279
Knowledge of barriers	< – >	Knowledge of practices	0.245
Knowledge of barriers	< – >	Awareness	0.344
Knowledge of barriers	< – >	Knowledge of benefits	0.237
Basic terminology	< – >	Knowledge of practices	0.457
Basic terminology	< – >	Awareness	0.586
Basic terminology	< – >	Knowledge of benefits	0.516
Knowledge of practices	< – >	Awareness	0.59
Knowledge of practices	< – >	Knowledge of benefits	0.46
Awareness	< – >	Knowledge of benefits	0.618

Table 13.8 Discriminant validity estimates.

	Composite reliability (CR)	Average variance extracted (AVE)	Maximum shared variance (MSV)
Awareness	0.888	0.666	0.382
Knowledge of barriers	0.915	0.685	0.118
Basic terminology	0.867	0.620	0.343
Knowledge of practices	0.863	0.613	0.348
Knowledge of benefits	0.843	0.643	0.382

implementation of E-waste management, and awareness about E-waste management from individual's perspective is moderately correlated in nature. In order to confirm the presence of discriminant validity, the estimated coefficient of correlation between different aspects of E-waste management should not be very high. It is found that the Pearson's correlation coefficients between different constructs are moderate, positive, and <0.7. This indicates that moderate level of correlations exists between different constructs of E-waste management. The presence of moderate correlation between different constructs by itself does not establish the presence or absence of discriminant validity in the developed scale. For this purpose, maximum variance shared by a dimension with any other dimension of the scale is to be calculated and compared to AVE of that factor. These indicators of validity measures are presented in Table 13.8.

As can be seen from Table 13.8, AVE of each factor is >0.5 as well as the value of MSV, thereby by establishing the discriminant validity. To further establish discriminant validity, square root of the AVE of each dimension is also compared with the correlation of each dimension with all other dimensions of E-waste management.

The results are presented in Table 13.9.

The results indicate that the square root of the AVE of each dimension of E-waste management is greater than the correlation coefficient. Hence, it can be concluded from this statistical analysis that the measurement model has discriminant validity.

To assess the overall fit of the model, various indices are calculated and are presented in Table 13.10.

As can be seen from Table 13.10, CMIN/df is 1.933 which is less than the maximum acceptable value of 3, GFI is 0.906 which is more than the requisite value of 0.8, augmented goodness-of-fit index (AGFI) is 0.876 which is more than the prescribed value of 0.8, comparative fit index (CFI) is 0.957 which is more than the expected value of 0.9, Tucker–Lewis coefficient (TLI) is 0.949 which is more than the required value of 0.9, and root mean square error approximation (RMSEA) is 0.057 which is less than the recommended value of 0.08. Thus it can be concluded that all the calculated values are within the acceptable range and hence overall statistical fit of the measurement model is acceptable.

Table 13.9 Square root of AVE and correlation estimates.

	Awareness	Knowledge of barriers	Basic terminology	Knowledge of practices	Knowledge of benefits
Awareness	0.816				
Knowledge of barriers	0.344	0.828			
Basic terminology	0.586	0.279	0.787		
Knowledge of practices	0.590	0.245	0.457	0.783	
Knowledge of benefits	0.618	0.237	0.516	0.460	0.802

Table 13.10 Statistical fitness of CFA model for individuals.

Goodness of fit index	CMIN/ df	Goodness of fit index (GFI)	Augmented goodness of fit index (AGFI)	Comparative fit index (CFI)	Tucker-Lewis index (TLI)	Root mean square error approximation (RMSEA)
Calculated value	1.933	0.906	0.876	0.957	0.949	0.057
Expected value	>3	<0.8	<0.8	<0.9	<0.9	>0.08

13.4.3 Concluding remarks—confirmatory factor analysis

The presence of convergent validity indicates that all the selected statements to represent/measure various factors of E-waste namely basic terminology, knowledge of practices, knowledge of barriers, knowledge of benefits, and overall awareness about E-waste are significantly representing it and each factor used to measure the variability of statements explains its substantial portion. On the other hand, discriminant validity indicates that the different factors used for explaining various aspects of overall awareness level of individuals about E-waste management included in the scale do not have significant correlation among them and each factor is measuring some unique dimension of E-waste awareness. The results of CFA showed that the questions in the questionnaire are valid and constructs are not significantly related to each other which are the basic need of such analysis. The result also indicated that the conducted survey was valid.

Since the CFA established the validity of various constructs and the measurement questions associated with each construct, the response of individuals can be used to measure average level of awareness of individuals about E-waste. Tables 13.11—13.15 present summery statistics for each measured item of all the constructs.

13.4.4 Concluding remarks—questionnaire

Tables 13.11—13.15 present the percentage of responses received for each option of various questions in the questionnaire, mean score for different questions, and mean score for each construct. Since the average score of all the constructs is low, there is a need to increase this awareness level. For this, it needs to be established if each of the construct significantly contributes to overall awareness of individuals and whether all the constructs are directly affecting overall awareness or these are some mediating effects present. Answer to these questions will help government to frame the appropriate awareness improvement strategy. The following analysis was carried out to answer these questions.

13.5 Structural equation Model A for organizations

The following hypothesis are proposed and need to be tested with the help of SEM (Fig. 13.4):

Hypothesis 1:
"There is no impact of awareness about basic terminology of E-waste management on overall awareness level of the organizations"

Table 13.11 Knowledge about basic terminology of individuals on E-waste disposal system.

S. No.	Basic terminology	No idea {1}	Little knowledge {2}	Average knowledge {3}	High knowledge {4}	Highly updated knowledge {5}	Mean	SD
1.	Do you have any idea of the term "E-waste"?	10.1%	40.1%	28.6%	14.3%	7.0%	2.68	1.1
2.	Do you have idea about how a common man can contribute towards managing E-waste efficiently?	7.0%	31.0%	27.9%	24.0%	10.1%	2.99	1.11
3.	Are you aware of hazardous materials present in E-waste?	7.0%	30.3%	29.3%	23.3%	10.1%	2.99	1.11
4.	Do you consider life of the electronic product as an important factor at the time of buying a new product for reducing E-waste generation?	10.1%	39.4%	25.8%	14.3%	10.5%	2.76	1.14
Mean total							2.86	

Table 13.12 Knowledge of practices of individuals about E-waste management.

S. No.	Knowledge of practices	No idea {1}	Little knowledge {2}	Average knowledge {3}	High knowledge {4}	Highly updated knowledge {5}	Mean	SD
1.	How do you rate official take back system used by some e-products manufacturing companies, as a good E-waste management practice	22.0%	29.3%	17.8%	17.8%	13.2%	2.71	1.34
2.	Do you know that there are better channels available for disposing E-waste rather than selling these to a normal rag picker?	27.2%	28.2%	13.9%	15.7%	15.0%	2.63	1.41
3.	In Indian perspective, do you think that easy access to informal collection and recycling channels is a barrier in adopting formal E-waste management practices?	25.1%	22.0%	19.9%	20.9%	12.2%	2.73	1.36
4.	Do you think that inadequate financial incentive offered by official take back channels of the companies is a barrier in their adoption in India?	19.9%	28.2%	20.9%	19.2%	11.8%	2.75	1.3
Mean total							2.71	

Table 13.13 Knowledge of benefits of E-waste management.

S. No.	Knowledge of benefits	No idea {1}	Little knowledge {2}	Average knowledge {3}	High knowledge {4}	Highly updated knowledge {5}	Mean	SD
1.	Do you think proper removal of hazardous materials from E-waste is beneficial in minimizing negative effect on human health and environment?	21.3%	43.6%	15.7%	17.4%	2.1%	2.36	1.1
2.	To what extent, do you think, formal E-waste management will reduce public health hazards?	24.7%	42.2%	14.6%	14.3%	4.2%	2.31	1.12
3.	Do you think formal recycling is helpful in reducing the final volume of E-waste (Recovery of components/parts)?	34.5%	23.7%	18.8%	17.8%	5.2%	2.36	1.26
Mean total							2.34	

Table 13.14 Knowledge of barrier of E-waste management.

S. No.	Knowledge of barrier	No idea {1}	Little knowledge {2}	Average knowledge {3}	High knowledge {4}	Highly updated knowledge {5}	Mean	SD
1.	Do you think, limited access to formal recycling facility is a barrier in formal disposal of E-waste?	39.0%	40.8%	7.3%	5.2%	7.7%	2.02	1.17
2.	Do you think, easy access to informal collection channels is a barrier in adoption of formal E-waste disposal methods by individuals?	41.8%	35.2%	8.7%	7.0%	7.3%	2.03	1.20
3.	Do you think lack of knowledge about hazardous effects of E-waste act as a barrier in adopting formal E-waste management practices?	46.7%	24.4%	11.8%	11.5%	5.6%	2.05	1.25
4.	To what extent do you think that inadequate legislations are resulting in nonadoption of formal E-waste recycling approach?	41.1%	28.6%	15.0%	8.0%	7.3%	2.12	1.24
5.	Do you think that inadequate collection efforts by recyclers/ Govt. agencies are resulting in piling up of E-waste at home?	50.2%	29.3%	13.2%	4.9%	2.4%	1.80	1.01
Mean total							2.0	

Table 13.15 Awareness of individuals on E-waste management.

S. No.	Awareness	No idea {1}	Little awareness {2}	Average awareness {3}	High awareness {4}	Highly updated awareness {5}	Mean	SD
1.	To what extent do you know how to dispose off E-waste safely for human beings as well as for the environment?	30.0%	44.9%	16.7%	4.9%	3.5%	2.07	0.99
2.	Are you aware of any electronic waste management policy currently implemented in India for safe disposal of electrical and electronic items?	53.7%	16.7%	17.8%	6.6%	5.2%	1.93	1.20
3.	What is your perspective towards health and environmental hazards associated with E-waste?	46.0%	30.3%	14.6%	4.2%	4.9%	1.92	1.10
4.	Do you think toxic/hazardous materials from discarded e-products require special treatment for environmentally sound disposal?	35.9%	33.8%	17.8%	9.1%	3.5%	2.10	1.1
Mean total							2.01	

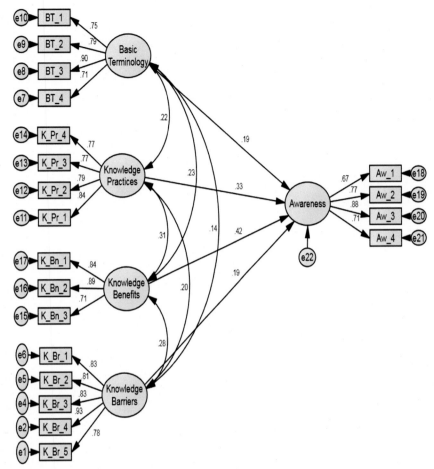

Figure 13.4 CFA model with awareness as mediating variable.

Hypothesis 2:
"There is no impact of knowledge of E-waste management practices on overall awareness level of the organizations"

Hypothesis 3:
"There is no impact of knowledge about benefits arising out of proper E-waste management on overall awareness level of the organizations"

Hypothesis 4:
"There is no impact of knowledge of barriers about E-waste management on overall awareness level of the organizations"

Table 13.16 presents the values of slope coefficients, regression weight, standard error, critical ratio, P value for each construct, and the R^2. The slope coefficient is used to examine the *cause-and-effect relationship* mentioned in the proposed hypothesis. The probability value of critical ratio in case of all the considered constructs of E-waste management is found to be <0.05. The conclusions made on the basis of SEM analysis are discussed below.

13.5.1 Hypothesis 1: "There is no impact of awareness about basic terminology of E-waste management on overall awareness level of the organizations"

Conclusion: The standardized slope coefficient indicating the *cause-and-effect relationship* between awareness of basic terminology of E-waste management and overall awareness level of organizations about E-waste management is 0.193 with P value $<.05$. Thus the hypothesis that *"There is no impact of awareness about basic terminology of E-waste management on overall awareness level of the organizations"* is rejected at 5% level of significance. The awareness about basic terminology of E-waste management has a significant positive impact on overall awareness level of organizations about E-waste management. Thus in framing overall viewpoint on E-waste management at organizational level, knowledge of basic terminology of E-waste management is important.

13.5.2 Hypothesis 2: "There is no impact of knowledge of E-waste management practices on overall awareness level of the organizations"

Conclusion: The standardized slope coefficient indicating the *cause-and-effect relationship* between knowledge of practices about E-waste management and overall awareness level of organizations about E-waste management is 0.327 with P value $<.05$. Thus the hypothesis that *"There is no impact of knowledge of E-waste management practices on overall awareness level of the organizations"* is rejected at 95% confidence level. The knowledge of E-waste management practices has a significant positive impact on overall awareness level of organizations about E-waste management.

13.5.3 Hypothesis 3: "There is no impact of knowledge about benefits arising out of proper E-waste management on overall awareness level of the organizations"

Conclusion: The standardized slope coefficient between knowledge of benefits about E-waste management on the overall awareness level of organizations about E-waste management is 0.419 with P value $<.05$. Thus the hypothesis that *"There is no impact of knowledge about benefits arising out of proper E-waste management on overall awareness level of the organizations"* is rejected at 5% level of significance. Knowledge of benefits of proper E-waste management will act as a driving force to take interest and enhance their knowledge on different dimensions of overall awareness about E-waste management.

Table 13.16 Regression weights for Model A.

		Standardized slope coefficient	Unstandardized regression weight	Standard error	Critical ratio	P	R^2
Awareness	<--	Basic terminology	0.193	0.157	0.074	2.124	.034	0.58
Awareness	<--	Knowledge of practices	0.327	0.21	0.064	3.299	***	
Awareness	<--	Knowledge of benefits	0.419	0.31	0.082	3.779	***	
Awareness	<--	Knowledge of barriers	0.187	0.16	0.076	2.122	.034	

***P-values are almost zero indicating that there is significant relationship between the constructs and the measured items.

13.5.4 Hypothesis 4: "There is no impact of knowledge of barriers in implementing proper E-waste management on overall awareness level of the organizations"

Conclusion: The standardized slope coefficient indicating the *cause-and-effect relationship* between knowledge of barriers in adoption of proper E-waste management and overall awareness of organizations about E-waste management is 0.187 with P value $<.05$. Thus the hypothesis that *"There is no impact of knowledge of barriers in implementing proper E-waste management on overall awareness level of the organizations"* is rejected at 95% confidence level. The knowledge of barriers about E-waste management has a significant positive impact on overall awareness level of organizations about E-waste management. Knowledge of barriers in implementing E-waste management practices and the methods to overcome those barriers are important determinants of overall awareness of organizations about E-waste management. For effective E-waste management system, government needs to take some effective steps to overcome the barriers.

The value of coefficient of determination (R^2) should be >0.39 (O'Rourke & Hatcher, 2013) for assessing the percentage of variance of dependent variable explained by independent variables.

The R^2 value of the SEM model explaining the overall awareness of the organizational users in E-waste management with the help of included factors is found to be 0.58. The R^2 of 0.58 means that 58% of variation of overall awareness is explained with the help of four exogenous constructs. The goodness–of–fit indices of the structural model are presented in Table 13.17.

The results for the measurement model indicating the different attributes of E-waste management have the following fit indices:

- calculated value of CMIN/df is 1.355 which is less than the maximum acceptable value of 3,
- calculated value of GFI is 0.835 which is more than the required minimum value of 0.8,
- calculated value of AGFI is 0.811 which is more than the minimum expected value of 0.8,

Table 13.17 Statistical fitness of Model A.

Goodness of fit index	CMIN/ df	Goodness of fit index (GFI)	Augmented goodness of fit index (AGFI)	Comparative fit index (CFI)	Tucker-Lewis index (TLI)	Root mean square error approximation (RMSEA)
Calculated value	1.355	0.835	0.811	0.951	0.944	0.059
Expected value	>3	<0.8	<0.8	<0.9	<0.9	>0.08

- calculated value of CFI is 0.951 which is more than the minimum requisite value of 0.9,
- calculated value of TLI is 0.944 which is more than the minimum expected value of 0.9, and
- calculated value RMSEA is 0.057 which is less than the maximum desired value of 0.08.

Thus it is concluded that overall statistical fitness of the measurement model is acceptable.

Through this model meets all the statistical requirements of a good fit model; yet there is no guarantee that it is the best fit model. Any model with a significantly lesser Chi-square value will be a better model. This possibility was explored by testing the possible SEM models.

13.5.5 Structural equation Model B for organizations

In this model, we are using knowledge of practices as a mediator between basic terminology, knowledge of benefits, and knowledge of barriers and overall awareness of E-waste management. This approach is based on the premise that an organization first becomes aware of basic terminology, benefits, and barriers of E-waste management. Once they gained the certain level, the organizational consumers try to learn about best practices for proper E-waste management and this knowledge completes his overall awareness level about E-waste management.

The hypothesis for comparison between Model A and proposed mediation model is

"Model with knowledge about practices acting as mediator between other three exogenous constructs and endogenous construct is a better model than Model A, i.e., the model without any mediation."

The structural model is developed to test this hypothesis and the same is shown in Fig. 13.5.

Various statistical fit indices for this model are presented in Table 13.18.

13.5.6 Criteria to compare different structural equation modeling models

- Models may be compared to each other w.r.t. goodness-of-fit indices for the values of CMIN/df, GFI, AGFI, CFI, etc. and the model with the superior values will be considered as better fitting model (Hair et al., 2010).
- Models may also be compared for the Chi-square values and the model with lower Chi-square value will be considered as model with better fitting (Hair et al., 2010).
- R^2 is also another useful parameter to compare the two models which defines the percentage of variance for dependent variables explained by independent variables.

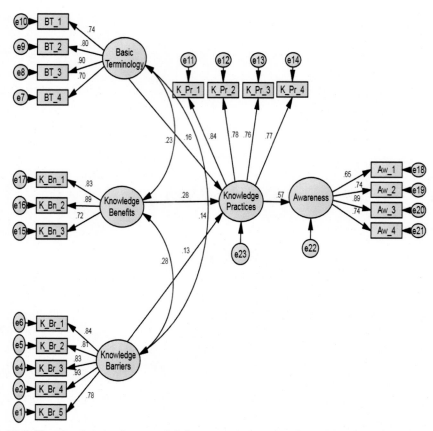

Figure 13.5 Structural equation model for organizations with mediation effect of practices (Model B).

The model having higher value of R^2 is taken as better fitting model under observations (Hair et al., 2010).
- Also standard slope coefficient for all the constructs under consideration should be significant at 95% level of confidence (Hair et al., 2010).

Although all the fit indices of this model are in the acceptable range, yet this model is inferior to Model A on account of the followings:

13.5.7 Comparison of structural equation modeling Model A with structural equation modeling Model B

1. The goodness–of–fit indices for Model A and Model B are within acceptable range, but if we compare the values of both, the values for Model A are statistically better as compared to Model B. So we can say that Model A is better than Model B as shown in Table 13.19.

Table 13.18 Statistical fitness of Model B.

Goodness of fit index	CMIN/ df	Goodness of fit index (GFI)	Augmented goodness of fit index (AGFI)	Comparative fit index (CFI)	Tucker-Lewis index (TLI)	Root mean square error approximation (RMSEA)
Calculated value	1.548	0.821	0.801	0.926	0.913	0.073
Expected value	>3	<0.8	<0.8	<0.9	<0.9	>0.08

2. Chi-square value of Model A is 216.829 whereas for Model B is 252.252. Chi-square value of Model A is less than Model B. This indicate that Model B is worse than Model A.
3. The explanatory power of Model B (R^2 0.33) is inferior to that of Model A (R^2 0.58). Therefore Model A is considered to be better fit.
4. The regression coefficient between knowledge of practices and awareness of basic terminology along with knowledge of practices and knowledge about barriers are not statistically significant at 5% level of significance.

13.5.8 Structural equation Model C for organizations

In this model, we are using knowledge of benefits as a mediator between the independent variables namely basic terminology, knowledge of practices, knowledge of barriers, and dependent variable namely overall awareness of E-waste management. This model that we tested for better fit is based on the presumption that an organization initially becomes aware of the basic terminology, practices, and barriers of E-waste management. Once they gain the knowledge about these three constructs up to satisfaction level, organizations try to learn about likely benefits to gain from proper E-waste management and this knowledge completes his overall knowledge about E-waste management.

The hypothesis for comparison between Model A and proposed mediation Model C is

"Model with knowledge about benefits acting as mediator between other three exogenous constructs and endogenous construct is a better model than Model A, i.e., the model without any mediation effect."

The structural model is developed to test this hypothesis and the same is shown in Fig. 13.6.

Various regression weights and their statistical significance are summarized in Table 13.20.

Statistical fit indices for this model are given in Table 13.21.

Although all the fit indices of this model are in the acceptable range, yet this model is inferior to Model A on account of the following:

Table 13.19 Regression weights for Model B.

			Standardized slope coefficient	Unstandardized regression weight	Standard error	Critical ratio	P	R^2
Knowledge of practices	<--	Basic terminology	0.163	0.211	0.141	1.495	.135	0.17
Knowledge of practices	<--	Knowledge of benefits	0.278	0.317	0.133	2.385	.017	
Knowledge of practices	<--	Knowledge of barriers	0.131	0.176	0.143	1.224	.221	
Awareness	<--	Knowledge of practices	0.57	0.357	0.078	4.574	***	0.33

***P-values are almost zero indicating that there is significant relationship between the constructs and the measured items.

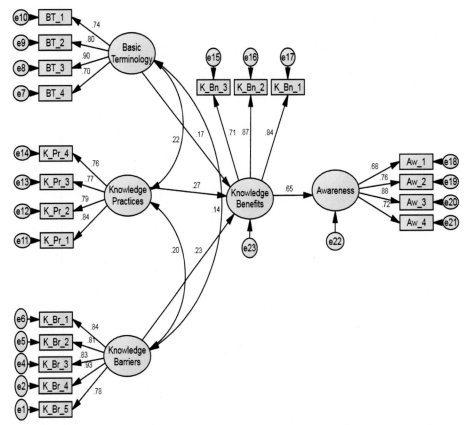

Figure 13.6 Structural equation model for organizations with mediation effect of benefits (Model C).

13.5.9 Comparison of structural equation modeling Model A with structural equation modeling Model C

1. The goodness-of-fit indices for Model A and Model C are within the acceptable range, but if we compare the values of both, the values for Model A are better as compare to Model C. So we can say that Model A is better than Model C.
2. Chi-square value of Model A is 216.829, whereas for Model C is 243.331. Chi-square value of Model A is less than Model C. This indicate that Model C is worse than Model A.
3. The explanatory power of Model C (R^2 0.42) is slightly inferior to that of Model A (R^2 0.58). The R^2 of 0.58 and 0.42 means that 58% of variance (Model A) and 42% of variance (Model C) of the awareness of E-waste management can be

Table 13.20 Regression weights for Model C.

			Standardized slope coefficient	Unstandardized regression weight	Standard error	Critical ratio	P	R^2
Knowledge of benefits	<-	Basic terminology	0.174	0.194	0.12	1.61	.107	0.21
Knowledge of benefits	<-	Knowledge of practices	0.272	0.233	0.096	2.422	.015	
Knowledge of benefits	<-	Knowledge of barriers	0.229	0.265	0.122	2.161	.031	
Awareness	<-	Knowledge of benefits	0.651	0.492	0.101	4.86	***	0.42

*** *P*-values are almost zero indicating that there is significant relationship between the constructs and the measured items.

Table 13.21 Statistical fitness of Model C.

Goodness of fit index	CMIN/ df	Goodness of fit index (GFI)	Augmented goodness of fit index (AGFI)	Comparative fit index (CFI)	Tucker-Lewis index (TLI)	Root mean square error approximation (RMSEA)
Calculated value	1.487	0.822	0.804	0.934	0.923	0.069
Expected value	>3	<0.8	<0.8	<0.9	<0.9	>0.08

explained with the help of included factors. Therefore Model A is considered the best fitting.

4. The regression coefficient between knowledge about benefits and awareness about basic terminology is not statistically significant.

13.5.10 Structural equation Model-D for organizations

This next model that we tested for better fit is based on the presumption that an organization initially becomes aware of the basic terminology of E-waste and once his level of interest about basic terminology has been aroused beyond a minimum threshold level, organizational users try to learn, in details, about suitable practices used to manage E-waste, likely benefits to gain from proper E-waste management, barriers in managing E-waste, etc. and knowledge about these completes his overall awareness level about E-waste management. Thus, in this model, we are using knowledge of benefits, knowledge of practices, and knowledge of barriers as a mediator between independent variable, basic terminology, and dependent variable and overall awareness of E-waste management. Knowledge of benefits, knowledge of practices, and knowledge of barriers are endogenous as well as exogenous constructs in this model (Table 13.22).

The hypothesis for comparison between Model A and proposed mediation Model D is

"Model with knowledge about benefits, barriers and practices acting as mediator between basics awareness and overall awareness is a better model than Model A, i.e., the model without any mediation."

The structural model is developed to test this hypothesis and the same is shown in Fig. 13.7.

All the fit indices of this model are in the acceptable range, yet this model too is inferior to Model A on account of following:

Table 13.22 Regression weights for Model D.

			Standardized slope coefficient	Unstandardized regression weight	Standard error	Critical ration	P	R^2
Knowledge of practices	<--	Basic terminology	0.245	0.317	0.146	2.175	.03	0.06
Knowledge of benefits	<--	Basic terminology	0.260	0.29	0.127	2.278	.023	0.07
Knowledge of barriers	<--	Basic terminology	0.152	0.147	0.105	1.398	.162	0.02
Awareness	<--	Knowledge of benefits	0.493	0.345	0.083	4.165	***	0.50
Awareness	<--	Knowledge of practices	0.400	0.241	0.064	3.793	***	
Awareness	<--	Knowledge of barriers	0.231	0.186	0.075	2.486	.013	

*** P-values are almost zero indicating that there is significant relationship between the constructs and the measured items.

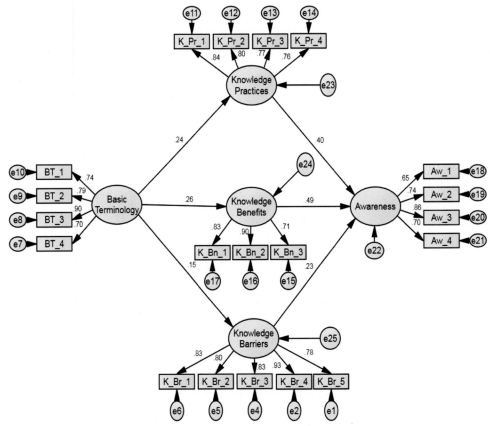

Figure 13.7 Structural equation model for organizations with parallel mediation (Model D).

13.5.11 Comparison of structural equation modeling Model A with structural equation modeling Model D

1. The goodness–of–fit indices for Model A and Model D are within the acceptable range, but if we compare the values of both, the values for Model A are better as compare to Model D. So we can say that Model A is better than Model D as shown in Table 13.23.

2. Chi-square value of Model A is 216.829 whereas for Model D is 233.533. Chi-square value of Model A is less than Model D. This indicates that Model D is worse than Model A.

3. The explanatory power of Model D (R^2 0.50) is inferior to that of Model A (R^2 0.58). The R^2 of 0.58 and 0.50 means that 58% of variance (Model A) and 50% of variance (Model D) of the awareness of E-waste management can be explained with the help of included factors. Therefore Model A is considered the best fitting.

Table 13.23 Statistical fitness of Model D.

Goodness of fit index	CMIN/ df	Goodness of fit index (GFI)	Augmented goodness of fit index (AGFI)	Comparative fit index (CFI)	Tucker-Lewis index (TLI)	Root mean square error approximation (RMSEA)
Calculated value	1.424	0.826	0.808	0.942	0.933	0.064
Expected value	>3	<0.8	<0.8	<0.9	<0.9	>0.08

4. The regression coefficient between knowledge of barriers and knowledge of basic terminology of E-waste management is not statistically significant.

13.5.12 Structural equation Model E for organizations

The only difference between Model D and Model E is that the basic terminology about E-waste management in Model E is also affecting the overall awareness level of the organizations directly along with knowledge of practices, knowledge of benefits, and knowledge of barriers in proper recycling of E-waste. We have considered a model in which basic awareness leads to search for knowledge about practices, knowledge of benefits, and knowledge of barriers of E-waste management. These constructs in turn determine overall awareness.

The structural model examined is shown in Fig. 13.8.

13.5.13 Comparison of structural equation modeling Model A with structural equation modeling Model E

1. The goodness-of-fit indices for Model A and Model E (as shown in Table 13.24) are within the acceptable range, but if we compare the values of both, the values for Model A are better as compare to Model E. So we can say that Model A is better than Model E.
2. Chi-square value of Model A is 216.829 whereas for Model E is 233.533. Chi-square value of Model A is less than Model E. This indicate that Model E is worse than Model A.
3. The explanatory power of Model E (R^2 0.53) as shown in Table 13.25 is slightly inferior to that of Model A (R^2 0.58). The R^2 of 0.58 and 0.53 means that 58% of variance (Model A) and 53% of variance (Model E) of the awareness of E-waste management can be explained with the help of included factors. Therefore Model A is considered the best fitting.
4. The regression coefficient between knowledge of barriers and knowledge of basic terminology of E-waste management is not statistically significant.

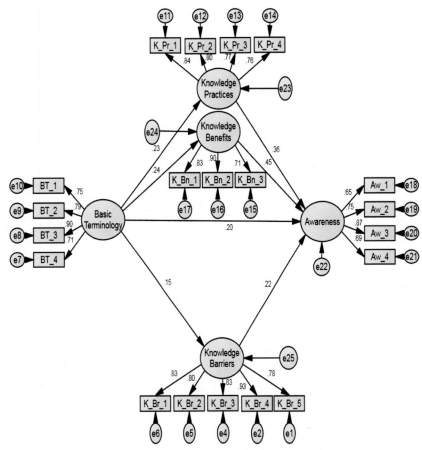

Figure 13.8 Structural equation model for organizations with mediators and direct effect of BT (Model E).

Table 13.24 Statistical Fitness of Model-E.

Goodness of fit index	CMIN/ df	Goodness of fit index (GFI)	Augmented goodness of fit index (AGFI)	Comparative fit index (CFI)	Tucker-Lewis index (TLI)	Root mean square error approximation (RMSEA)
Calculated value	1.406	0.827	0.808	0.945	0.936	0.063
Expected value	>3	<0.8	<0.8	<0.9	<0.9	>0.08

Table 13.25 Regression weights for Model E.

			Standardized slope coefficient	Unstandardized regression weight	Standard error	Critical ration	P	R²
Knowledge of practices	<--	Basic terminology	0.229	0.292	0.143	2.038	.042	0.05
Knowledge of benefits	<--	Basic terminology	0.242	0.266	0.125	2.125	.034	0.06
Knowledge of barriers	<--	Basic terminology	0.147	0.141	0.104	1.357	.175	0.20
Awareness	<--	Knowledge of benefits	0.448	0.315	0.08	3.94	***	0.53
Awareness	<--	Knowledge of practices	0.359	0.218	0.062	3.503	***	
Awareness	<--	Knowledge of barriers	0.216	0.175	0.074	2.382	.017	
Awareness	<--	Basic terminology	0.197	0.152	0.076	1.996	.046	

***P-values are almost zero indicating that there is significant relationship between the constructs and the measured items.

13.6 Conclusion

Although the original first-order SEM model was statistically significant, the possibility of mediation role played by one or more exogenous construct was explored. Four different possible SEMs each with a different types of mediation have been hypothesized and compared with a model having no mediation effect. From this analysis, it can be concluded that although all these models have acceptable fit values, yet the first model, without any mediation, is the best among all the compared models. The implications of the current work are that if we want to increase the awareness level of end users (individuals/organizational) about negative impacts of E-waste on human health and on surroundings, then all the constructs that define overall awareness need to be taken care of simultaneously.

References

Dangi, H. K., & Dewen, S. (2016). *Business research methods*. Cenage.

Garlapati, V. K. (2016). E-waste in India and developed countries: Management, recycling, business and biotechnological initiatives. *Renewable and Sustainable Energy Reviews, 54*, 874—881.

Gupta, T. K., & Singh, V. (2015). A systematic approach to evaluate supply chain management environment index using graph theoretic approach. *International Journal of Logistics Systems and Management, 21*(1), 1—45.

Hair, J. F., Jr., Anderson, R. E., Tatham, R. L., & Black, W. C. (2010). *Multivariate data analysis* (7th ed.). Pearson Education.

Khan, M. (2017). Environment and health issues associated with E-waste. *International Journal of Advanced Research, 5*(1), 1425—1430.

O'Rourke, N., & Hatcher, L. (2013). *A step-by-step approach to using SAS for factor analysis and structural equation modeling* (2nd ed.). SAS Institute Inc.

Saoji. (2012). E-waste management: An emerging environmental and health issue in India. *National Journal of Medical Research, 2*(1), 107—110.

Trochim, W., Donnelly, J. P., & Arora, K. (2015). *Research methods: The essential knowledge base*. Nelson Education.

CHAPTER 14

Global E-waste management: consolidated information showcasing best available practices

Dayanand Sharma[1,2], Anudeep Nema[3], Rajnikant Prasad[4], Kumari Sweta[1], Dipeshkumar R. Sonaviya[5] and Sandip Karmakar[1]

[1]Department of Civil Engineering, National Institute of Technology, Patna, Bihar, India
[2]Department of Civil Engineering, Sharda University, Greater Noida, Uttar Pradesh, India
[3]Department of Civil Engineering, School of Engineering, Eklavya University, Damoh, Madhya Pradesh, India
[4]Civil Engineering Department, G. H. Raisoni Institute of Engineering and Business Management, Jalgaon, Maharashtra, India
[5]M.S. Patel Department of Civil Engineering, C. S. Patel Institute of Technology, Charusat, Anand, Gujarat, India

14.1 Introduction

E-waste can be defined as unused/broken waste from electronic and electrical equipment (WEEE) waste that cannot perform its intended function and is redirected to recycle, reuse, and recovery (RRR) or disposal. The use of electronic equipment has increased tremendously as the IT and communication sectors have grown. Consumers are forced to discard obsolete electronic items more quickly due to faster electronic product upgrades, increasing E-waste loading to junk. WEEE is a growing problem that requires more attention toward recycling and better waste management. This rapid growth of WEEE is attributable to industrial revolution, lifestyle changes, and urbanization (Shittu et al., 2021). When WEEE becomes unsuitable for its intended persistence or has reached beyond its expiration date, it is referred to as E-waste. E-waste includes wastes from computer parts, server wastes, mainframes, monitor screens, compact discs, digital versatile discs, waste parts from printers, scanners, photocopiers, calculators, fax machines, batteries, mobile phones, transceivers, televisions, medical apparatus, laundry machines, fridges, and ACs (when unfit for use). Due to quick industrialization breakthroughs and the development of newer electronic apparatus, electronic apparatus is quickly exchanged with a new one. As a result, the amount of E-waste created has skyrocketed. The public is more likely to change to newer versions, and thus, product lifespans have shortened.

Numerous initiatives, such as life cycle assessment (LCA) and extended producer responsibility (EPR), have been implemented in conjunction with targeted administrations and legislation to limit E-waste output and recover important metals. Globally, E-waste is disposed of mainly through recycling and landfilling. India is currently dealing with its own and imported E-waste from other countries. The use of inappropriate

Global E-waste Management Strategies and Future Implications
DOI: https://doi.org/10.1016/B978-0-323-99919-9.00002-7
289

and unsafe techniques to recycle and dispose of E-waste in India has adversely impacted the health and safety of the workers. This chapter elaborates on global E-waste management and best available practices adopted in Germany, Switzerland, Japan, and Bangalore (India). The chapter also presents the current status of global E-waste management, existing methods, associated challenges, and opportunities and strategies undertaken to manage E-waste globally and locally (in India). It also discusses the deleterious effects of improper management of E-waste.

Worldwide, E-waste is a large sector increasing at a rate of around 2 million tons (Mt) per year due to increasing sales and E-waste due to the variety and affordability of technological innovation. Due to the lack of awareness, most end-users dispose of E-waste as household rubbish. Electronic trash is high in hazardous metals and other contaminants, endangering life and ecosystems. As a result, most governments have developed rules to regulate producers, consumers, and recyclers. However, in a globalized society, the efficiency of these local rules is restricted. Biohydrometallurgy and hydrometallurgy appear to be the most promising formal recycling technologies compared to pyrometallurgy (Arya & Kumar, 2020a). Consumers are apathetic toward E-waste disposal and proper recycling (Nithya et al., 2021). E-waste recycling should be practiced without the emission of harmful pollutants. However, the lack of environmentally sound standard operating procedures has led to improper recycling of E-waste, resulting in the emission of pollutants that do not have a proper remedy. E-waste has enormous metal content in it, which is then converted to metallic oxides, nitrates, and sulfates during the treatment process. These secondary forms of metals are easily soluble in the environment. Thus, it is important to devise proper treatment procedures. Once the treatment process is confirmed, the source of treatment has to be concentrated. Plastics may release toxic elements when mixed with acid, so they must be segregated before the treatment. Furthermore, it is important to concentrate on workability and maintain an accident-free environment. A recycler must practice good health and safety practices. The health hazard can be monitored only when the exposed workers are tested for metal accumulation in blood, with which lots of health and safety equipment can be mounted in the environment. The occupational health hazard not only damages a specific individual but also damages the entire generation by bioaccumulation and gene damages, and thus, it is very important to focus on safety.

Metals, polymers, cathode ray tubes (CRTs), microchips, wires, and other materials commonly make up E-waste. If E-waste is handled correctly, costly metals such as gold, silver, copper, and platinum can be recycled. If E-waste is deconstructed and managed in a basic way using rudimentary processes, dangerous compounds such as liquid crystal, lithium, mercury, nickel, polychlorinated biphenyls (PCBs), selenium, arsenic, barium, brominated flame retardants, cadmium, chromium, cobalt, copper, and lead can be extracted. Humans, animals, and the ecosystem are all at risk from E-waste. A very small amount of heavy metals and extremely poisonous compounds such as lead, mercury, cadmium, and beryllium represent a severe environmental hazard.

Consumers have the key to effective E-waste management. EPR, design for the environment (DfE), and 3Rs (Reduce, Reuse, Recycle) are the technological platforms that connect markets and promote a circular economy. These technologies aim to incentivize users to manage their E-waste properly with the recycle and reuse principle to accept workable user habits. E-waste management is deemed highly important in developed nations. Still, it is aggravated in developing countries by completely accepting or repeating developed country E-waste management and numerous associated issues such as the lack of assets and technically expert personnel.

In addition, the obligations and roles of parties and organizations involved in the management of E-waste are not well described. India has successfully amended rules from year to year to execute effective E-waste management. The new E-waste management guidelines were issued in the year 2016 by MoEF & CC (Ministry of Environment, Forest, and Climate Change), which replaced the older E-waste management and handling rules proposed in 2011. Despite an increase in E-waste research, the number of publications by most countries remained limited. Few factors have to be overcome by investigators and researchers, such as epidemiology data and associations that may result in biomagnification, cross findings, and understanding of biological mechanisms. Some precautionary actions must be taken to reduce the adverse effects of the exposure of metals and metallic compounds among children. Changes in subject selection and analytic qualities have also impacted the breadth and limitations of current studies. The expanding number of newly produced procedures and the lack of extensive and up-to-date investigations in this research domain have hampered the analysis of numerous methodological concerns.

14.2 E-waste generation and current E-waste market

E-waste has become indispensable in modern society. It is one of the rapidly increasing waste streams worldwide due to its increased generation rate. There is a high human reliance on EEE to live a luxurious life, and as a result, consumption and demand for EEE goods have increased (Balde et al., 2017). It includes a wide variety of products used in day-to-day life such as phones, televisions, fridges, air conditioners, laptops, computers, and televisions. The most common type of E-waste is small equipment such as microwaves, electric kettles, and cameras. E-waste is categorized under EU-6 classification depending upon waste management category as:

1. Temperature exchange equipment
2. Screens and monitors
3. Lamps
4. Large equipment
5. Small equipment
6. Small IT and telecommunication equipment

In 2016, the global E-waste waste generation was reported to be 44.7 metric tons (MT) (Initiative Step, 2016). Projected E-waste generation by Balde et al. (2017) for the year 2018 was 50 Mt. As per the latest data of 2019, the global E-waste generation is approximately 53.6 MT, with an average per capita generation rate is 7.3 kg. The continent-wise global waste generation data are shown in Fig. 14.1. However, the E-waste generation growth rate is about 2 MT yearly and is expected to exceed 74.7 Mt by 2030. Based on the estimation, the annual growth rate of E-waste generation will be around 3% from the year 2018 to 2047 (Islam et al., 2019). Increased E-waste generation rate can be attributed to the increasing consumption rate and increased buying potential. Any country's average waste generation rate is directly proportional to the average income per capita. High-income countries have an average household size of 2.8 with an average purchasing power of USD 51.58/capita. Similarly, high-to-middle, middle, middle-to-low, and low-income countries have an average household size of 3.6, 4.5, 5.3, and 5.0 with the purchasing power of USD 21.7, USD 9.87, USD 3.50, and USD 1.26, respectively (Forti et al., 2020).

The electronic equipment generally has a short lifespan with limited reusability or repair options and thus end up in the waste stream. The increase in waste generation can also be attributed to the fact that people tend to purchase new products instead of repairing and reusing older ones. The repair cost is generally high compared to its after repair life with a low functioning warranty after repair. With the current generation rate, the global documented collection and recycling rate was 9.3 Mt, that is, only 17.4% of the E-waste generation. In 2019, 10.8 Mt of temperature exchange

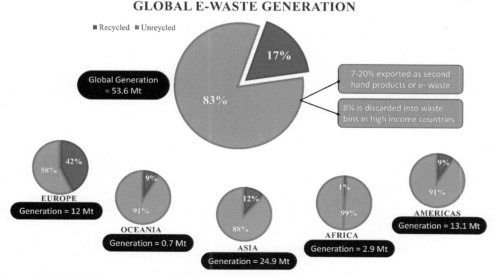

Figure 14.1 Global E-waste generation with the percentage of recycled and unrecycled (Forti et al., 2020).

equipment, 6.7 Mt of screens and monitors, 0.9 Mt of lamps, 13.1 Mt of large equipment, 17.4 Mt of small equipment, and 4.7 Mt of small IT and telecommunication equipment were generated globally (Forti et al., 2020). Studies have reported a 7% increase in temperature exchange equipment waste generation, 5% increase in large equipment, 4% increase in lamps and small equipment, and 2% increase in small IT and telecommunication equipment since 2014. However, a reduction of 1% was observed for screens and monitors. This reduction can be due to increased life or replacement of heavy monitors and screens with lighter screens and monitors compared to other categories of E-waste. The increased generation rate of other categories may be due to the increasing consumption in lower-income countries.

For continent-wise E-waste generation, Asia ranks on top with a total of 24.9 Mt E-waste, followed by America (13.1 Mt), Europe (12 Mt), Africa (2.9 Mt), and Oceania (0.7 Mt). The continent-wise top three Asian countries with the highest E-waste generation in descending order were China, India, and Japan, with generation rate of 10,129, 3230, and 2569 kt, of E-waste, respectively. The United States, Brazil, and Mexico had E-waste generation rates of 6918, 2143, and 1220 kt, respectively. In Europe, Russian Federation, Germany, and the UK of Great Britain and Northern Ireland had E-waste generation rates of 1631, 1607, and 1598 kt, respectively. For Africa, Egypt, Nigeria, and South Africa, had E-waste generation rates of 586, 461, and 416 kt, respectively. For Oceania, Australia, New Zealand, and Papua New Guinea, the generation rates of 554, 96, and 9 kt, respectively, were reported (Forti et al., 2020).

The highest per capita generation rate was in Europe with 16.2 kg. Moreover, Europe recorded the highest collection and recycling rate of 42.5%. However, the fate of global E-waste generated is uncertain due to a lack of proper collection and treatment facilities. Out of 53.6 Mt global E-waste generation, 44.3 Mt is unrecycled. Europe leads the continent in collection and recycling rates, with 45.5%, followed by Asia (11.7%), America (9.4%), Oceania (8.8%), and Africa (0.9%). The unrecycled waste is exported as second-hand products, accounting for 7%—20% of unrecycled waste, and 8% are discarded in the bins that find its final disposal in the landfills.

Currently, India, the world's second-most populous country, ranks third in the world in terms of E-waste production, trailing only China and the United States, with 3230 kt of E-waste (Arya & Kumar, 2020a, 2020b) as shown in Fig. 14.2. In India, Mumbai produces the highest quantity of E-waste, followed by New Delhi, Bangalore, and Chennai (Imran et al., 2017).

Around 82.6% of the total E-waste generated in 2019 was handled by informal sectors. It was usually not handled in an environmentally sound manner and was not documented. The recycled E-waste (17.4%) contributes to the equivalent of 4 Mt of raw material and USD10 million of valuable material. The value of global raw materials present in E-waste is estimated to be around USD57 billion. The current rate of

Figure 14.2 World's top 12 countries for E-waste generation (Forti et al., 2020).

generating is expected to increase to 74.4 Mt of E-waste globally by 2030, which is approximately 9 kg of waste per capita (Forti et al., 2020), an increase of 1.7 kg/capita. To handle the increasing E-waste stream, there is a need to increase the pace of formal collection systems and develop technology and infrastructure to handle the same. The main constraint in E-waste recycling and treatment is the associated cost (Yang et al., 2021). The cost of recycled materials is much lower than the cost involved in the recovery. The high recycling cost is accounted for infrastructure establishment, collection, storage, treatment facility, and disposal of waste generated during the recycling process (Ghimire & Ariya, 2020).

For example, the cost of E-waste recycling per ton of waste was around USD450 to USD1000 compared to USD150 to USD250 for landfills (Sun et al., 2017). Proper treatment technology is needed in the recovery and recycling technology to reduce the waste reaching landfill sites. The energy generated from E-waste can be used for power generation provided availability of cost-efficient technology (Lin et al., 2019). Despite the presence of precious metals in E-waste, recovery is based on precious metal quantity, wastage level (approximately 7%), and recovery rate (approximately 93%—97%) (Andeobu et al., 2021).

There is a lack of money for automated machines and proper collection facilities to meet the cost incurred in the recycling of waste. Moreover, the inadequate quantity of waste collection and the cost of skilled manpower have restricted formal recycling facilities (Lundgren, 2012). This has resulted in the major fraction of E-waste being handled by the informal sector, which can generate profit from the waste at the cost of polluting our environment. Because these informal sectors lack the infrastructure to handle such waste, the waste is treated or handled using obsolete technologies. Recycling methods employed by informal sectors are expected to release toxic additives and hazardous substances contained or produced from E-waste treatment into the environment, posing a significant threat to the environment (Ahirwar & Tripathi, 2021).

A considerable amount of natural resources are present in the E-waste that includes rare earth elements. According to Forti et al. (2020), the estimated amount of raw materials that could be recovered from the E-waste generated in 2019 was USD57 billion. The availability of a proper recycling facility would have saved a substantial amount of natural resources required from exploiting our natural resources. Improving our current E-waste collection and treatment facility will reduce the import of crucial raw materials such as cobalt, copper, and nickel, thus avoiding over-extraction and dependency on the import. This increases the potential of job creation, especially in developing countries and will also reduce the dependency on the need for mining to extract these metals (Zeng et al., 2018).

14.3 E-waste management, treatment, and disposal

E-waste treatment and recovery from E-waste includes collection, preprocessing, and end-treatment processes. As far as global E-waste management is concerned, Switzerland is the first country to implement the organized E-waste management system in the world (Sinha-Khetriwal et al., 2005). In India, the Ministry of Environment and Forests is in charge of hazardous waste management. Effective government laws, awareness programs, and public waste collection facilities all expedite E-waste collection. In a collection center, recyclable and nonrecyclable components are safely separated from the E-waste. After sorting at the collecting center, usable electronic components are returned to consumer supply chains (Khaliq et al., 2014). Nonrecyclable E-waste is processed, stored, and dumped in both developed and developing countries. Fig. 14.3 shows a typical picture of E-waste treatment. The majority of developed countries have progressed their manual and semi-automated disassembly and separation methods to manage the formal collecting industry. On the other hand, developing countries collect E-waste from informal industries. Certain underdeveloped countries do not collect E-waste properly. However, some backyard techniques (such as open incineration, rudimentary smelters, and cyanide leaching) are

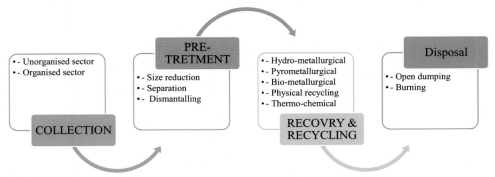

Figure 14.3 Flowchart of the E-waste treatment system.

employed directly to recover some precious metals and dispose of the remainder in open dumps alongside city solid garbage (Mmereki et al., 2016).

14.3.1 Informal treatment or recovery process

The informal sector in developing countries processes a large volume of E-waste. The informal procedure of recovering metals in a nontraditional and unscientific manner. The informal method of recycling comprises no safeguards and little control over the material's recovery (Fu et al., 2013). Informal recycling implies lesser automated procedures and health-protective measures, as well as much more scattered components. For informal recycling, the most typical techniques include manual disassembling, sorting, open burning, and de-soldering printed circuit boards. Because of the health repercussions, these activities have a disproportionate impact on disadvantaged people. The informal recovery method pollutes the surrounding environment and natural resources and adversely impacts human health. In general, important metals such as Al, Ag, Cu, Pb, Hg, Pt, and Cd are recovered by burning E-waste in an open yard, which emits a significant quantity of hydrocarbon and other hazardous substances (Song & Li, 2014). Other metals such as Au, Pd, and Ag are also chemically recovered using highly toxic acids like HCl, HNO_3, H_2SO_4, and cyanide, which heavily contaminate the atmosphere. Additionally, informal treatment sectors lack control over treatment leftovers. At the conclusion of the process, all residues are disposed of directly into the surrounding environment (land or water). These harmful metal ions are found in amounts greater than those permitted by environmental protection organizations and the World Health Organization. Diverse experts from around the world have discovered various harmful substances in natural land and water sources as a result of unethical E-waste management (Pradhan et al., 2014). In addition, the fertility of soil and groundwater are affected by improper landfilling of E-waste by the informal treatment process. The informal E-waste processing efficiency is less than 80%, which can increase up to 95% by formal recycling. To save the ecosystem and its inhabitants, appropriate E-waste recovery and recycling solutions must be developed.

14.3.2 Formal treatment or recovery process

E-waste management is now regulated in most developed countries (such as the United States, Japan, Canada, and Sweden) as well as a few developing countries (such as Colombia and China). Numerous formal recycling facilities act as distributors for third-party material recovery of plastics, glass, and metals. A typical formal E-waste recycling process begins with shorting, pretreatment, physical separation, and metal recovery activities such as pyrometallurgy, hydrometallurgy, and biohydrometallurgy. Systematic and different formal treatment processes are as follows:

14.3.2.1 Pyrometallurgy method

Pyrometallurgy is a high-energy technology used to treat or extract nonferrous metals from metallurgical materials at elevated temperatures. Pyrometallurgical methods are commonly used worldwide to recover metals from E-waste. When metals are combined with various nonmetals and ceramics, typical recycling procedures may prove difficult to use. The choice of several critical processes is heavily impacted by the type and quantity of E-waste and the requirement for smelting activities. Pyrometallurgical processes may be used to recover metals such as Cu, Ag, Au, Pd, Ni, Se, Zn, and Pb from E-waste (Nithya et al., 2018). The primary disadvantage of this procedure is the high temperature, which requires a delicate balance between the heat and the substance being treated. Toxic gases such as dibenzo-p-dioxin, biphenyl, anthracene, polybrominated dibenzofurans, and polybrominated dibenzodioxins were emitted during the pyrometallurgical process. Additionally, slag formation limits recovery yield due to the inevitable discharge of high-temperature dust and smoke, which have a negative impact on human health and the surrounding environment. Natural air or pure oxygen is used as fuel in the combustion process. Natural air includes 79% nitrogen dioxide, which reduces thermal efficiency by evaporating heat into the environment. On the other hand, employing pure oxygen may improve efficacy while reducing flue gas output. Due to the rapidity with which metallurgical processes approach equilibrium, it is impossible to predict the mechanical parts of the recovery process. Additional studies are required to optimize the recovery process and eliminate dangerous gas emissions.

14.3.2.2 Biohydrometallurgy method

Biohydrometallurgy processing of E-waste is a technique in which microorganisms help dissolve elements from solid objects into liquids that may subsequently be recovered by separation. Biometallurgy offers a wealth of possibilities in terms of research and business due to the less energy-intensive procedures and minimal use of chemicals that have made this field new and promising. Furthermore, from a business perspective, the technology is referred to as clean and green due to its sustainability. This method has been used to recover elements such as copper, silver, and aluminum from Rio Tinto mines in southwest Spain since the pre-Roman period. However, it was commercially adopted only a few decades ago. It is classified into biosorption and bioleaching. Biosorption is the process of adsorbing metals using adsorbents made from waste biomass or abundant biomass. The biosorption process utilizes *Aspergillus niger* (fungi), *Chlorella vulgaris* (algae), *Penicillium chrysogenum* (bacteria), hen eggshell membrane, ovalbumin, and alfalfa (Debnath et al., 2018). The factors affecting biosorption include the type of biological ligands, the type of biosorbent, chemical and stereochemical, characteristics of the metal to be recovered, and characteristics of the metal solution (Tsezos & Remoudaki, 2006). On the other hand, bioleaching refers to the mobilization of metal cations from insoluble materials through biological oxidation and other complexation processes (Ilyas & Jae, 2014). Bioleaching includes four typical steps, which are acidolysis, complex analysis, redoxolysis, and bioaccumulation.

Bioleaching includes three groups of bacteria, namely, autotrophic bacteria, heterotrophic bacteria, and heterotrophic fungi (Arya & Kumar, 2020b). The most known autotrophic bacteria in the leaching process are *Acidobacillus thiooxidans*, *Acidobacillus ferrooxidans*, and *Sulfolobus* sp. (Natarajan & Ting, 2015). These bacteria generate energy by oxidizing ferrous ions and reducing sulfur-containing compounds. The heterotrophic bacteria (*Pseudomonas* sp. and *Bacillus* sp.) and fungi (*Aspergillus* sp. and *Penicillium* sp.) can translocate metals from waste. The mechanism used for translocation is facilitated through acidolysis, metal complexation, redox reaction, or bioaccumulation. Using thermophilic microorganisms, more than 80% recovery of copper—nickel and zinc was observed by Ilyas et al. (2007), whereas using *A. thiooxidans* and *A. ferroxidans* consortium, more than 90% recovery of copper, nickel, zinc, and lead was observed. Several metals have been reported to recover from E-waste efficiently through this process, such as gold 8%—85% (Sheel & Pant, 2018), copper 23%—99% (Jagannath et al., 2017), silver 12%—41% (Ruan et al., 2014), nickel 38%—96% (Horeh et al., 2016), zinc 64%—95% (Ilyas et al., 2010).

14.3.2.3 Hydrometallurgy method

The hydrometallurgy method incorporates mechanical shredding as a pretreatment, followed by a sorting procedure before the leaching phase. To remove metals from a dispersion media, a leaching process involves a reaction between a solid material and a chemical reagent such as extractant or lixiviant. Low solute investment and high recovery rate are the primary benefits that have made them a superior procedure to the previous operation (Yazici & Deveci, 2014). There are four types of common leaching processes: cyanide leaching, halide leaching, thiourea leaching, and thiosulfate leaching (Debnath et al., 2018). In hydrometallurgy, solute transfer occurs in a variety of liquid mediums, such as halide, thiourea, thiosulfate cyanide, HCl, H_2SO_4, and HNO_3 (Nithya et al., 2021). The required metals are precipitated due to gravity settlement or centrifugal forces and then are progressed to the metal recovery process. Design parameters such as the particle size, operating temperature, nature of lixiviants, and its sample ratio with the leaching time are considered in this intermediate process. The metals from this suspension are recovered downstream using solvent extraction, electrodeposition, ion-exchange, and adsorption. Final forms of metals are achieved through electro-refining or chemical reduction processes. The downstream process is dependent on the type of metal to be recovered and a level of purity is desired. The advantage of this method includes a high recovery rate, moderate temperature, less energy consumption, zero toxic gas emission, and minimal secondary waste generation. Various metals have been reported to recover from E-waste efficiently via this process such as gold 12%—98% (Quinet et al., 2005), copper 92%—96% (Cui & Anderson, 2020), silver 4%—100% (Oh et al., 2003), nickel 95% (Cui & Anderson, 2020), palladium 58%—98% (Ha et al., 2014). Researchers are now focusing on antimony and tin recycling from E-waste (Barragan et al., 2020). The pyrometallurgy process has a few limitations, such as high time requirement, shredded particle requirement, required toxic, and expensive lixiviant with high safety standards. Moreover, it requires corrosion-resistive equipment for the leaching process and has the possibility of metal loss through many downstream processes.

14.3.2.4 Physical recycling methods

E-waste is often treated with physical recycling processes, which assist in liberating the embedded metals and nonmetals. After exploring numerous E-waste recycling facilities worldwide, a previous study concluded that physical recycling is the most systematic approach. This is one of the most influential metal recovery methods; it is often called a pretreatment step before further processing. The various processes include dismantling, shredding, chopping, and disassembling, among others. To separate metals and nonmetals, various mechanical machines such as shredders, pregranulators, and granulators have been used to accomplish these stages. Magnetic separation, eddy current separation, corona discharge method, density-based separation, milling, froth floatation, and density separation are some of the methods used. It is possible to recover metal fractions containing more than 50% copper, 24% tin, and 8% lead using a combination of electrostatic and magnetic separation, which separates the metallic components from the nonmetallic ones (Veit, 2005).

14.3.2.5 Thermo-chemical methods

Pyrolysis is a necessary thermo–chemical process that ensures the thermal degradation of a material in the absence of air. Different kinds of pyrolysis, such as vacuum pyrolysis, microwave-induced pyrolysis, catalytic pyrolysis, and co-pyrolysis, have been documented (Debnath & Ghosh, 2018). E-waste pyrolysis is now restricted to the laboratory, although Jectec (a Japanese company) has previously deployed pyrolysis at their facilities. The plasma method for E-waste treatment and metal recovery has recently garnered massive attention. Plasma technology is a high-temperature, low-impact technology. E-waste has been processed using a high-enthalpy plasma jet, a plasma reactor, and a plasma torch (Arya, Patel, et al., 2021; Arya, Rautela, et al., 2021). PyroGenesis Canada Inc. is already putting it to good use despite the lack of information on the subject.

The story of current Indian E-waste management differs from global practices. In India, rag pickers pay a fee to the customer from whom they collect waste, whereas in developed countries, a recycling fee is charged to customers to manage waste effectively. The informal sector performs the majority of the activities such as collection, transportation, segregation, dismantling, recycling, and disposal. Rag pickers (also known as *kabadiwala*) collect all types of waste, such as papers, books, newspapers, plastic, cardboard, polythene, and metals and make a living by selling it to intermediaries or scrap dealers. This is a great way to make money for rag pickers, intermediaries, and scrap dealers. E-waste is mainly handled by unskilled workers who do not take adequate safety precautions.

Furthermore, no proper location is used for E-waste disposal. The operations to treat E-waste are carried out within the cities and slums. In some places, operations are carried out without adequate ventilation or lighting. Due to a lack of appropriate technology, recycling and disposal are not carried out properly. Also, very few companies have implemented a "take back" system voluntarily.

14.4 Problems and opportunities associated with E-waste

The Basel Convention, for example, aims to reduce and regulate the mobility of harmful waste between countries. Despite this convention, unlawful shipping and disposal of E-waste continues. It was projected in 2018 that 50 Mt of E-waste are being produced worldwide. Personal gadgets, such as computers, displays, cellphones, tablets, and televisions, account for half of this, with bigger household appliances and apparatus such as cooling and heating, accounting for the rest. Even though E-waste law covers 66% of the world's population, only 20% of global E-waste gets reused every year, implying that 40 Mt of E-waste are either burnt for resource retrieval or unlawfully sold and processed in a substandard manner. More than 100 million computers are discarded in the United States alone, with only around 20% of them being properly recycled. A total of 160 million electronic gadgets are discarded in China every year. China was formerly thought to be the world's largest E-waste dump. Separating E-waste is a skill that hundreds of thousands of individuals have.

The volume of E-waste is growing at a pace of 5%–10% annually worldwide. Every year, India creates 146,000 tons of E-waste (Mitra, 2013). These figures, however, only reflect E-waste created in the United States and do not account for garbage imports, including both lawful and unlawful, which are significant in growing economies like India and China. This is because India imports a considerable EEE quantity from other nations. Switzerland is world's first country to create and operate a formal E-waste management system, recycling 11 kg of E-waste per capita, compared to the European Union (EU)'s aim of 4 kg/capita.

EU's directive establishes specific gathering, retrieval, and reprocessing objectives for EU associate nations. As a result, all member nations must meet a least gathering objective of 4 kg per capita each year. Such gathering- and weight-based reprocessing objectives aim to limit the number of harmful chemicals dumped into landfills in addition to increasing the availability of recyclable resources, which indirectly encourages new product manufacturers to use fewer virgin materials.

In the EU, one-third of E-waste is collected separately and processed properly. In South Korea, implementing the EPR system in 2003 was the most significant move, with producers collecting almost 70% of E-waste. Also, 12% and 69% of E-waste were recycled and reprocessed during the same period. The rest was disposed of in landfills or incinerator plants, accounting for 19% of the total.

The EU and Japan are well-developed initiatives at all levels to change consumer behavior. E-waste management in developing nations is impaired by nonenforcement of the existing regulatory authority, lack of consciousness and sensitization, and scarce professional protection for persons involved in these processes. As a result, developing nations must implement efficient programs to encourage E-waste reuse, refurbishment, or recycling in particular amenities to evade environment and earth's fragile ecosystem.

Developing nations such as India, China, Ghana, Peru, Nigeria, and Pakistan are top receivers of the E-waste (Singh & Seth, 2013). The Basel Action Network (BAN) guarantees that E-waste is handled sustainably. It protects the environment from harmful trade-in waste. A linked network of environmental advocacy NGOs in the US includes BAN, Silicon Valley Toxic Coalition (SVTC), and Electronics Take-Back Coalition (ETBC). The shared goal of the three groups is to advocate country-level solutions for management of the hazardous waste. Electronic Stewards, a mechanism for monitoring and certifying recyclers and takeback programs so that careful customers know which ones satisfy high standards, is a recent project.

14.4.1 Challenges and opportunities associated in India

India ranks 177 out of 180 nations according to a study issued by the World Economic Forum (WEF) and is among the worst five countries in terms of the Environmental Performance Index (EPI) in 2018. This is attributed to low environmental health policy performance and mortality from air pollution categories. In addition, after the United States, China, Japan, and Germany, India is rated fifth in the world among major E-waste-generating countries, recycling less than 2% of the entire E-waste it creates yearly. Since 2018, India has produced over 2 Mt of E-waste per year and imports huge E-waste from other nations. Open dumpsites are a regular sight, resulting in difficulties including groundwater poisoning and bad health, among others. According to the Associated Chambers of Commerce and Industry of India (ASSOCHAM) and KPMG, the management of E-waste in India, would be found nearly 70% from computer apparatus accounts, 12% from telecommunication apparatus phones, 8% from electrical apparatus, and 7% from medical apparatus, with the remainder coming from domestic E-waste.

The informal sector controls E-waste gathering, conveyance, handling, and recovering. The industry is well connected but uncontrolled. The materials that are retrieved are not valuable. Furthermore, severe concerns about pollutants are leaked with environment and health of employees.

India has its largest E-waste disposal center at Seelampur, Delhi. Adults and progenies give 8—10 h every day to collect recyclable materials, valuable metals, and numerous functioning pieces from electronic equipment. E-waste recycler's practice methods include open incineration and acid-leeching. This issue might be rectified by raising consciousness and strengthening recycling unit infrastructure and existing legislation. An unorganized sector manages the majority of the E-waste collected in India.

In addition, informal electronics recycling/reuse channels such as restoration factories, used-goods dealers, and e-commerce site merchants acquire a considerable share of abandoned electronics for reuse and component cannibalism.

Many developing countries, such as India, face significant challenges in managing E-waste. Processes for the safe and sustainable management and disposal of E-waste must be implemented by relevant authorities in developing and transition nations. The E-waste Management Rules were established in India in 2016 to minimize E-waste output and boost recycling. Studies have shown that in accordance with these rules, the government implemented EPR, which requires manufacturers to collect around 30%–70% of the E-waste generate during seven years.

Integrating the informal sector into a transparent recycling system is critical for improved environmental and human health impact control. A few attempts have been made to integrate the current informal sector into the growing environment. German Society for International Co-operation (GIZ), for example, has established alternative business models for guiding informal sector associations toward authorization. These business models advocate for a city-wide collecting system that feeds the manual dismantling facility and a plan based on the best available technological facilities to generate more income from printed circuit boards. Safer practices and more income per unit of E-waste collected are achieved by substituting the traditional wet chemical leaching procedure for gold recovery with export to integrated smelters and refineries.

E-waste contains a high concentration of metals such as gold, silver, and copper, which may be recovered and reintroduced into the manufacturing cycle. The effective recovery of precious elements in E-waste has enormous economic potential and can create income-generating opportunities for people and businesses. The Indian government modified the E-waste Management Rules of 2016, in March 2018, to facilitate and successfully execute ecologically sound E-waste management in India. With effect from October 1, 2017, the updated rules changed the collection objectives under the provisions of EPR. Effective and enhanced E-waste management will be assured through updated objectives and monitoring under the Central Pollution Control Board (CPCB).

Nokia is one of the few businesses that appears to have made a concerted effort in this area since 2008. In line with an EPR authorization plan authorized by CPCB in India, the firms were deemed liable for building routes for appropriate management of E-waste. Recently, the import licenses of several large corporations were suspended due to violations of E-waste regulations. The policies have a significant impact on India's efficient execution of E-waste management. Any activity can have its unique set of incentives to attract contributors.

14.4.2 Influence of recycling E-waste in developing countries

Most of the E-waste includes some reusable material, such as metal, glass, and metals. However, owing to incorrect discarding procedures and processes, these materials cannot be recovered for further purposes. If E-waste is deconstructed and treated improperly, its hazardous elements may harm the environment. Disassembling components, incineration, and wet type biochemical process are utilized to clear off the garbage, resulting in direct exposure and inhalation of dangerous chemicals. Face masks and gloves are not regularly utilized due to employees frequently lacking the expertise. Therefore, they need experience to do their duties effectively. Furthermore, hand extraction of hazardous metals results in the introduction of harmful material into the bloodstream of the person doing the extraction. The risks of hazardous metals include kidney damage, liver damage, and neurological issues. The method of recovering of E-waste produce different forms of pollution, such as water pollution, soil pollution, and air pollution. The incineration process is used to extract metals from the E-waste. This process releases brominated dioxins, chlorinated dioxins, and carcinogens, which can cause cancer in human beings. During the recycling process, poisonous substances with little commercial use are discarded. These harmful compounds drain into the underground aquifer, eroding the quality of the local groundwater and leaving it unsuitable for human consumption and agricultural reasons. E-waste placed in the ground contains metals like a cadmium, lead, arsenic which can contaminate the soil and render it unsuitable for farming use. Studies on E-waste recycling have found higher concentrations of heavy metals on the ground surface. In India, E-waste is processed by the informal sectors in cities including New Delhi, Kolkata, Mumbai, and Chennai (Cairns, 2005). According to the findings of such investigations, the primary locations for such persistent hazardous chemicals are those involved in metal recovery operations. According to the same researchers, the semi-volatile nature of the determined organic pollutants created or released during the recovering process allows them to escape into the ambient air.

14.4.3 How can governments, city administration, and citizens help?

According to the Associated Chambers of Commerce and Industry of India report (2017), the government should consider cooperating with the sector to develop norms regarding operating processes and a phased approach to decrease E-waste to the lowermost possible level. Alternatively, the authorities may refer to practices used by other nations for effective E-waste gathering and reprocessing. According to a previous study, South Korea is a leading electronics maker and has recycled 21% of the total 0.8 Mt of electric waste created in 2015.

The government should promote new entrepreneurs by providing the necessary financial assistance and technological advice due to the adverse effects of unprocessed

E-waste on surface of ground, air, and water. New start-ups related to E-waste reprocessing and dumping should be established and promoted through specific discounts. The unorganized sector has a well-established collecting network. However, it is capital-intensive in the case of the organized industry. As a result, if both sectors cooperate and work in harmony, the materials gathered by the unorganized sector may be given over to the organized sector to be processed in an environmentally beneficial manner. In this case, the government may play a critical role in bridging the gap between the two sectors and ensuring the successful processing of E-waste. It is time for the government to take a proactive approach to recycle and dispose of E-waste safely to safeguard the environment and ensure the well-being of the general population and other living species.

The EPR concept is rapidly being used for E-waste handling in many nations, and its relative efficacy and success have been established in EU countries. Instruments for implementing EPR can be a combination of economic, regulatory, and informational. The authorities such as a state government, municipalities, and NGOs should collectively facilitate E-waste collection, facilitation, and infrastructure creation to ensure the success of E-waste management. DfE is currently gaining popularity worldwide as a novel approach to addressing environmental contamination. The DfE concept in product design is a technique that drastically reduces the environmental effect of items before they are placed on the market. It is frequently observed that India's strict standards are rendered ineffective owing to lax execution.

Consumers have a critical part in E-waste management. Most consumers casually trash numerous little devices together with dumped rubbish, and many individuals publicly burn that collected waste. Various dangerous compounds, such as dioxins and furans, are produced throughout our breathing process. This is a dangerous behavior that must be discontinued immediately. Some of the most advanced resident welfare associations (RWAs) have clearly labeled containers for collecting E-waste. All such residential communities should follow this approach. Students and women self help groups (SHGs) in their respective RWAs might be mobilized for this exercise.

14.5 International policies, regulations, conventions, and initiatives related to E-waste

Many countries and unions worldwide have established various approaches to managing E-waste under legislative scope and instrument effectiveness. However, there are still many countries such as European countries, China, Taiwan, and Korea, where illegal migration of E-waste is observed. In this context, many countries have addressed their E-waste problems, and accordingly, their comprehensive and directive legislation has been approved and progressed.

In European countries, the government had regularized the first E-waste regulation in 2007 and named that regulation Waste Shipment Regulation. This regulation emphasized that no EU member could export hazardous E-waste to non-OECD (Organization for Economic Cooperation and Development) countries. Furthermore, the EU had passed E-waste directive legislation in 2003, which is mainly aimed at changing products' design and enhancing recycling rates of WEEE. Furthermore, the EU implemented the restriction of hazardous substances (RoHS) to eliminate toxic substances such as mercury, lead, and other heavy metals from electrical and E-waste products. These regulations are primarily designed for facilitating 3R policy for WEEE by using the EPR theory, which deals with the financial responsibilities for collecting and managing WEEE.

On the other hand, individual producer responsibility primarily deals with managing new products launched in the market. Also, the European Commission proposed a revision in 2008 to deal with the ineffective and deficient rule of the directive legislation. They proposed to perform several modifications in the directive legislation to reduce unlawful export of E-waste to the OECD countries. Furthermore, the provision dealing with packaging directives was implemented in UK laws.

In the United States, federal regulations have been divided into two major types, namely, the Resource Conservation and Recovery Act of 1976 (RCRA) and the Environmental Protection Agency (EPA)'s CRT rule (Kumar & Singh, 2013). RCRA is a tracking system that is mainly designed to control hazardous wastes. This rule requires permission from the importing countries while handing and disposing of E-wastes. However, RCRA has two loopholes. First, RCRA will be effective only when the material comes under the definition of toxic waste. Second, the waste products generated from houses and businesses (up to 220 pounds of hazardous waste per month) can be buried in landfills. This shows that RCRA is not working effectively to control the disposal of wastes in the United States. Currently, CRT is more effective in the United States. The rule deals with the exportation of arranged CRT glass and their recycling. The United States has also initiated its legislation and regulation to make an effective management system. Nearly 23 states of the United States passed their legislation and regulation to control the disposal of E-waste.

Furthermore, the Chinese government has gone through various environmental laws and regulations to form an effective E-waste management system (Brigden et al., 2005). In February 2000, the Chinese government passed the regulation of "Catalogue for managing the import of E-waste," which primarily bans the import of E-waste. In April 2006, "Technical Policy on Pollution Prevention and Control of WEEE" was passed to set a principle of 3R and "polluter pays principle." In March 2007, the rules and regulations regarding controlling pollution from E-waste were employed to inhibit the use of toxic materials. Furthermore, in February 2008, administrative measures on WEEE pollution prevention were primarily developed to prevent pollution caused by E-waste disassembly, recycling, and disposal. The latest regulation was formed in January 2011, which is used to manage WEEE recycling and disposal. This regulation makes the recycling of E-waste compulsory.

Furthermore, Thailand has developed "The Enhancement and Conservation of the National Environmental Quality Act (NEQA) 1992," which is directed toward enhancing the quality of the environment (Kumar & Singh, 2013). The Factory Act 1992 points toward the supervision of factory activities. This act was classified into Classes 1, 2, and 3 (as per the directive). These classes were formed to control the wastes generated from the factories. Factories were classified as Class 1 if they manufactured the products without official request and permission. On the other hand, factories were classified as Class 2 and Class 3 if they manufactured the products with official request and official permission, respectively. Moreover, the Substance Act 1992 was declared to regulate dangerous substances that may harm living beings such as animals, human, plants, etc. Furthermore, Thailand introduced the latest strategic plan in 2007, which deals with environmental issues and has an E-waste effect.

Japan is one of the largest electronic product manufacturers. However, it has a very limited number of regulations regarding E-waste. Japan took its first initiative in 1998, where the regulation "Specified Home Appliance Recycling Law (SHARL)" was formed (Kumar & Singh, 2013). The Home Appliance Recycling Law was amended in 2001, which mainly deals with the proper treatment of waste home appliances.

E-waste management is a major challenge for India. In this context, India has formed several waste control rules or regulations. First, The Environmental Protection Act was formed in 1986. This regulation defined hazardous waste for the first time. Furthermore, two special regulations were made under this provision, namely, the Hazardous Waste Rules (Management and Handling) and the Rules Batteries (Management and Handling) (Alexeew et al., 2009). However, the Indian Municipal Solid Wastes Rules (Management and Handling) 2000 does not cover E-waste. The Hazardous Waste (Management and Handling) Rules 2008 replaced the previous hazardous waste management (HWM) rules to add more provisions in E-waste handling. The Batteries (Management and Handling) Act covers the management of lead—acid batteries. Under this rule, the manufacturer and importers are responsible for taking back the system for batteries. The Central Pollution Control Board (Gaela, 2010) further issued the guidelines on E-waste management in 2008. This guideline was developed to restrict the use of hazardous substances in manufacturing electronic appliances and the acceptance of eco-friendly technologies in recycling E-waste. The most recent regulation formed by the Indian government was "E-waste (Management and Handling) 2011," which involves all the stakeholders in handling E-waste, EEE production, dealers, refurbishers, etc. This rule states that E-waste producers should confirm that their waste should not cause any harm before their disposal, and their product should be as per the rule of RoHS in the manufacturing of EEE. In addition, this rule mainly deals with the total ban on illegal imports. Consequently, it has become an important part of regulation for dealing of E-waste in India.

14.6 Case studies on issues of best practice of E-waste management

It is becoming more difficult to dispose of electronic trash responsibly due to the urgent threat of environmental contamination from dangerous compounds found in these items. Hence, managing these materials might help reduce environmental impact in addition to generating economic benefits. This management is based on triple R, that is, reduce, reuse, and recycle. However, the global living standards are changing drastically, which is directly proportional use of electronic goods. As a result, a reduction in the production of E-waste is not feasible. Hence, the management of E-waste completely depends on reuse and recycling. Different countries and institutions follow several practices to manage E-waste. In this section, several effective practices of E-waste management around the globe have been discussed.

14.6.1 E-waste management scenario in India

Until 2008, there were no laws to govern the disposal of E-waste. As per Schedule IV of the hazardous waste rules (2008), registration in CPCB for E-waste recyclers is necessary.
 According to a 2020 report by the Central Pollution Control Board (India), India generated 1,014,961 tons of E-waste in 2019—2020, which is 32% more than 2018—2019. This report also concluded that only 3.6% and 10% of these were collected in 2018 and 2019, respectively. It was found that the concentration of metals in E-waste is more than the concentration present in ores. This has created a golden opportunity in the field of E-waste management (Debnath & Ghosh, 2018).
   Formal and informal recyclers currently handle the collection, segregation, and recycling of E-waste in India. Informal recyclers include unapproved and unregistered units. These informal recyclers collect E-waste directly from people's houses and recycle it in an unprocessed manner. On the other hand, the formal recyclers are experienced, authorized, and registered units. These units are doing a business out of managing the E-waste by recovering the metals and polymers (Ghosh et al., 2014). By using mechanical dismantling, the formal sectors separate metals, polymers, electronic circuits, and glass (Ghosh et al., 2014). The high content metal units are further sent to the third parties who extract metals from the printed circuit boards using metallurgical techniques (Debnath et al., 2018). The various methods used for metal recovery are incineration or open burning (used by informal sector), mechanical dismantling (informal and formal), acid leaching (informal), electro-chemical process (formal), and metallurgical process (formal). Out of these methods, the metallurgical process has high metal recovery efficiency compared with others. In India, supply chain network is facing many issues. As the generated E-waste has a heterogeneous nature, from collection to the recovery of metals ends up with high processing costs, making it costlier than virgin metals. In India, there is common thinking that recycled product is of lower grade. This belief creates a hurdle in business sustainability in E-waste recycling.

Nowadays, most of the countries are adopting and emphasizing the concept of 4R principles. The E-waste plastics among such are used in the construction of the roads, building block materials, etc. To utilize such E-plastic or thermosetting wastes (e.g., printed writing boards, IC packages), full-scale performance studies were carried out in several sectors, including road construction Kumar & Singh, 2013). However, the recycling of cited waste is extremely difficult because existing epoxy resins in them are rarely remelted and hard to pulverize, including inorganic filler, such as glass fibers. Additionally, the outcomes of these studies have revealed that the use of modified bitumen in the construction/maintenance of bituminous roads significantly improved the pavement performance and its cost-effectiveness, considering the life cycle cost as per the standard. Meanwhile, Guo et al. (2009) grounded such wastes and tried to improve the properties of bitumen by blending in the same. Their results showed that the viscosity of the E-waste modified bitumen increased with the presence of nonmetals in such wastes. The penetration index (PI) of that modified bitumen was observed to be higher than the required values, which anticipated the use of the modified bitumen in high-temperature areas. The smaller particle size in such grounded waste had more surface area per unit volume, which led to greater bitumen—particle interactions. Furthermore, reduced ductility of the modified bitumen depicted its inconspicuous low-temperature stress concentration withstanding capability. Eventually, it was documented that nonmetals in that modified bitumen with the smallest particle size (i.e., 0.07—0.09, 0.09—0.15, 0.15—0.30, and 0.30—0.45 mm) had the best overall performance. Subsequently, the rheological analysis conducted by Colbert (2012) explored that the E-waste modified bituminous binders had a higher mixing and compaction temperature than the virgin bitumen and increased with greater percentages of grounded E-waste blended into such bitumen.

Low-temperature performance by creep stiffness analysis of the modified bitumen showed that the stiffness of the modified asphalt binder increased with every incremental frequency in bending beam rheometer characterization. Again, lower percentages (approximately 5% by weight of bitumen) of the E-waste yielded superior creep stiffness to resist low-temperature cracking. On the contrary, more amount (approximately 15% by weight of bitumen) of such waste was required to resist rutting at pavement temperature. A diverse range of E-waste percentages have dwindled its applicability in tropical regions like India. Thus Shahane et al. (2021) initiated an investigation to explore the rheological compatibility of the E-waste to the virgin bitumen. Outcomes of that study have revealed that 5% of E-waste plastic powder by weight of bitumen met the Indian specifications at high temperatures. Besides, the phase angle was found smaller than the virgin bitumen, which indicated its good elastic behavior even at smaller frequencies. The aforementioned modified binder enriched bituminous mix sustained maximum resistance to fatigue cracking at 10°C and rutting resistance at 40°C temperature significantly. However, structural evaluation or performance analysis of E-waste modified bitumen enriched bituminous mix is required to judge its rutting and moisture damage resistivity. Thus rigorous study on all those points should be conducted.

14.6.2 E-waste management scenario in Bangalore, India

E-waste recycling in Bengaluru, the silicon city of India, is a multicrore market, with E-waste management at Gowripalya and Nayandahalli. The E-waste recyclers make around Rs. 2—3 lakhs every month from the sale of the disassembled E-waste they collect (Prakash & Shaikh Mubarak, 2019). There are a few recycling centers in Karnataka, including e- Wardd, e-Parisara, K.G. Nandini Recyclers, Ash Recyclers, New Port Computer Services India Pvt. Ltd. Recyclers, and E-R3 Solutions Pvt. Ltd. In Bangalore, the unit e-friendly E-waste recycling has a daily capacity of 1 ton to store and dismantle E-waste. The most consistent effort is being made to ensure that E-waste dismantling is handled by licensed professional recyclers rather than unlicensed local merchants. Waste recyclers have unique capacity to separate waste into several categories and improve scrap so that it may be reused and sold. It contributes to recycling by guiding the technical parts of waste treatment with guidelines on safety and safeguards. Electronic merchants and service facilities now have containers specifically designed for their recyclables. The hydrometallurgical process does the further separation of complicated and rejected E-waste. Plastic, polymers, and noxious gas, among other non–biodegradable and hazardous substances, were also salvaged and disposed of properly (Prakash & Shaikh Mubarak, 2019).

The Eco Birdd E-waste recycling system is another recycling unit in Bangalore that has a daily capacity of 2 tons of E-waste. It provides an end-to-end solution for all electronic trash created in an environment-friendly and sustainable manner. The E-waste is collected from several sectors, including households, schools, hospitals, businesses, bulk generators, manufacturers, consumers, and NGOs. The disassembly is carried out in accordance with all applicable regulations and safety procedures. The free-hand approach separated metallic compounds from nonmetallic substances in the separations. The chemical treatment method is used for the metal and other products recovery. The main objective of the treatment is to get the metal out of E-waste (Prakash & Shaikh Mubarak, 2019)

14.6.3 E-waste management scenario in Germany

In Germany, E-waste management is regulated by the Electrical and Electronic Equipment Act (ElekrtoG). As per ElekrtoG, the Public Waste Management Authority (PuWaMa) is in charge for E-waste collection. The PuWaMa is responsible for setting up places in their districts where people can bring E-waste back or collect E-waste from people's homes. It also separates the E-waste into five different containers based on the five categories mentioned in the act and gives them to the manufacturers free of charge. The ElektroG adheres to EPR in its operations. Individual brand-selective takeback schemes (IBTS), individual nonselective takeback schemes (INTS), and collective takeback schemes (CTS) are defined by the ElektroG as three methods through which producers might collect E-waste. In IBTS and INTS arrangements, manufacturers often contract with end-of-life service providers (ESPs) to handle the logistics, treatment, and disposal of E-waste generated during the manufacturing process. In CTS, producers get E-waste in proportion to their market share, regardless of their brand. When ESPs collect E-waste, they process it in an ecologically friendly manner and provide the producer with information about this treatment. On the other hand, the producer is required to transmit the information to the Elektro-Altgerate Register (EAR) (ElekrtoG, 2005).

14.6.4 E-waste management scenario in Switzerland

The Swiss Association for Information, Communication and Organizational Technology (SWICO) and the Stiftung für Entsorgung Schweiz (SENS) are the two bodies in Switzerland accountable for the management of E-waste. The SENS. created a vignette system in 1991 to compensate for the inadequacy of freezers and air conditioners when they were disposed of properly. After 2003, the SENS proposed the concept of the advanced recycling fee (ARF) on all electrical items, which is now in effect on all electrical products. The SWICO organized a feasibility study committee in 1989 to examine the viability of a takeback and recycling scheme for IT and office equipment. SWICO Recycling Guarantee was established in 1994 to assist in collecting, transporting, and financing IT and office equipment recycling. In consultation with industry groups, the government considers numerous policy and regulatory issues (Sinha, 2004). In Switzerland, material flows are divided into two types: those that occur before the consumption (preconsumption) and those that occur after consumption (postconsumption). The preconsumption flow includes all of the flows directed from the manufacturer/importer through the distributor and finally to the consumer. Postconsumption flows are when the consumer returns E-waste to stores or authorized collection stations after they have purchased something. After the collecting phase, the waste is sent to facilities for disassembly or recycling in an environmentally sound and sustainable system. Material recovery occurs throughout the recycling process, and the remainder is disposed of in a landfill or incinerator. When a new product is sold, the ARF is collected by distributors and retailers from the people who purchased the product. All the expenses related to E-waste collection, logistics, and recycling are met by these funds. The government acts as a monitor and does not become involved in day-to-day operations in the E-waste management system.

14.6.5 E-waste management scenario in Japan

There have been several laws established in Japan to promote more efficient resource utilization; for example, in 2001, the Japanese Home Appliance Recycling Law and

Table 14.1 Comparison of the E-waste management system of India, Germany, Switzerland, and Japan (Debnath & Ghosh, 2018; Kumar & Singh, 2013).

Indicative parameters	Country			
	India	Germany	Switzerland	Japan
Laws, legislation, and regulations	Ineffective	Effective	Very effective	Effective
Collection system	2% (very ineffective)	50% (average)	62% (effective)	30% (below average)
Recycling and recovery	2% (very ineffective)	45% (average)	61% (effective)	80% (very effective)
Infrastructure	Ineffective	Effective	Very effective	Effective
Consumer participation	Very ineffective	Effective	Effective	Effective
Data availability	Very ineffective (no data available)	Very effective	Very effective	Very effective

the Law for Promotion of Effective Utilization of Resources (LPEUR) were passed. The two rules are designed to reduce the amount of electronic trash generated and maximize the utilization of recycled resources. The Japanese Home Appliance Recycling Law is based on EPR. It applies to four major household equipment (refrigerators, huge television sets, air conditioners, and washing machines). Later, the regulation was expanded to include personal computers and copiers. Instead of focusing on collection objectives, this regulation focuses on recycling goals. As a result of the revision of the law in 2003, PC manufacturers were forced to implement a take-back system to encourage the efficient usage of resources required by PC makers. Fees earned from consumers are used to fund recycling activities (Table 14.1).

14.7 Conclusions

Generation of E-waste is rapidly becoming a major public health concern, and it is just getting worse. It is important for an E-waste recycler to recover maximum resources out of E-waste with least or no emission of toxic gases by adopting cleaner production. To collect, process, and manage E-waste independently, as well as divert it from traditional dumps and waste incineration, it is important to unite the informal and formal sectors. It is necessary to increase evidence drives, building capacity, and consciousness to support environment-friendly programs of E-waste management. There is a large gap in the physical, mental, and educational health and behavioral outcomes due to the exposure to various chemical compounds due to the recycling of E-waste. Formal E-waste recyclers must offer a safe working environment. Although informal recyclers work hard in hazardous conditions, the health dangers are widespread. To prevent illicit E-waste trading, increased efforts are urgently needed to enhance present processes such as collection schemes and management methods.

Optimizing the number of dangerous compounds in e-products will also help with dealing with certain E-waste streams because it will aid in the preventative process. In the sphere of E-waste management, the government must offer encouragements, such as tax breaks or refunds, to guarantee compliance throughout the electronics sector. Furthermore, E-waste collection objectives must be evaluated and revised on a regular basis to maintain compliance with E-waste collection.

References

Ahirwar, R., & Tripathi, A. K. (2021). E-waste management: A review of recycling process, environmental and occupational health hazards, and potential solutions. *Environmental Nanotechnology, Monitoring and Management, 15*(December 2020), 100409. Available from https://doi.org/10.1016/j.enmm.2020.100409.

Alexeew, et al. (2009). *E-waste handling practices in Europe & India: Lessons learned from both sides.* Berlin.

Andeobu, L., Wibowo, S., & Grandhi, S. (2021). An assessment of E-waste generation and environmental management of selected countries in Africa, Europe and North America: A systematic review. *Science of the Total Environment, 792*, 148078. Available from https://doi.org/10.1016/j.scitotenv.2021.148078.

Arya, S., & Kumar, S. (2020a). E-waste in India at a glance: Current trends, regulations, challenges and management strategies. *Journal of Cleaner Production, 271*, 122707. Available from https://doi.org/10.1016/j.jclepro.2020.122707.

Arya, S., & Kumar, S. (2020b). Bioleaching: Urban mining option to curb the menace of E-waste challenge. *Bioengineered, 11*(1), 640−660. Available from https://doi.org/10.1080/21655979.2020.1775988.

Arya, S., Patel, A., & Kumar, S. (2021). Urban mining of obsolete computers by manual dismantling and waste printed circuit boards by chemical leaching and toxicity assessment of its waste residues. *Environmental Pollution, 283*, 117033. Available from https://doi.org/10.1016/j.envpol.2021.117033.

Arya, S., Rautela, R., Chavan, D., & Kumar, S. (2021). Evaluation of soil contamination due to crude E-waste recycling activities in the capital city of India. *Process Safety and Environmental Protection, 152*, 641−653. Available from https://doi.org/10.1016/j.psep.2021.07.001.

Balde, C. P., Forti, V., Gray, V., Kuehr, R., & Stegmann, P. (2017). *The global E-waste monitor 2017. Quantities, Flows, and Resources.*

Barragan, J. A., Carlos Ponce, D. L., Juan Roberto, A., Castro, A., Peregrina-Lucano, F., Gómez-Zamudio., & Larios-Durán, E. R. (2020). Copper and antimony recovery from E-waste by hydro-metallurgical and electrochemical techniques. *ACS Omega, 5*(21), 12355−12363. Available from https://doi.org/10.1021/acsomega.0c01100.

Brigden, K., Labunska, I., Santillo, D., & Allsopp, M. (2005). Recycling of E-wastes in China & India: *Workplace & environmental contamination E-wastes* contamination.

Cairns, C. N. (2005). E-waste and the consumer: Improving options to reduce, reuse and recycle. *Proceedings of the 2005 IEEE International Symposium on Electronics and the Environment.* Available from https://doi.org/10.1109/ISEE.2005.1437033.

Colbert, B. W. (2012). *The performance and modification of recycled electronic waste plastics for the improvement of asphalt pavement materials.* Michigan Technological University.

Cui, H., & Anderson, C. (2020). "Hydrometallurgical treatment of waste printed circuit boards: Bromine leaching.". *Metals, 10*(4), 1−18. Available from https://doi.org/10.3390/met10040462.

Debnath, B., Chowdhury, R., & Ghosh, S. K. (2018). Sustainability of metal recovery from E-waste. *Frontiers of Environmental Science and Engineering, 12*(6), 1−12. Available from https://doi.org/10.1007/s11783-018-1044-9.

Debnath, B., & Ghosh, S.K. (2018). E-waste recycling in India: A case study.

Forti, V., Baldé, C. P., Kuehr, R., & Bel, G. (2020). *The global E-waste monitor 2020: Quantities, flows, and the circular economy potential.* United Nations University (UNU)/United Nations Institute for Training and Research (UNITAR) − Co-Hosted SCYCLE Programme, International Telecommunication Union (ITU).

Fu, J., Zhang, A., Wang, T., Qu, G., Shao, J., Yuan, B., Wang, Y., & Jiang, G. (2013). Influence of E-waste dismantling and its regulations: Temporal trend, spatial distribution of heavy metals in rice grains, and its potential health risk. *Environmental Science and Technology*, *47*(13), 7437–7445. Available from https://doi.org/10.1021/es304903b.

Gaela. (2010). *Waste regulation in India: An overview*. CPPR.

Ghimire, H., & Ariya, P. A. (2020). E-wastes: Bridging the knowledge gaps in global production budgets, composition, recycling and sustainability implications. *Sustainable Chemistry*, *1*(2), 154–182. Available from https://doi.org/10.3390/suschem1020012.

Ghosh, S. K., Baidya, R., Debnath, B., Biswas, N. T., & Lokeswari, M. (2014). E-waste supply chain issues and challenges in India using QFD as analytical tool. *International Conference on Computing, Communication & Manufacturing*, 287–291.

Guo, Y., Huang, C., Zhang, H., & Dong, Q. (2009). Heavy metal contamination from electronic waste recycling at Guiyu, Southeastern China. *Journal of Environmental Quality*, *38*(4), 1617–1626.

Ha, V. H., Lee, J. C., Huynh, T. H., Jeong, J., & Pandey, B. D. (2014). Optimizing the thiosulfate leaching of gold from printed circuit boards of discarded mobile phone. *Hydrometallurgy*, *149*, 118–126. Available from https://doi.org/10.1016/j.hydromet.2014.07.007.

Horeh, N., Bahaloo, S. M., Mousavi., & Shojaosadati, S. A. (2016). Bioleaching of valuable metals from spent lithium-ion mobile phone batteries using *Aspergillus niger*. *Journal of Power Sources*, *320*, 257–266. Available from https://doi.org/10.1016/j.jpowsour.2016.04.104.

Ilyas, S., & Jae, chun L. (2014). Biometallurgical recovery of metals from waste electrical and electronic equipment: A review. *ChemBioEng Reviews*, *1*(4), 148–169. Available from https://doi.org/10.1002/cben.201400001.

Ilyas, S., Munir, A. A., Niazi, S. B., & Afzal Ghauri, M. (2007). Bioleaching of metals from electronic scrap by moderately thermophilic acidophilic bacteria. *Hydrometallurgy*, *88*(1–4), 180–188. Available from https://doi.org/10.1016/j.hydromet.2007.04.007.

Ilyas, S., Ruan, C., Bhatti, H. N., Ghauri, M. A., & Anwar, M. A. (2010). Hydrometallurgy column bioleaching of metals from electronic scrap. *Hydrometallurgy*, *101*(3–4), 135–140. Available from https://doi.org/10.1016/j.hydromet.2009.12.007.

Imran, M., Haydar, S., Kim, J., Awan, M. R., & Bhatti, A. A. (2017). E-waste flows, resource recovery and improvement of legal framework in Pakistan. *Resources, Conservation and Recycling*, *125*, 131–138. Available from https://doi.org/10.1016/j.resconrec.2017.06.015.

Initiative Step. (2016). *Guiding principles to develop E-waste management systems and legislation*.

Islam, M. T., & Huda, N. (2019). E-waste in Australia: Generation estimation and untapped material recovery and revenue potential. *Journal of Cleaner Production*, *237*, 117787. Available from https://doi.org/10.1016/j.jclepro.2019.117787.

Jagannath, A., Shetty, K. V., & Saidutta, M. B. (2017). Bioleaching of copper from E-waste using *Acinetobacter* sp. Cr B2 in a pulsed plate column operated in batch and sequential batch mode. *Journal of Environmental Chemical Engineering*, *5*(2), 1599–1607. Available from https://doi.org/10.1016/j.jece.2017.02.023.

Khaliq, A., Rhamdhani, M. A., Geoffrey, B., & Syed, M. (2014). Metal extraction processes for E-waste and existing industrial routes: A review and Australian perspective. *Resources*, *3*(1), 152–179. Available from https://doi.org/10.3390/resources3010152.

Kumar, U., & Singh, D. N. (2013). E-waste management through regulations. *International Journal of Engineering Inventions*, *3*(2), 6–14.

Lin, H., Tien, E., Yamasue, K. N., Ishihara., & Okumura, H. (2019). Waste shipments for energy recovery as a waste treatment strategy for small islands: The case of Kinmen, Taiwan. *Journal of Material Cycles and Waste Management*, *21*(1), 44–56. Available from https://doi.org/10.1007/s10163-018-0760-3.

Lundgren, K. (2012). *The global impact of E-waste: Addressing the challenge*.

Mitra, A. (2013). E-waste management—Indian context. *International Journal of Managment, IT and Engineering*, *3*(1), 358–366.

Mmereki, D., Li, B., Baldwin, A., & Hong, L. (2016). *The generation, composition, collection, treatment and disposal system, and impact of E-waste. E-waste in transition — From pollution to resource* (p. p. 13) InTech.

Natarajan, G., & Ting, Y. P. (2015). Gold biorecovery from E-waste: An improved strategy through spent medium leaching with pH modification. *Chemosphere*, *136*, 232–238. Available from https://doi.org/10.1016/j.chemosphere.2015.05.046.

Nithya, R., Sivasankari, C., & Thirunavukkarasu, A. (2021). E-waste generation, regulation and metal recovery: A review. *Environmental Chemistry Letters*, *19*(2), 1347—1368. Available from https://doi.org/10.1007/s10311-020-01111-9.

Nithya, R., Sivasankari, C., Thirunavukkarasu, A., & Selvasembian, R. (2018). *Novel adsorbent prepared from bio-hydrometallurgical leachate from waste printed circuit board used for the removal of methylene blue from aqueous solution* (Vol. 142). Elsevier B.V.

Oh, C., Jung, S. O., Lee, H. S., Yang, T. J., Ha., & Kim, M. J. (2003). Selective leaching of valuable metals from waste printed circuit boards. *Journal of the Air and Waste Management Association*, *53*(7), 897—902. Available from https://doi.org/10.1080/10473289.2003.10466230.

Pradhan, J., Kumar., & Kumar, S. (2014). Informal E-waste recycling: Environmental risk assessment of heavy metal contamination in Mandoli Industrial Area, Delhi, India. *Environmental Science and Pollution Research*, *21*(13), 7913—7928. Available from https://doi.org/10.1007/s11356-014-2713-2.

Prakash, K. L., & Shaikh Mubarak, N. (2019). E-waste handling — A case study of E-friendly and eco-Birdd recycling units, Bangalore. *Research Journal of Life Sciences, Bioinformatics, Pharmaceutical and Chemical Sciences*, *5*(660), 660—679. Available from https://doi.org/10.26479/2019.0503.54.

Quinet, P., Proost, J., & Lierde, A. V. (2005). Recovery of precious metals from electronic scrap by hydrometallurgical processing routes. *Minerals & Metallurgical Processing*, *22*(01), 17—22.

Ruan, J., Zhu, X., Qian, Y., & Jian, H. (2014). A new strain for recovering precious metals from waste printed circuit boards. *Waste Management*, *34*(5), 901—907. Available from https://doi.org/10.1016/j.wasman.2014.02.014.

Shahane, H. A., & Bhosale, S. S. (2021). E-Waste plastic powder modified bitumen: Rheological properties and performance study of bituminous concrete. *Road Materials and Pavement Design*, *22*(3), 682—702.

Sheel, A., & Pant, D. (2018). Recovery of gold from E-waste using chemical assisted microbial biosorption (hybrid) technique. *Bioresource Technology*, *247*, 1189—1192. Available from https://doi.org/10.1016/j.biortech.2017.08.212.

Shittu, O. S., Ian, D. W., & Peter, J. S. (2021). Global E-waste management: Can WEEE make a difference? A review of E-waste trends, legislation, contemporary issues and future challenges. *Waste Management*, *120*, 549—563. Available from https://doi.org/10.1016/j.wasman.2020.10.016.

Singh, R. P., & Seth, S. K. (2013). India: A matter of E-waste; the government initiatives. *Journal of Business Management & Social Sciences Research*, *2*(15).

Sinha-Khetriwal, D., Kraeuchi, P., & Schwaninger, M. (2005). A comparison of E-waste recycling in Switzerland and in India. *Environmental Impact Assessment Review*, *25*(5 spec. Iss.), 492—504. Available from https://doi.org/10.1016/j.eiar.2005.04.006.

Song, Q., & Li, J. (2014). Environmental effects of heavy metals derived from the E-waste recycling activities in China: A systematic review. *Waste Management*, *34*(12), 2587—2594. Available from https://doi.org/10.1016/j.wasman.2014.08.012.

Sun, Z., Cao, H., Xiao, Y., Sietsma, J., Jin, W., Agterhuis, H., & Yang, Y. (2017). Toward sustainability for recovery of critical metals from E-waste: The hydrochemistry processes. *ACS Sustainable Chemistry and Engineering*, *5*(1), 21—40. Available from https://doi.org/10.1021/acssuschemeng.6b00841.

Tsezos, M., Remoudaki, E., & Hatzikioseyian, A. (2006). Biosorption-principles and applications for metal immobilization from wastewater streams. *Workshop on Clean Production and Nano Technologies, Seoul, Korea*.

Veit, H. M. (2005). Utilization of magnetic and electrostatic separation in the recycling of printed circuit boards scrap. *Waste Management*, *25*, 67—74. Available from https://doi.org/10.1016/j.wasman.2004.09.009.

Yang, W., Dong, Q., Sun., & Ni, H. G. (2021). Cost-benefit analysis of metal recovery from E-waste: Implications for international policy. *Waste Management*, *123*, 42—47. Available from https://doi.org/10.1016/j.wasman.2021.01.023.

Yazici, E. Y., & Deveci, H. (2014). Ferric sulphate leaching of metals from waste printed circuit boards. *International Journal of Mineral Processing*, *133*, 39—45. Available from https://doi.org/10.1016/j.minpro.2014.09.015.

Zeng, X., John, A. M., & Li, J. (2018). Urban mining of E-waste is becoming more cost-effective than virgin mining. *Environmental Science and Technology*, *52*(8), 4835—4841. Available from https://doi.org/10.1021/acs.est.7b04909.

CHAPTER 15

An overview of E-waste generation and management strategies in metro cities of India

Dolly Kumari, Gunjan Singh and Radhika Singh
Department of Chemistry, Dayalbagh Educational Institute, Agra, Uttar Pradesh, India

15.1 Introduction

The use of electronic gadgets by humans is increasing rapidly as the result of day-by-day development and innovations throughout the world. Presently, electronic gadgets are used in most sectors, which leads to the production of a large proportion of global waste, and the exploitation of electrical equipment has doubled equally over the several years. The rapid need for waste and consumption has expanded the opportunities for trade of electrical and electronic equipment (EEE) and created electronic web (E-Web) all over the world (Arya et al., 2021a; Arya & Kumar, 2020a). Electronic devices, such as cellphones, computers, and diversion hardware, require proper disposal, which is rapidly evolving all over the world. As a result of good developments in equipment skills, features, and cost, as well as improved internet use, the manufacturing and use of electronics have substantially improved in the last two decades. In addition, most countries have largely overlooked this rising challenge in the rush for new developments. People in all regions of the world, for example, rely on cellphones as an essential part of their daily lives, and in most circumstances, life cannot be imagined without these gadgets. To fulfill the needs of today's consumers, end-to-end companies are developing cellphones to meet the needs of consumers, which has also allowed consumers to replace their existing cellphones for upgraded versions, which is one of the reasons for E-waste generation
(Sahu et al., 2018). Annually, E-waste output is expected to be 40 million tons, accounting for 5% of all global spending. E-waste is made up of around 9 million tons of abandoned cellphones, televisions, and computers in the European Union. According to the United Nations Environment Program (UNEP), E-waste from personal computers (PCs) is the most common, and dumping of televisions (TVs) has increased several times recently (Islam et al., 2020). Over two decades ago, the Indian economy announced major changes as the norm for the economy along the way. India's

Global E-waste Management Strategies and Future Implications
DOI: https://doi.org/10.1016/B978-0-323-99919-9.00015-5

electronics industry has emerged as a rapidly evolving area in terms of the point of creation, internal use, and transportation costs. The propagation of technological processes and high degrees of aging have triggered the development of E-waste. The current E-waste preparation in India is largely responsible for the planned environment, where the removal and reuse of personal computer (PC) waste are common and extremely dangerous to the environment (Dwivedy & Mittal, 2010). Because of its high cost of production, hazardous nature, and lack of removal strategies to deal with it, trash is constantly a source of concern. This has aggravated electronics product integration, which has in turn led to a significant increase in hazardous electrical trash. Toxic and other rare earth metals are present in this hazardous waste, which are harmful to human health and ecosystems. Unnecessary treatment of waste promotes material shortages just as it promotes major environmental and financial issues (Kumar & Dixit, 2018). The immense generation of E-waste is directing the latest ecological problems. Different bibliometric observations of research in miscellaneous parts of the E-waste board that was newly distributed, could deliver strong proof of the significance of E-waste test controlling. It has gained attention of experts worldwide. E-waste is inevitable and widely produced worldwide, and high-income countries, such as Japan, European Union nations, the United States, and Australia are reflected to be the main causes of E-wastes (Hossain et al., 2015). Thus there is a need to evaluate the sources of E-waste, how it is treated, how much quantity of E-waste is generated, and how to recycle it without impacting the environment.

15.2 Global generation of E-waste

The global E-waste generation was reported to be 53.6 million metric tons (Mt) in 2019 and has been predicted to amplify approximately to 74 Mt by 2030. Developed countries such as the United States, Europe, and Australia are the major producers of E-waste. These wastes are then transported to developing countries such as India, China, Ghana, Pakistan, Nepal, Bhutan, and Vietnam due to easy availability of open space for dumping and low-cost labor for recycling purposes (Dutta et al., 2021). In 2019, worldwide waste production included 17.4 million tons of small equipment, 13.1 million tons of large equipment, 10.8 million tons of commercial equipment, 6.7 million tons of monitors and screens, 4.7 million tons of small IT and telecom equipment, and 0.9 million tons of lighting. In the publication *Worldwide E-squander Monitor 2020*, the set amount of E-waste produced by different countries for 2019 is specified as the starting point for collecting of EEE points under the previously indicated categories. Asia and its 17 nations have produced the most significant portion of E-waste in 2019. China, India, Japan, and Indonesia have produced the majority of E-waste among Asian nations (Ahirwar & Tripathi, 2021). According to a collaborative study *Rethinking Waste Scaling Opportunity in India*, China (6.1 million Mt), Japan (2.2 million Mt), and India (3.2 million Mt) are the top three

Asian countries with the maximum E-waste generation (Garg & Adhana, 2019). Brazil's largest city produces an estimated 534,000 tons of E-waste every year and all emerging countries produce a substantial amount of E-waste. Brazil's largest city produced 6.5 kg of waste per person in 2015 and 8 kg of waste per person in 2017, that is, 65.4 million tons of waste. This is not the same as Mexico's average (8.2 kg per person) from 2014, but it is greater than that of other BRICS countries, specifically Brazil, Russia, India, China, and South Africa, and with the exemption of Brazil, the rest of the world. Russia, India, China, and South Africa generate approximately 8.7 kg, 1.3 kg, 4.4 kg, and 6.6 kg of E-waste per person (Azevedo et al., 2017; Rodrigues et al., 2020). The Malaysian government is currently undergoing a challenge to address the growing E-waste issues. Based on the study by Perunding Good Earth (PGE), the total burden of E-waste in 2008 was about 688,066 tons, but it was estimated to reach 1.119 million tons by 2020. Conferring to the challenge, the E-waste guideline has been recognized in Malaysia since 2005, with the enforcement of the Environmental Quality (Scheduled Wastes) Regulations 2005 permitted on August 15, 2005. According to this regulation, dumping E-waste at landfills is strictly prohibited. This regulation has been widely followed by the industrial and other sectors, but it is still an issue at the household level (Tiep et al., 2015). As most Pakistanis have limited access to new and upgraded electrical gadgets, they must rely on the use of second-hand products. The current situation has sparked demand in the market for lower-cost second-hand products or the elimination of the imported equipment. In 2012, the timing of E-waste testing in Pakistan was estimated to be 316 kt, when UNU estimated it to be 266 kt, that is, 1.4 kg per person in 2014 (Iqbal et al., 2015). As previously stated, approximately 80% of E-waste produced in industrialized countries is exported and shipped to Asian countries, particularly Pakistan, China, and India for further processing and alternative uses. The cheap cost of labor in Pakistan, which is less than USD2.00 reuse per unit PC compared to USD20.00 in industrialized countries, may be a significant consequence of E-waste imports. Furthermore, more stringent laws and guidelines are being formed in countries, mainly China and India, and it is projected that due to noncompliance with environmental legislation, mass E-waste would locate its way to Pakistan (Sajid et al., 2019).

15.3 E-waste management in India

After 1990, E-waste-related concerns in India began to surface after the first financial period, owing to stiff competition in the market for quality, brand, services, and price supplied by many industries, including foreign and Indian. India's electrical and consumer industries expanded during this time. Similarly, due to lower prices and increased purchasing power, the Indian commercial industry, particularly in-home equipment, such as ovens, TVs, refrigerators, laundry, air conditioning, and PCs, among others, exploded in the predevelopment era. Furthermore, the IT revolution in India is being ignored due to fundamental changes and e-administration with the spectacular utilization of new data in all domains (Wath et al., 2011). India is the fifth leading producer of E-waste in the world. The volume coming from OECD countries includes half of the 60% of the total E-waste in India, which makes its direct testing a challenging task. Government, civil society, and NGOs produce about 75% of E-waste, while the contribution of individual household is only 16%. Studies have shown that the rapid financial development with an improved purchasing limit of the expanded middle-income urban society provides a dramatic increase in the production of E-waste, and the two best-selling items in the Indian market are cellphones and PCs. The volume of PCs will keep increasing until 2022, subsequently steadily reaching a steeping point by 2028. On the other hand, no immersion site can detect E-waste from cellphones (Borthakur & Singh, 2016; Pathak & Srivastava, 2017). India's gadget industry has emerged as a rapidly evolving sector in terms of production and internal use and transportation costs. As indicated by the Electronic Industries Association of India, the service/business area denotes 80% of all IT tools bazaar access in India. The financial performance of the E-waste recovery framework depends entirely on the mix of the return item (Vishwakarma et al., 2022). However, laptops have a high product value area, but they have low mass limit, making them more difficult to recover than PCs. Contrary to the worldwide rate of 27 PCs per 1000 persons and more than 500 PCs per 1000 persons in the United States in 2004, India had one entry level PC to at least nine PCs for every 1000 persons. However, the size of Indian markets is larger than the majority of high-income countries (Dwivedy & Mittal, 2012). New Delhi, a major metro region in India, generates about 98,000 tons E-waste, Bangalore generates about 92,000 tons, Chennai generates 67,000 tons, Kolkata generates about 50,000 tons, Ahmedabad generates about 36,000 tons, Hyderabad about 32,000 tons, and Pune generates about 25,000 tons E-waste, while Mumbai reaches 1.2 Mt is top in the list. City-wise, Mumbai hit the masses when it brought electricity, followed by New Delhi, Bangalore, and Chennai (Garg & Adhana, 2019).

15.3.1 Status of E-waste in Mumbai

Mumbai, India's main financial center and port city, is the origin of the IT industry, producing over 20,270 tons of E-waste per year. In India, Mumbai is ranked first among the top 10 cities that generate the huge amount of E-waste. The intricate industrial belt between Mumbai and Pune is one of the world's major technology manufacturers. Mumbai is indeed more than just a port for importing new and used gear. It is also the largest customer and manufacturer base, both of which generate large amount of E-waste (Kalana, 2010; Srivastava & Pathak, 2020).

15.3.2 Status of E-waste in Delhi

Without a solid institutional structure, Delhi, the capital of India, has emerged as the world center of E-waste reuse. According to the management report, E-waste reuse continues in at least 20 areas in Delhi, including Gandhi Nagar, Kirti Nagar, Mayapuri, Mandoli, Mustafabad, Seelampur, Shastri Park, Turkman Gate, and Wazirabad. However, according to current information, there were only eight recyclers in Delhi with a combined capacity of 20,300 Mt per year, implying that Delhi produced around 30,000 Mt of E-waste yearly before 2015, a figure that is projected to rise dramatically in the future years. Many registered and unlicensed recyclers in Delhi import E-waste from abroad due to a lack of E-waste in the city and storing it in various urban areas across the country, as a result impacting worker. Delhi is a money magnet due to the high demand for the country. India imports roughly 50,000 Mt of E-waste each year, but only 11% of it has been lawfully reused (Das, 2011; Sinha, 2013).

15.3.3 Status of E-waste in Bangalore

Bangalore is one of India's main media outlets, covering the Indian IT industry and consumer markets. It was reported in 2013 that 20,000 tons of E-waste was produced, with an annual growth rate of 20%. The study also predicted that by the end of 2020, the amount of personal computer (PC) waste would be increased by roughly 500%. A few middle-class members, such as retailers, peddlers, residual retailers, and special authorities, are an important part of the WEEE administered by a unique area. Only 5% of E-waste is treated in a formal setting, with the residual 95% going to either a second-hand market or a home workshop (informal sector) for reuse (Awasthi & Li, 2018). E-waste sellers ships separate and demolish sections of E-waste to Delhi and Mumbai every day. Also, E-waste recyclers purchase roughly 2—3 lakh INR from the sale of used E-waste to Delhi every month (Das, 2011; Sinha, 2013).

15.3.4 Status of E-waste in Hyderabad

In India, Hyderabad was recognized as the Silicon Valley capital. The annual E-waste is predicted to reach 3,263,994 Mt from various sorts of hardware, such as PCs (3111.25 Mt), TVs (61.0 Mt), printers (86.46 Mt), and cellphones (5.284 Mt). Hyderabad's total E-waste production was 98,163 kg in 2010, with a population of 74.42 lakhs. It contains 42,869 kg of PCs, 53,581 kg of TVs, and 1713 kg of cellphones. For a population of 81.8 lakhs, the overall volume of E-waste was estimated to exceed 107,886 kg in 2013, with 47,117 kg of PCs, 58,890 kg of TVs, and 1881 kg cellphones. The most common size used by E-waste authorities and recyclers is the size which simply decreases (destroys) and separates (Gaidajis et al., 2010).

15.4 Initiatives and regulations

Except for the enormous E-waste produced in many countries, electronic equipment prices are not modified or reused in the countries where they were produced. Overall, the development of the E-waste limit was preexisting and waste collected in developed countries such as Australia, Japan, the United States, Canada, Korea, and Europe could be generally exported to Asian territories, such as China and India, where inexpensive disposal amenities and environmental strategies with less complicated laws exist. However, stronger E-waste legislation in certain countries has made finding alternate e-trans-border channels a priority (e.g., countries in West Africa). Despite the fact that accurate statistics and specifics on unlawful WEEE exports are unavailable, improvements in existing regulations are critical to the productive management of E-waste (Ilankoon et al., 2018). Provincial and public law can work with eco-plans all over the world, helping to save resources, reduce emissions, eliminate toxicity, etc. (Borthakur & Govind, 2017). The Department of Environment, Forestry, and Climate Change issued the National Environmental Management Policy (NEMP) in 2006, emphasizing the need for material recovery and reuse to reduce E-waste. As a result, garbage could be used as a commodity. The NEMP's main goal was to develop a framework for resource integration. Furthermore, the NEMP emphasized the importance of strengthening small areas by providing people with legal recognition, institutional funding, and significant development so that they can be associated with the normal activities of a reusable business that protects and safeguards both human and environmental health. The Central Pollution Control Board (CPCB) released Guidelines for the Strict Management of Natural E-Waste in April 2008, which included recommendations for the identification of various E-waste sources and recommended procedures for E-waste treatment. These rules recommend liability stages and harmful substance discounts. They have also enhanced the informal sector's participation in the proper collection and distribution of E-waste (Kaur & Goel, 2016). E-waste was not seen as a severe impediment to progress in the first two decades of India's financial development. This phase has seen an intensifying dissemination of E-waste into the market due to the extension of the gadget and IT zones. For the most part, E-waste was a hidden and unnoticed problem that coincided with the country's tremendous expansion (Borthakur & Govind, 2017). In India, strict waste control is unavoidably tough. The Control of Disposal Regulations/Rules and Regulations are clearly labeled by Kumar and Singh (2013).

- The Environmental Protection Rules, 1986
- Bio-Medical Waste (Management and Handling) Rules, 1998
- The Batteries (Management and Handling) Rules
- The Hazardous Waste (Management and Handling) Rules
- The Noise Pollution (Regulation and Control) Rules, 2000

- The Ozone Depleting Substances (Regulation and Control) Rules, 2000
- The Hazardous Wastes (Management, Handling and Trans-boundary Movement) Rules, 2008
- The Plastics (Manufacture, Usage and Waste Management) Rules, 2009
- The E-Waste (Management and Handling) Rules, 2011

15.5 E-waste recycling and management

Consumer awareness is one of the biggest distractions behind destroying the management of E-waste recycling. Organization and recycling of E-waste is a part of the varied spectrum discussed as waste management. Therefore, researchers are still drawn to gaining a better grasp of the prime challenges raised by consumer expectations and their associated behavior (Skinner et al., 2010). At the national and/or regional levels, global noise management necessitates a core focus on collection centers, transportation, treatment, investment, recovery, and disposal of E-waste as well as a better exposition of these facilities (Jayapradha, 2015). The government must empower NGOs and manufacturers to establish facilities for E-waste collection, commercialization, and recycling (Cucchiella et al., 2015). Exporting E-waste to poor and developed countries is complicating E-waste management both in those cultures (because to environmental and health issues) and around the world, as some contaminants associated with reprocessing may be re-exported. It includes ways for producing recyclable resources like metals, which can also be recycled as nonrecyclable materials (Rautela et al., 2021). However, dangerous compounds such as metals (e.g., As, Ba, Be, Cr, Cd, Ni and Pb), brominated regular mixes, and polychlorinated biphenyls (PCBs) are plentiful in E-waste. Various metals, such as polyhalogenated hydrocarbons, polycyclic aromatic hydrocarbons, dibenzofurans, and polychlorinated dibenzo-p-dioxins (PCDDs) are formed by low-temperature combustion of E-waste. When these compounds can be formally removed, their distribution may harm the environment and public health (Taghipour et al., 2012). For example, India recycled 5% of the E-waste it handled, which is remarkable for local governments working in hazardous and polluted conditions. In this way nonindustrial countries compare to nations in terms of E-waste development. E-waste development rates are higher due to faster ICT development rate, while legal recycling structures are ineffective in this regard. As a result, E-waste is a more visible problem in the global South than in the global North, although prior experience is the most significant knowledge gap (Heeks et al., 2015). Older limits should be supplied in accordance with unfriendly environmental requirements to avoid the effects of hostile environmental consequences on recycling workers. Air pollution control systems for evacuation and emissions are required in reusable offices. Nowadays, the private sector and various exploration organizations are approaching each other with the aim of E-waste exploration because it is a source of money for the private association due to the availability of valuable resources (Pandey & Govind, 2014).

15.6 Informal or nonformal sector

Small units that work on the development of items and services that can be physically tested but not recorded, certified, or overseen by community specialists are referred to as the informal sector. These units are household enterprises that operate at a low level of integration and are not established as direct legal entities. Personnel in this segment have been depicted as unorganized local workforces who are unable to categorize themselves in search of their shared benefits due to various constraints, including business freedom and ignorance with a small and dispersed domain (Kolli, 2011). People who manage and eliminate scroungers in India are increasingly subjected to social discrimination, restrictions, and low standards. Members of lower castes are forced to migrate to cities where they undertake hazardous jobs, such as rag picking, recycling, and cleaning works due to a lack of alternative job opportunities and embarrassment to the community who have joined their traditional positions in their villages. Conferring to a government statement, the greater part (55.5%) of rag pickers in the New Okhla Industrial Development Authority (NOIDA) area were scheduled caste teenagers (Sekar, 2004). As a result, embracing an informal sector is critical to accomplish a successful transformation of active E-waste recycling and handling operations. Unfortunately, the draught legislation does not effectively address this challenge. Official controllers and dismantlers may be operative, but there is an encouragement for market contributors to circumvent anomalies and illegally get E-waste as long as reusable users can pay more for it. The ability of particular organizations to communicate the proper risk of usage will determine the effectiveness of the guideline in limiting the role of recyclables (Arya et al., 2021b). The use of hazardous waste disposal techniques such as open burning for the recovery of designated metals (e.g., copper, steel, aluminum, etc.), acid leaching for the recovery of metal from PCBs and mother boards (e.g., copper and precious metals) leaves all hazardous materials such as Pb, Hg, Cd, and other metals in the open treatment areas causing weathering explosions (Skinner et al., 2010; Vats & Singh, 2014). In India, 95% of E-waste is reprocessed in an unofficial sector with only 5% of E-waste being examined by a proper specialist. There are presently about 2000 units in India running in the nonformal subdivision for E-waste recycling, and nonformal E-waste recyclers can be found in virtually every state, predominantly in Delhi, Gujarat, Karnataka, Maharashtra, and West Bengal (Greenpeace, 2008). In India, there are reputable collectors/sellers, waste disposal workers, and recyclers, most of whom work in the informal sector. Each of these units functions on a small scale in association to authorities/door-to-door sellers, known as "*kabariwalas*" (Chatterjee, 2012). Informal units generally use methods such integrating E-waste from fabric collectors, dismantling the components of their functional parts, components, and modules by calculating exchange prices. The remaining material is chemically processed to recover important metals that have been discharged into the soil, air, and water. The expense of such a recycling technique is minimal, and recovery is mostly confined to precious metals such as gold, silver, copper, aluminum, etc. and different materials such as zinc, tantalum, cadmium, palladium, and other materials that cannot be restored. In order to offer an integrated system support framework, the activities now operating in the informal sector should be eliminated. This will allow them to mainstream the informal sector and cooperate with it to ensure environmental compatibility (Manomaivibool & Panate, 2009). Disassembling, separating, categorizing, lowering size, and re-splitting emergent ingredients and cataloging are all common E-waste recycling processes used worldwide. A portion of these cycles is being rebuilt to increase functional separation. With the exception of waste streams, office and excellent management have tremendous value in recycling (Öztürk, 2015). Fig. 15.1 depicts E-waste recycling process in terms of four levels.

15.6.1 Step: 1—Collection, disassembly, and segregation

The initial step in E-waste recycling is collection, disassembly, and segregation where E-waste is collected from various customers such as factories, workplaces, households, private and public organizations by unorganized E-waste collectors (*kabaries*). The collected E-waste is separated into several groups using disassembled and segregated procedures and differentiated by E-waste according to their market interest. Metal objects, glass parts, screws, transformers, heat sinks, plastic enclosed, batteries, connectors (PCBs and connectors are the most important components because they contain gold, copper, silver, and other precious metals) and other items are removed from concentrated E-waste and *kabaries* are able to sell recovered items at a fair market cost when one can be certain of giving a feasible return on the collection and segregation of necessities including gold-rich PCBs. Treatment of PCBs with raw materials recycling by unorganized recyclers can be poor. In a subsequent cycle, high-density PCBs including connections, chips, and other components are separated to get precious metals such as Cu, Ag, Au, Pd, Ta, and other elements (Gao et al., 2002; Li et al., 2004).

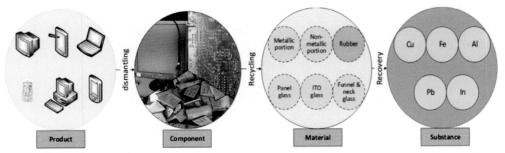

Figure 15.1 E-waste recycling process in terms of four levels, namely, product, component, material, and substance (Zeng et al., 2017).

15.6.2 Step: 2—Shredding, crushing, and pulverization

The second stage involves shredding and crushing of E-waste. After these complete cycles, shredders destroy the E-waste to reducing its size. Various forms of isolation cycles are employed in the scattering of streams to organize plastic, metal, and other materials, based on this simple combination. Appealing forces, size separation, current vortex separation, and electrostatic separation are the most common separation methods, and parts of these cycles can be utilized sequentially in waste flows as unique with an attractive splitting strategy. Separation is offered for ferrous metals from several metals and materials by means of magnets for this purpose, and the size separation approach varies on the size division. In the streams flowing, the heavy parts are separated as light without any problem. Current vortex separation process is used to separate nonalloys such as Al and Cu from different materials. The flow of electricity density with the electrostatic separation mechanism conduction is a basic limitation (Wen et al., 2005).

15.6.3 Step: 3—Refining

Refining is the third step in recycling E-waste. Discrimination of resources in E-waste can be considered, and technical solutions exist to recover the contaminants with negligible environmental impact. The majority of the fragments must be distinguished or molded to be marketed as resulting raw constituents or to be removed independently on the ultimate ejection site. Through the refinement process, three improvements are kept to a least amount: metals, plastics, and glass (Anam & Syed, 2013). Technical system of E-waste treatment and its exterior societal borderline for employment is represented in Fig. 15.2.

15.7 The hazardous impact of E-waste

E-waste imitates the atmosphere as much as human lifespan. It also has an adverse influence on the country's monetary sector. Due to the generation of a large-scale E-waste, the expenditure rate is equally increased, which has a significant impact on public performance (Masud et al., 2019). The following subsections discuss the unsafe impacts of E-waste on diverse resources.

15.7.1 Impact on soil

In recent years, the expansion of human activities has resulted in various stages of toxic contamination in the soil as well as pollution of farms, while landfill owners claim that overloaded dumping sites are capable of safely isolating contaminants, contained in electronics from the environment (Arya et al., 2021a). Thousands of old, unregulated landfills including a mix of putrescible and E-waste are a major source of concern

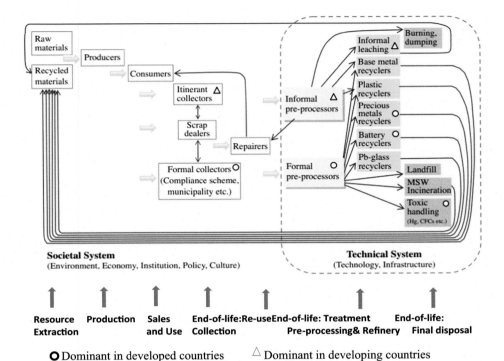

Figure 15.2 Technical system of E-waste treatment and its external societal boundary for implementation (Wang et al., 2012).

from which pollutants may seep into the earth and groundwater. Carbon-based and putrescible constituents in landfills depreciate and block the ground like landfill leachate. High volumes of dispersants can be found in leachates. On the other hand, poisonous composites in leachate are reliant on the assets of the waste and the phases of waste dumping in a distinctive landfill (Tanwani, 2021). Opening the passage of heavy metals inside a specific soil—vegetable or grain system and humans is largely reliant on food types. The ability of plant-based foundations to digest hazardous substances (especially large metals) from the soil through the stem and leaf can also be used. For instance, *Waste Soil Journal* has verified 15 recycling slums in Bangalore, India, and found to have up to 2850 mg/kg Pb, 957 mg/kg Sn, 180 mg/kg Sb, 49 mg/kg Hg, 39 mg/kg Cd, 4.6 mg/kg In, and 2.7 mg/kg Bi in soil. Compared with control areas near the same city, these attributes were doubled to hundredfold (Li & Achal, 2020).

15.7.2 Impact on health

During the disassembly, shredding, acid baths, and incineration processes, there is potential for mistakes such as cuts and burns. Similarly, exposure to the following substances has certain harmful effects on human health. For example, PCBs are incorporated into fish and other organisms and pass through collections that add high value to high meat diets as humans. PCBs can also be absorbed through the skin and inhaled or ingested, resulting in neurotoxicity, immunosuppression, liver trauma, cancer, behavioral changes, and reproductive disorders (Arya & Kumar, 2020b; Sivaramanan, 2013). Children living in networks that use E-waste are more likely to be exposed to harmful compounds at a greater level throughout their lives. Neurodevelopmental insufficiency is a genuine issue when it comes to E-waste exposure. Infants and young children have a lot more moderate body weight than adults, but because they are frequently underweight, their toxic burden may grow. This is the basic window of neural development, division, mobility, synaptogenesis, and myelination in newborns and young children. End-to-end neurodevelopmental testing focuses on input (IQ), memory, language, large and positive motor skills, cognition, and leadership potential and behavior (Chen et al., 2011).

15.7.3 Impact on environment

The majority of EEE shipments are sold from African countries to connect the computerized segmentation, but a considerable portion of the E-waste explodes and is burned in landfills, polluting the environment. Researchers have discovered that E-waste consumers in Africa are unaware of the dangers of open E-waste burning, which results in the release of harmful hazardous metals such as As, Pb, Hg, and Cd, which have damaged the environment and human wellbeing. To reduce E-waste accumulation and environmental degradation, most countries are moving to open burning and incineration, resulting in environmental degradation. Conferring to several reports, unacceptable use of E-waste recovering observers and connectors lead to the liberation of hazardous chemicals into the surroundings. In Ghana, incorrect E-waste recycling has resulted in pollution and contaminated streams due to the release of arsenic, mercury, and lead, among other things, affecting ordinary workers and local communities. Following the investigation, it was discovered that E-waste was responsible for more than 75% of the heavy metals such as polyvinyl chloride (PVC) and brominated flame retardant (BFRs) found in landfills, polluting the ecosystem (Maphosa & Mfowabo, 2020).

15.8 Future recommendations

E-waste comprises crucial as well as dangerous elements that can influence human fitness and the environment. Unfortunately, because they are outmoded and no longer useful, a considerable amount of E-waste is accumulating in households and workplaces. Storage and removal challenges arise as a result of this increased accumulation space. A few experiments have shown the value of E-waste information in the productive management of this waste pollution (Miner et al., 2020). Households and institutional consumers are less inclined to reimburse their losses in the formal sector because of a lack of awareness about E-waste and the cost of recovering end-of-life equipment in organized depletion (Gupta, 2011). In particular, the distinctive atmosphere combined with the ease of integrating home and financial facilitators (regardless of the possibility) encourages customers to reclaim their waste, thereby diversifying and managing resources. This presents an ethical claim that can take the form of social solutions, and the success of any E-waste management system will be determined by the ability to identify this challenge (Turaga et al., 2019). Many international tests have demonstrated the value of personal recycling ethics and trash disposal in steering waste management activities in this way. According to Bhat and Patil, "customer attentiveness plays an expressive role in producing E-waste in permissible collection amenities and recycling reserves for safe ejection" (Bhat & Patil, 2014). According to Shah (2014), a major societal notion is required for strong E-waste assortment and improved reuse levels to occur. Furthermore, several researchers have linked missing elements on E-waste management to poor recycling attitudes, implying that the present level of awareness is not always sufficient for formal E-waste management. The database of natural data for the consumer was derived from exposure to WEEE

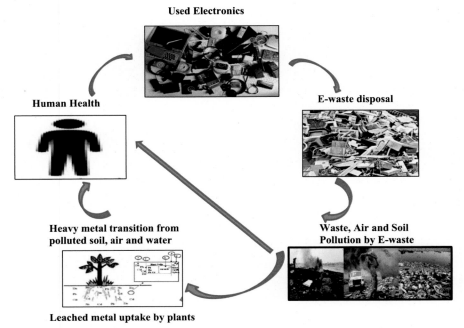

Figure 15.3 A representation of E-waste cycle and its impact on the environment and human health (Kwatra et al., 2014).

contamination hazards and resistance from TV and newspaper (39%), governmental propaganda (32%), environmental awareness activities (22%), family and friend (7%), that is, consumer to obtain more information about awareness activities from either print or online media (Shittu et al., 2020). Most consumers agree that the informal sector provides outstanding service and does it at a lower cost than the formal sector. The majority of users sell their EEEs second-hand or restrict them to *kabariwalas*. Only 9% of residents in this unusual situation classified these workers as illegal, while 11% of the respondents were unaware of them. These data indicate that conducting aberrant E-waste activities is not a good idea. Specialists from municipal authorities, such as the State Pollution Control Board and the local public corporate city should be mandated to join a consistent public site in this regard (Jamshidi et al., 2011). In-depth regional studies are expected to identify customer perceptions of management misconduct as well as communicate with consumers about E-waste potential training boards. Examining consumer's current attitudes, understanding, and awareness of E-waste, as well as its collection and recycling techniques should provide some useful information for experts looking to improve the savings process and encourage sustainable E-waste management practices among consumers (Masud et al., 2019). Fig. 15.3 depicts E-waste cycle and its impact on the environment and human health.

15.9 Conclusion

This chapter summarizes the present state of knowledge regarding E-waste activities. Electronic trash is a genuine and growing concern to human health worldwide, even in the formal sector. E-waste knowledge and awareness is critical for safe disposal, reuse, and recycling as well as reducing exposure to dangerous components. Recycling E-waste is vital, but it must be done in a safe and consistent manner. The profitability of E-waste recycling determines its success. If the amount of recycling (comprising foundation, workforce, and power) is greater than the cost of mining/construction of virgin raw resources, requirement for reprocessing draw stuffs may drop. When manufacturers antagonize an economic or physical constraint of recycling their electronics after use, eco-design for more supportable, less dangerous, and easily biodegradable electronics can be reinforced in reaction to extended producer responsibility (EPR) legislation. At least two stages in the life cycle of electronics can recover co-design from qualitative to enumerative by computing recyclability and recycling complications. In the early stages of production, manufacturers can use the newly developed measuring method to quantitatively design the materials and packaging. Instead of being destroyed, E-waste should be reconditioned and reused in its whole wherever possible. E-waste should be demolished by skilled, protected, and well-compensated personnel in technologically advanced areas when restoration is not viable. The manufacturer's obligation may be systematically evaluated during the end of life (EoL) segment using the capacity method, making the financial change of accountability from manufacturer to recycler extensively more considerable. To accomplish effective and lucrative E-waste recycling, new technologies and innovations are required to overcome difficulties such as separate E-waste collection, disassembly, and material segregation. Systematizing E-waste reprocessing, increasing output and dropping energy convention will be life-threatening in the future for sustainable E-waste recycling and governments should encourage such developments through financing and incentives. There are a few key concepts that should underpin all E-waste regulations. Countries that already have E-waste legislation must reinforce it through systematic assessments and modifications on a regular basis. However, enacting E-waste legislation in every country will not be enough to solve the global E-waste problem. An international council is needed to oversee and coordinate E-waste management around the world. In spite of the circumstance that the government regulates the ratio of E-waste recycling, the global finances of recycled supplies govern the E-waste reutilizing industry. The poll also highlighted some of the elements that may help respondents modify their E-waste knowledge and awareness as well as their readiness to assist in its management and probable recycling. According to the assumptions developed for this survey, respondents' levels of awareness and knowledge were unaffected by any of their socio-demographic variables. Damage to existing products, theft, product upgrades, and the launch of newer models were the most common reasons for respondents to purchase more electronic gadgets. Poor practices in disposing of E-waste in respondents' neighborhoods are linked to a number of environmental issues, the most severe of which are the adverse consequences on esthetics, ambient air attribute, and public management's capability to proficiently agree with this E-waste flood.

References

Ahirwar, R., & Tripathi, A. K. (2021). E-waste management: A review of recycling process, environmental and occupational health hazards, and potential solutions. *Environmental Nanotechnology, Monitoring & Management, 15*, 100409.

Anam, A., & Syed, A. (2013). Green computing: E-waste management through recycling. *International Journal of Scientific & Engineering Research, 4*(5), 1103.

Arya, S., & Kumar, S. (2020a). E-waste in India at a glance: Current trends, regulations, challenges and management strategies. *Journal of Cleaner Production, 271*, 122707. Available from https://doi.org/10.1016/j.jclepro.2020.122707.

Arya, S., & Kumar, S. (2020b). Bioleaching: Urban mining option to curb the menace of E-waste challenge. *Bioengineered*, 640−660. Available from https://doi.org/10.1080/21655979.2020.1775988.

Arya, S., Patel, A., Kumar, S., & Pau Loke, S. (2021b). Urban mining of obsolete computers by manual dismantling and waste printed circuit boards by chemical leaching and toxicity assessment of its waste residues. *Environmental Pollution, 283*, 117033. Available from https://doi.org/10.1016/j.envpol.2021.117033.

Arya, S., Rautela, R., Chavan, D., & Kumar, S. (2021a). Evaluation of soil contamination due to crude E-waste recycling activities in the capital city of India. *Process Safety and Environmental Protection, 152*, 641−653. Available from https://doi.org/10.1016/j.psep.2021.07.001.

Awasthi, A. K., & Li, J. (2018). Assessing resident awareness on e-waste management in Bangalore, India: A preliminary case study. *Environmental Science and Pollution Research, 25*(1), 11163−11172.

Azevedo, L. P., da Silva Araújo, F. G., Lagarinhos, C. A. F., Tenório, J. A. S., & Espinosa, D. C. R. (2017). E-waste management and sustainability: A case study in Brazil. *Environmental Science and Pollution Research, 24*(32), 25221–25232.

Bhat, V., & Patil, Y. (2014). E-waste consciousness and disposal practices among residents of Pune city. *Procedia-Social and Behavioral Sciences, 133*, 491–498.

Borthakur, A., & Govind, M. (2017). How well are we managing E-waste in India: Evidences from the city of Bangalore, Energy. *Ecology and Environment, 2*(4), 225–235.

Borthakur, A., & Singh, P. (2016). Researches on informal E-waste recycling sector: It's time for a 'Lab to Land' approach. *Journal of Hazardous Materials, 323(Pt B)*, 730–732.

Chatterjee, S. (2012). Sustainable electronic waste management and recycling process. *American Journal of Environmental Engineering, 2*(1), 23–33.

Chen, A., Dietrich, K. N., Huo, X., & Ho, S. M. (2011). Developmental neurotoxicants in e-waste: An emerging health concern. *Environmental Health Perspectives, 119*(4), 431–438.

Cucchiella, F., D'Adamo, I., Koh, S. L., & Rosa, P. (2015). Recycling of WEEEs: An economic assessment of present and future e-waste streams. *Renewable and Sustainable Energy Reviews, 51*, 263–272.

Das, D. (2011). Ewaste rule puts onus on producers. *Times of India*, 12 June.

Dutta, D., Arya, S., Kumar, S., & Lichtfouse, E. (2021). Electronic waste pollution and the COVID-19 pandemic. *Environmental Chemistry Letters*, 1–4. Available from https://doi.org/10.1007/s10311-021-01286-9.

Dwivedy, M., & Mittal, R. K. (2010). Estimation of future outflows of e-waste in India. *Waste Management, 30*(3), 483–491.

Dwivedy, M., & Mittal, R. K. (2012). An investigation into e-waste flows in India. *Journal of Cleaner Production, 37*, 229–242.

Gaidajis, G., Angelakoglou, K., & Aktsoglou, D. (2010). *Journal of Engineering Science and Technology, 3*(1), 193–199.

Gao, Z., Li, J., & Zhang, H. C. (2002). Electronics and the environment. *IEEE International Symposium*, 234–241.

Garg, N., & Adhana, D. (2019). E-waste management in India: A study of current scenario. *International Journal of Management, Technology and Engineering, 9*(1).

Greenpeace. (2008). *Take back blues: An assessment of e-waste takeback in India.* www.greenpeaceindia.org.

Gupta, S. (2011). E-waste management: teaching how to reduce, reuse and recycle for sustainable development-need of some educational strategies. *Journal of Education and Practice, 2*(3).

Heeks, R., Subramanian, L., & Jones, C. (2015). Understanding e-waste management in developing countries: Strategies, determinants, and policy implications in the Indian ICT sector. *Information Technology for Development, 21*(4), 653–667.

Hossain, M. S., Al-Hamadani, S. M., & Rahman, M. T. (2015). E-waste: A challenge for sustainable development. *Journal of Health and Pollution, 5*(9), 3–11.

Ilankoon, I. M. S. K., Ghorbani, Y., Chong, M. N., Herath, G., Moyo, T., & Petersen, J. (2018). E-waste in the international context: A review of trade flows, regulations, hazards, waste management strategies and technologies for value recovery. *Waste Management, 82*, 258–275.

Iqbal, M., Breivik, K., Syed, J. H., Malik, R. N., Li, J., Zhang, G., & Jones, K. C. (2015). Emerging issue of e-waste in Pakistan: A review of status, research needs and data gaps. *Environmental Pollution, 207*, 308–318.

Islam, A., Ahmed, T., Awual, M. R., Rahman, A., Sultana, M., Abd Aziz, A., & Hasan, M. (2020). Advances in sustainable approaches to recover metals from e-waste: A review. *Journal of Cleaner Production, 244*, 118815.

Jamshidi, A., Taghizadeh, F., & Ata, D. (2011). Sustainable municipal solid waste management (case study: Sarab County, Iran). *Annals of Environmental Science*, 5.

Jayapradha, A. (2015). Scenario of E-waste in India and application of new recycling approaches for E-waste management. *Journal of Chemical and Pharmaceutical Research, 7*(3), 232–238.

Kalana, J. A. (2010). Electrical and electronic waste management practice by households in Shah Alam, Selangor, Malaysia. *International Journal of Environmental Sciences, 1*(2), 132–144.

Kaur, H., & Goel, S. (2016). E-waste legislations in India—A critical review. *Management and Labour Studies, 41*(1), 63–69.

Kolli, R. (2011). Measuring the informal economy: Case study of India. In: *15th conference of commonwealth statisticians*. New Delhi, Conference of Commonwealth Statisticians. Online access: http://cwsc2011. gov.in/papers/sna/Paper_4.pdf. (accessed 24 June 2013).

Kumar, A., & Dixit, G. (2018). An analysis of barriers affecting the implementation of e-waste management practices in India: A novel ISM-DEMATEL approach. *Sustainable Production and Consumption*, *14*, 36–52.

Kumar, U., & Singh, D. N. (2013). E-waste management through regulations. *International Journal of Engineering Inventions*, *3*(2), 6–14.

Kwatra, S., Pandey, S., & Sharma, S. (2014). Understanding public knowledge and awareness on e-waste in an urban setting in India: A case study for Delhi. *Management of Environmental Quality: An International Journal*, *25*(6), 752–765.

Li, J., Shrivastava, P., Gao, Z., & Zhang, H. C. (2004). Transactions on electronics packaging manufacturing. *IEEE*, *27*(1), 33–42.

Li, W., & Achal, V. (2020). Environmental and health impacts due to e-waste disposal in China: A review. *Science of the Total Environment*, *737*, 139745.

Manomaivibool, P., & Panate. (2009). Extended producer responsibility in a non-OECD context: The management of waste electrical and electronic equipment in India. *Resources, Conservation and Recycling*, *53*(3), 136–144.

Maphosa, V., & Mfowabo, M. (2020). E-waste management in Sub-Saharan Africa: A systematic literature review. *Cogent Business & Management*, *7*(1), 1814503.

Masud, M. H., Akram, W., Ahmed, A., Ananno, A. A., Mourshed, M., Hasan, M., & Joardder, M. U. H. (2019). Towards the effective E-waste management in Bangladesh: A review. *Environmental Science and Pollution Research*, *26*(2), 1250–1276.

Miner, K. J., Rampedi, I. T., Ifegbesan, A. P., & Machete, F. (2020). Survey on household awareness and willingness to participate in e-waste management in Jos, Plateau State, Nigeria. *Sustainability*, *12*(3), 1047.

Öztürk, T. (2015). Generation and management of electrical–electronic waste (e-waste) in Turkey. *Journal of Material Cycles and Waste Management*, *17*(3), 411–421.

Pandey, P., & Govind, M. (2014). Social repercussions of e-waste management in India: A study of three informal recycling sites in Delhi. *International Journal of Environmental Studies*, *71*(3), 241–260.

Pathak, P., & Srivastava, R. R. (2017). Assessment of legislation and practices for the sustainable management of waste electrical and electronic equipment in India. *Renewable and Sustainable Energy Reviews*, *78*, 220–232.

Rautela, R., Arya, S., Vishwakarma, S., Lee, J., Ki-Hyun, K., & Kumar, S. (2021). E-waste management and its effects on the environment and human health. *Science of the Total Environment*, *773*, 145623. Available from https://doi.org/10.1016/j.scitotenV.2021.145623.

Rodrigues, A. C., Boscov, M. E., & Günther, W. M. (2020). Domestic flow of e-waste in São Paulo, Brazil: Characterization to support public policies. *Waste Management*, *102*, 474–485.

Sahu, A. K., Narang, H. K., & Rajput, M. S. (2018). A Grey-DEMATEL approach for implicating e-waste management practice: Modeling in context of Indian scenario. *Grey Systems: Theory and Application8*, *1*, 84–99.

Sajid, M., Syed, J. H., Iqbal, M., Abbas, Z., Hussain, I., & Baig, M. A. (2019). Assessing the generation, recycling and disposal practices of electronic/electrical-waste (E-Waste) from major cities in Pakistan. *Waste Management*, *84*, 394–401.

Sekar, H.R. (2004). *Child labour in urban informal sector: A study of ragpickers in NOIDA* (No. 2004-2055). VV Giri National Labour Institute.

Shah, A. (2014). An assessment of public awareness regarding E-waste hazards and management strategies. *Independent Study Project Collection 1820*.

Sinha, A. (2013). Delhi NCR likely to generate 50,000 metric tonnes of e-waste by 2015: Assocham, *Times of India*, 31 August.

Shittu, O. S., Williams, I. D., & Shaw, P. J. (2020). Global E-waste management: Can WEEE make a difference? A review of e-waste trends, legislation, contemporary issues and future challenges. *Waste Management*, *120*, 549–563.

Sivaramanan, S. (2013). E-waste management, disposal and its impacts on the environment. *Universal Journal of Environmental Research & Technology, 3*(5).

Skinner, A., Dinter, Y., Lloyd, A., & Strothmann, P. (2010). The challenges of E-waste management in India: Can India draw lessons from the EU and the USA. *Asien, 117*(7), 26.

Srivastava, R.R., & Pathak, P. (2020). Policy issues for efficient management of E-waste in developing countries. In *Handbook of electronic waste management, international best practices and case studies*, (pp. 81–99).

Taghipour, H., Nowrouz, P., Jafarabadi, M. A., Nazari, J., Hashemi, A. A., Mosaferi, M., & Dehghanzadeh, R. (2012). E-waste management challenges in Iran: Presenting some strategies for improvement of current conditions. *Waste Management & Research, 30*(11), 1138–1144.

Tanwani, L. (2021). E-waste: A global hazard and management techniques. *SPAST Abstracts, 1*(01).

Tiep, H. S., Kin, T. D. Y., Ahmed, E. M., & Teck, L. C. (2015). E-waste management practices of households in Melaka. *International Journal of Environmental Science and Development, 6*(1), 1811.

Turaga, R. M. R., Bhaskar, K., Sinha, S., Hinchliffe, D., Hemkhaus, M., Arora, R., Chatterjee, S., Khetriwal, D. S., et al. (2019). E-waste management in India: Issues and strategies. *Vikalpa, 44*(3), 127–162.

Vats, M. C., & Singh, S. K. (2014). Status of e-waste in India: A review. *International Journal of Innovative Research in Science, Engineering and Technology, 3*(10), 16917–16931.

Vishwakarma, S., Kumar, V., Arya, S., Tembhare, M., Rahul., Dutta, D., & Kumar, S. (2022). E-waste in information and communication technology sector: Existing scenario, management schemes and initiatives. *Environmental Technology & Innovation*, 102797. Available from https://doi.org/10.1016/j.eti.2022.102797.

Wang, F., Huisman, J., Meskers, C. E., Schluep, M., Stevels, A., & Hagelüken, C. (2012). The Best-of-2-worlds philosophy: Developing local dismantling and global infrastructure network for sustainable e-waste treatment in emerging economies. *Waste Management, 32*(11), 2134–2146.

Wath, S. B., Dutt, P. S., & Chakrabarti, T. (2011). E-waste scenario in India, its management and implications. *Environmental Monitoring and Assessment, 172*(1), 249–262.

Wen, X., Zhao, Y., Duan, C., Zhou, X., Jiao, H., & Song, S. (2005). Study on metals recovery from discarded printed circuit boards by physical methods. *Proceedings of the 2005 IEEE international symposium on electronics and the environment*, 121–128.

Zeng, X., Yang, C., Chiang, J. F., & Li, J. (2017). Innovating e-waste management: From macroscopic to microscopic scales. *Science of the Total Environment, 575*, 1–5.

CHAPTER 16

Transitions toward sustainable E-waste management plans

Simran Sahota[1], Maneesh Kumar Poddar[1] and Rumi Narzari[2]
[1]Department of Chemical Engineering, National Institute of Technology Karnataka, Surathkal, Karnataka, India
[2]Department of Energy, Tezpur University, Tezpur, Assam, India

16.1 Introduction

Electronic waste (E-waste) pollutants can diffuse into the air via dust or fume, and dominate the exposure pathways among open burning areas for humans through inhalation, ingestion, and skin absorption (Mielkel & Reagan, 1998). During combustion, E-waste creates fine particulate matter (PM), which leads to pulmonary and cardiovascular disease. PM is also known as particle pollution. It gives a major indication regarding the rise in air pollution levels. Recently, waste recycling has been identified as a major factor for the deterioration of air quality. When people are exposed to such hazardous substances, it includes many life-threatening diseases that lead to lung damage by inhaling fumes of major heavy metals which are Pb and Cd (Awasthi et al., 2016b). A work carried out by Gangwar et al. (2019) showed the high occurrence of hypertension in the E-waste-dominated area, which proves the link between exposure to E-waste and exponential increase of cardiovascular risk factors. It is found that more than the optimum limit of Cr, Ni, Cu, and Zn in the residential site is a very serious issue of public health. E-waste recycling regions in China, especially the ones that carried out improper recycling, depicted significantly high mean concentrations of PM10 ($100-243.310 \pm 22.729 \ \mu g/m^3$) and heavy metals were found in the air of E-waste burning centers, responsible for the higher exposure to the residents near those sites (Gangwar et al., 2019; Qin et al., 2019).

Similarly, water is one of the most valuable natural resources that are constantly under threat due to major increments in information and communication technology and Electronic and Electrical Equipment (EEE) instruments that generate a lot of E-waste. All the perilous materials that are present in the E-waste can permeate from the landfill sites into groundwater, thus resulting in the toxicity of groundwater. This detestable liquid is known as leachate (Fatta et al., 1999). In a study carried out by Panwar and Ahmed (2018), heavy metals have been reported to be the cause for possible health risks due to direct exposure and consumption of water in a specific area due to toxicity of water, acid fumes, and toxic gases on account of E-waste recycling

Global E-waste Management Strategies and Future Implications
DOI: https://doi.org/10.1016/B978-0-323-99919-9.00001-5

333

practices carried out in the specific area (Panwar & Ahmed, 2018). Heavy metals present in E-waste contribute to major contamination of the marine ecosystem and are a major source of oxidative stress to aquatic organisms (Jakimska et al., 2011). Likewise, the blooming of human activities is resulting in an increase in the release of hazardous contaminants generated by E-waste into the soil. The E-waste recycling centers are the major source of soil contamination as they slowly release metals and other toxic contaminants at a rapid pace which is evolved due to the primitive way of disposing of E-waste (Awasthi et al., 2016a). The soil can be contaminated mainly by two sources, either aerial contamination or indirect sources like irrigation processes. It is found that Pb, Ca, and Ba have near to zero recycling efficiency which contaminates the soil on an exceptional scale nearby the recycling area (Fu et al., 2008). Ca coming out from E-waste drastically change the pH of the soil which leads to a nutritional reduction in E-waste-affected area, Moreover, it is found that the toxic elements lead to depletion of nitrogen and phosphorus which play a very important role in plant growth (Dharini et al., 2017). Luo et al. (2011) found a high concentration of Cd and Pb in the majority of the paddy and vegetable samples, which indicate that E-waste can be a true menace in operational agriculture (Luo et al., 2011). A large number of heavy metals in the local environment can cause serious health problems to residents who are consuming the toxic agriculture products growing in that area. For instance, soils obtained from an E-waste recycling slum in Bangalore, India showed up to 2850 mg/kg Pb, 39 mg/kg Cd, 4.6 mg/kg In, 180 mg/kg Sb, 957 mg/kg Sn, 49 mg/kg Hg, and 2.7 mg/kg Bi (Ngoc et al., 2009).

16.2 Influence of human lifestyle on E-waste

The human lifestyle is at one of the most critical junctures since its inception and this negative impact has worked as the catalyst for highlighting issues like E-waste. People residing near E-waste processing areas are readily exposed to the hazardous substances released and transported via all types of natural pathways (namely, air, water, soil) (Awasthi et al., 2018). Toxic chemical constituents are released from E-waste like polycyclic aromatic hydrocarbons, perfluoroalkyls, polychlorinated dibenzofurans, polybrominated diphenyl ethers, polychlorinated biphenyls (dioxin-related polychlorinated biphenyls), and some other chemical elements; for instance, chromium, manganese, nickel, and lead have played an important role in deteriorating human health (Hashmi & Varma, 2019). It includes altercations to normal cell functioning, thyroid function, lung function, growth, and reproductive health concerns (Grant et al., 2013). In accordance with the work carried out by Everett et al. (2011), E-waste leads to the release of dioxin, lead, and possibly cadmium that increase the probability for chronic diseases like obesity, type 2 diabetes, hypertension, and cardiovascular diseases for the prolonged period of time (Everett et al., 2011). E-waste is now a major public health issue as it has negative effects on human health, both children and adults, collectively by impacting the soil, aquatic ecosystem, and the atmosphere (Weerasundara et al., 2020). Several investigations have been previously reported to understand the correlations between exposure pathways to heavy metals from E-waste and human health outcomes, mainly about neonate's health, children's health, and the changes in cellular expression.

16.3 Initiative toward sustainable transitions: exploring technologies

16.3.1 Legal and authorized

This sector is still in its initial stages in the country, India in particular. As per the data published in the Central Pollution Control Board of India (CPCB) E-waste Report 2016: 178 dismantling and recycling units were registered in 2016 with approximately 44,000 Mt/annum capacities. Now, this has increased to 312 units over a recycling capacity of 78 Lakhs Mt/annum (CPCB, list of registered E-waste dismantlers and recycler in India 2019). MeitY, a government sector, has developed two different prototypes of circuit board (PCB) recycling technologies, that is, 1000 kg/day capacity (continuous process) and 100 kg/batch, that work in agreement with the environmental norms. Both the types are designed suitable to Indian conditions (MeitY, 2019). The processing methods involve the use of various techniques, namely, automatic mechanized collection and separation (Chao et al., 2011), hydrometallurgical, and pyrometallurgical (Tuncuk et al., 2012). These are described in the following paragraphs.

16.3.1.1 Automatic separation

It is the initial step of E-waste treatment for metal recovery. It involves dismantling and crushing E-waste with bare hands utilizing some manual tools, like screwdrivers, pliers, chisels, and hammers (Arya & Kumar, 2020). Further, magnetic, screening, eddy current, and density separation techniques are employed for the separation of metals and nonmetals (Zhou & Qiu, 2010).

16.3.1.2 Solvent leaching

The segregated E-waste scrap is exposed to leaching either in acid or caustic material (Khaliq et al., 2014). Further, the leachate is subjected to an isolation or concentration process to extricate the metal with the assistance of the process like solvent extraction, distillation, precipitation, cementation, filtration and ion exchange methods, and particle trade strategies (Arya & Kumar, 2020). The process is highly in practice in most of the registered recycling centers across the country. Hydrometallurgy possesses numerous advantages such as the reduced risk of toxic emissions (Tuncuk et al., 2012), effective metal separation (Cui & Zhang, 2008), and negligible left residue. However, it has a few limitations as well, like time-consuming processes, large consumption of chemicals and water leading to water pollution, and adverse effects to the environment as well as human health (Iannicelli-zubiani et al., 2017).

16.3.1.3 Pyrometallurgy

It involves heating the waste sample in an inert gas atmosphere. In this process, the organic materials, like paper, wood, rubber, plastics, and so on are decomposed at higher temperatures and form volatile substances that can be used as a high-value chemical product like oil or source of energy generation (Gramatyka et al., 2007). However, the release of secondary pollutants, such as dioxin and furan and disposing of the crude and unwanted materials in open fields and nearby water banks (Wong et al., 2007), is considered to be unsustainable. These generate corrosive residues, consume a large amount of energy, and involve a high capital expenditure and operational expenses (Argumedo-Delira et al., 2019).

16.3.2 Illegal or backyard recycling

Informal recycling is majorly practiced in India (Kannan et al., 2016) employing numerous people living in the urban and semiurban regions in the country. It is also called "Backyard Recycling." It is involved in the collection, processing, segregation, refurbishing, and dismantling in informal ways. The collection of waste materials is carried out by the waste collectors from the households which are further sold to contractors and traders. The E-waste is handled by using random methods of which approximately 25%–30% is reused and the remaining is discharged into streams, channels, waterways, and open spaces. These methods are responsible for the air, water, and soil contamination crisis (Vetrivel & Devi, 2012). Due to poor awareness, lack of stringent regulations, and nonappearance of the cutting edge innovation and technology (Jayapradha, 2015), informal groups continue to practice recycling in small congested and unsafe areas using rudimentary techniques like open burning, manual stripping of cables, acid leaching and smelting of the PCBs.

16.4 A step toward successful transitions: global perspective

According to a study by the United Nations, it is estimated that approximately 20–50 million metric tons of E-waste are generated in global scenario each year, out of which 20%–25% of E-waste generated in the world are recycled in third world continents like Asia and Africa (Lee et al., 2017). The major chunk of E-waste is exported to China and India for remediation purposes. Some other players like Pakistan, Bangladesh, Ghana, Nigeria, and Kenya are also important countries that hold E-waste (Ahsan et al., 2016). With rapid growth and employable facets, slowly and steadily formal sector for E-waste treatment has been achieved (Zeng & Li, 2016).

The United Kingdom became the first country to draft the law in 2005 and made sure that products that were manufactured and tried to establish business activities in the European market should not contain any hazardous substances and that they comply with the requirements of the directive [Restriction of Hazardous Substances (RoHS) Regulations, 2007, Directive 2002/95/EC]. On a similar level, Switzerland became one of the earliest countries in the EU to introduce extended producer responsibility (EPR) that follows the take-back approach. On the other side, Germany is also following the best strategic management approach that complies Act Governing the Sale, Return and environmentally sound disposal of electrically sound disposal of EEE (the ElektroG) in 2005. In addition to the previously described methods, which are in extensive use currently to deal with the E-waste issues globally, there are a few tools that have been developed for application in E-waste recycling. Some of them have been described as follows.

16.4.1 Life cycle assessment

It is a powerful tool for identifying the possible environmental impacts to develop eco-design products. It also determines carcinogens, climate change, ozone layer, eco-toxicity, acidification, eutrophication and land use, to improve the environmental performance of products (Bakri et al., 2008; Duan et al., 2008; Hischier & Baudin, 2010; Kiddee et al., 2013). It is widely used for E-waste management. In Europe, extensive research has been conducted to evaluate the environmental impacts of the end of life (EoL) treatment of E-waste. Life cycle assessment has also been in wide application in Asia and South America.

16.4.2 Material flow analysis

It is a tool used to study the route of material (E-waste) flowing into recycling sites, or disposal areas. It interconnects the sources, pathways, and intermediate and final destinations of the material. It is a decision support tool for waste management (Kiddee et al., 2013). Shinkuma et al. (2009) determined that second-hand electronic devices from Japan are reused in Southeast Asia (e.g., Vietnam and Cambodia). In addition, Yoshida et al. (2009) have found that the proportion of personal computers sent for domestic disposal and recycling decreased to 37% in the fiscal year 2004, while the proportion of domestic reuse and exports increased to 37% and 26%, respectively, in Japan.

16.4.3 Multicriteria analysis

It is a decision-making tool for evaluating strategic decisions and solving complex multicriteria problems (Garfi et al., 2009). Multicriteria analysis (MCA) models have been applied to E-waste management. For instance, Hula et al. (2003) used

MCA decision-making methodologies to determine the trade-offs between the environmental benefits and economic profit of the EoL processing of coffee makers (Hula et al., 2003).

16.4.4 Extended producer responsibility

It is an environmental policy approach that allocates responsibility to manufacturers to take back products after use and is based on polluter-pays principles (Widmer et al., 2005). The developed nations are the leaders of EPR programs for E-waste management including the European Union (EU), Switzerland, Japan, and some parts of the United States and Canada.

16.5 Sustainable waste management and its characteristics

Promoting waste minimization is one of the most primitive ways of waste management. It can be initiated from the household level and can be easily implemented in other spheres of life. Minimal change in lifestyle is required to take up recycling instead of direct disposal which ultimately reduces waste in long term (Waite, 2013). Waste can be easily reduced by mainly two ways: first is introducing more efficient processes and investment in research & development so that it leads to unit waste production reduction, and the second is reducing demand so that fewer units are produced in the value chain (Cooper, 2020). Among the classical sustainable waste management theories and comprehensive construction and demolition waste management model (C&DW), the reduction is rated to be best due to the least minimalistic damaging effects on our environment; that is why reduction is the first choice while developing 3R (Reduce, Reuse, Recycle) or C&DW models (Huang et al., 2018). There are several benefits of reducing waste such as generating extra income by collecting useful materials and reducing the costs by purchasing less material, which automatically lead to reduced CO_2 emissions. Additionally, this also reduces the cost of transportation of wastes to landfills, etc. It is significant to conclude that the best environmental and cost-effective solution is to reduce the amount of waste produced in construction activities as one should consider using standard sizes and quantities of materials according to industry standards; thereby it automatically minimizes the rework from errors and reduces offcuts (Ding et al., 2016; Llatas & Osmani, 2016). The major hindrance in the proper implementation of waste reduction strategy occurs during reduction, as stakeholders do not have a common understanding and effective communication among themselves regarding management strategies. Operational workers working on specific sites are not able to take advantage of all aspects of reduction strategy because when reduction strategy is included in the C&DW management cycle for the purpose of waste minimization, workers start giving extra attention to the execution of the reduction strategy which makes it difficult to follow that every time (Esa et al., 2017).

16.6 Scale of E-waste management

With the global target to transition toward sustainable E-waste management, it has become an essential topic of investigation owing to varied aspects of E-waste management around the globe. Due to the complexity and staggering amount of E-waste extensive studies have been conducted to understand and churn out a management system that is effective, sustainable, and environmentally friendly and can address the maximum issues associated (Ilankoon et al., 2018; Kumar et al., 2017). To formulate a management system that deals with the humongous heaps of E-waste generated in a technologically efficient and cost- effective way with minimum impact on the environment is a herculean task. So, there are some major points which must be taken into consideration before planning and developing a sound management system:

- Ensuring a special logistic dedicated to collecting the E-waste from its source to its disposal or treatment/recycling/recovery/reuse site is a prerequisite.
- Identification and proper disposal of the hazardous substance in E-waste have a detrimental effect on health and the environment.
- Recovery and reuse of precious metals present in E-waste such as gold, silver, and copper into the production cycle.

Depending on the scale of waste generated and its management practices it can be divided into three categories (Kumar et al., 2017) as follows.

16.6.1 Macroscopic (products and components)

World communities have agreed to follow legislation on E-waste management as it was potent to trim down the detrimental impact of human health and the environment (Veit & Bernardes, 2015). Countries and regions like China, Japan, EU, etc. have adopted various international and regional E-waste. In this regard, countries such as Japan, Taiwan, South Korea, and the EU have even implemented remarkably sophisticated measures on the production and management of E-waste. Despite this, an anomaly in the global E-waste regulation is evident (Hsu et al., 2019; Li, Zeng, Chen, et al., 2015).

While regions and countries like the EU, Japan, and South Korea have sophisticated and stringent laws, countries such as China and the United States have introduced some new directives and regulations to manage E-waste. But countries like Brazil and India have regulations and laws but their implementation is still a challenge.

In this regard, the EU has launched and adopted various directives/guidelines to regulate E-waste such as Directive on Waste Electrical and Electronic Equipment (WEEE), RoHS, Directive on Registration, Authorization and Restriction of

Chemical Substances (Zeng et al., 2013). However, in the United States there is a lack of any federal regulation on E-waste, though, in certain states policies like EPR have been implemented (Ogunseitan et al., 2009). Regulation/legislation like Waste Management and Public Cleansing Law, Small Appliance Recycling, etc. was established in Japan to resolve the issues of E-waste (Zeng et al., 2017). China has also established laws and policies on E-waste management, such as the Law of the Peoples of China on Cleaner Production Promotion, Law of the People's Republic of China on Circular Economy Promotion, Environmental Protection Law of the People's Republic of China, etc. Apart from these various other regulations have been formulated, including Recycling of WEEE, Opinions on Strengthening the Prevention and Control of Pollution from E-waste, and Administrative Measures for the Prevention and Control of Environmental Pollution (Wang et al., 2010; Zhou & Xu, 2012).

After critical evaluation of regulations/legislation prevalent in various countries, it was observed that there are two distinctive characteristics in them. They are as follows:

- There is a discrepancy in various countries in their laws (applied or proposed) regarding the introduced E-waste generation, collection, recycling, and related legal liability; and
- Substantial differences in E-waste regulated list in different regions of the world.

16.6.2 Mesoscopic (material)

At this scale, the E-waste is directly or indirectly managed via scientific responses, viz. material compatibility during the design stage, material fatigue during the consumption stage, and material reclaiming in the EoL stage.

16.6.2.1 Material compatibility

The compatibility of production chemicals with the construction materials used for chemical storage, delivery, and production systems is evaluated (metallic and nonmetallic materials). However, it is not feasible to choose material solely based on its performance in the processing system. Various studies conducted reveals that chief design rules for material compatibility should consider application of the compatibility rules for a wide range of materials such as metals, plastics, and glass; avoidance of combinations of material that are not suitable for recycling; encouragement of mono-material plastics making recycling more effective; marking all plastic parts, avoidance of contamination by stickers and wire fixtures. All these measures helps to ensures the easy glass removal from other materials and enhances recycling yields (Stevels, 2007).

16.6.2.2 Material fatigue

Fatigue is a phenomenon that is responsible for the majority of failures in electrical equipment annually which might be catastrophic considering the casualties involved (Castillo & Fernández-Canteli, 2009; Nový et al., 2007). Repetitive application of loads during the operation of electronics sometimes causes early failure of materials. The major cause of obsolescence of electronics is a long-time or high-tensile operation which is responsible for a shorter obsolescent duration than the material fatigue cycle in practical terms. The four basic approaches which can help in the mitigation of E-waste are reuse, remanufacturing, recycling, and recovery (Gharfalkar et al., 2016). It covers a wide range of electronics such as: direct reuse of equipment without remanufacturing of equipment is possible before the onset of fatigue, appropriate recycling is possible if the electronics is moderately used and most materials are fatigued, and it can be recovered if the electronics was extensively used and all its materials are fatigued.

16.6.2.3 Material reclaiming

Valuable materials utilized in the equipment can be recovered during the EoL stage from E-waste. Mechanical treatment is one of the most widely accepted methods of recycling due to the complicated design and structure of E-waste. The majority of metals, plastics, and glass can be easily recovered through simple dismantling (Li, Zeng, & Stevels, 2015; Zeng & Li, 2016). The recycling process currently under operations can effectively reclaim the majority of the valuable components which aims at a closed-loop supply chain. However, a small amount of fraction is lost during recycling due to product design, social behavior, recycling technologies, and separation thermodynamics (Reck & Graedel, 2012). Owing to its low value the lost parts are a matter of least concern in contrast to the recovered materials.

16.6.3 Microscopic (substance)

Through the recycling process at a microscopic scale, some substances can be recovered but many times hazardous substances such as heavy metals penetrate new production. The entrance of toxic metals such as brominated flame retardants and lead is inevitable and tends to transfer into lower-grade products. Apart from this, recycled glass from the cathode ray tube monitor contains over 20% lead. The transfer of hazardous substances through the recycling of E-waste or its components (television, printer, personal computers, and funnel and neck glass) is a major concern as some products have entered the market for civil use. Due to the use of such new products both humans and the environment are being exposed to toxic substances which should be taken into consideration keeping in mind the life cycle of such products. Improvisation in the dissipation of information on the kind of chemical used in EEE, how they are used, handled, and eventually recycled or disposed of would facilitate in churning outlaws or regulations which would ensure the protection of public health and the environment (Scruggs et al., 2016). Although various measures have been taken by the electronics industry, still the issue with the growing quantity persists. The use of clean energy for the manufacture of electronics has yet to be addressed. Though the big leading brands have eliminated the use of the worst hazardous substances from products, however, a majority of companies are still lagging and is going to remain in E-waste for many years (Mddy & Dowdall, 2014).

16.7 Policy and government interventions

A regulatory framework is an essential component in achieving the goal of sustainable transition of waste into a resource and its management. To resolve the mammoth task of incorporation of sustainability in E-waste management it needs to facilitate proper collection and recycling and formulation of well-defined policies and regulations. Currently, approximately 66% of the world's population has been secured the national E-waste executive law from 61 nations (Baldé et al., 2016). This would ensure adequate investment by responsible partners, health, and safety of the people involved, sustainable management of waste and emission, and restructuring of the existing setups. To ensure proper functioning and viability of the collection and recycling system government should provide incentives in the form of subsidies, land, etc. As the organized sector is the prime source of E-waste, it is vital to formalize these channels along with strict enforcement to ensure a viable and sustainable system.

In an attempt to overcome issues of E-waste various measures have been taken up and EPR is one of them. EPR means that the manufacturers are responsible to take back the final disposal product which will reduce the quantity of waste and lower utilization of natural resources. By using the gravity/mechanical separation technique electronic waste are separated into metals (aluminum, iron, copper, gold, palladium, copper, silver, etc.) and plastic parts. The plastic parts can be recycled and converted into plastic bottles and energy through various thermochemical interventions (incineration, gasification, and pyrolysis). Also, deposit-refund schemes like advanced recycling fee (ARF) and advanced disposal fee (ADF) may be incorporated into E-waste management. ARF generally imposes a tax on the sale of E-waste to cover the cost of recycling, and ADF is the fund generated to pay for the cost of disposing of E-waste (Zarei et al., 2018). There are three indispensable steps to ensure complete E-waste treatments, viz. conversion of E-waste to raw through the rigorous process of refining, disassembling, and separation for the subsequent step. The second step involves all possible separation methods. In the final step recovery of valuable metals and their recycling is the concern.

Increasing impact of the rising heaps of E-waste has triggered the global communities; world leaders and thinkers to formulate international treaties like Basel Convention, EU Directives (WEEE and RoHS), and Stockholm Conference to restrict the transboundary movement of prohibited hazardous waste. The WEEE directives have laid down some vital methods to manage the threats of E-waste and have imposed strict guidelines on waste material disposed of in the landfill. The limit of E-waste for collection, recovery, and recycling was fixed to 4 kg/capita for the member states (Król et al., 2016). Adoption of E-waste management strategies and its implementation is one of the major objectives which makes it mandatory to bear some or entire costs to offer take-back, free return of E-waste by the consumer. Apart from this, business user have to pay from collection to recycling, and treatment in accordance with the set standards and collection of WEEE separately (Waste Management Licensing Regulations Directive 2002/96/EC). In 2005 United Kingdom formulated laws to restrict the entry of hazardous substances along with the product in the European market and abide by regulations (RoHS Directive 2002/95/EC). Switzerland can be considered as the pioneer in formulating EPR which prioritizes resource recovery which guarantees that no obsolete devices are mixed and disposed of with other solid waste as mentioned in Ordinance on the Return schemes and the Disposal of 302 EEE in 1998. Swiss Information and Communication Technology Industry Association charges an ARF ranging from 1 to 20 euros while purchasing any new E-gadgets, whereas in the United Kingdom the consumers take back the waste to the designated locations. Germany has implemented various strategies to curb the menace of E-waste by-laws which monitor the sale, return, and safe disposal of E-waste (the ElektroG) in 2005. In Germany the EPR can be divided into two categories: (1) the responsibility of resource recovery and final disposal is taken up solely by the producer, and (2) the producer shares the responsibility of resource recycling and recovery jointly with the Producer Responsibility Organization (Arya & Kumar, 2020). China has passed the ordinance to reduce the number of hazardous substances and to mitigate their effect and has also introduced the EPR system in the regulations (European Commission, 2017). Administrative of the Recovery and Disposal of WEEE was enacted by the state council in 2009, in the context of the EPR to ensure to establish financial support for resource recovery processing (Wang et al., 2017). When it comes to the framing of the legislature to regulate E-waste recycling; Japan is one of the first countries to do so under Specified Home Appliance Recycling Law (SHARL) and revised law to promote effective utilization of resources. These laws ensure that the consumer pays regarding the collection and recycling of the products. They introduced a coupon system to recycle household appliances where the consumers pay a fund for the scrap. Norway has also set a role model for WEEE Directive 2002/96/EU. The Norwegian legislation (1998) made it mandatory to receive all the types of electronic waste free of cost by the distributor or the urban local body, whereas, in countries such as Korea and Taiwan, the recycling is supported by the producer. The United States launched the National Strategy for Electronics Stewardship (2011) intending to improve the design and management of obsolete electronic items with features of the green design, stringent enforcement of legislation, safer and effective E-waste management strategies, and regulated E-waste trading (Shamim & K, 2015). Global e-sustainability Initiative regulation by the United States focusing on collaborative and innovative approach toward sustainable E-waste management comprises a diverse global member base and partnership (Alghazo & Ouda, 2016). Besides these global directives and regulatory frameworks, every nation has its policy and legislature to regulate and monitor E-waste based on its society, culture, production, and utilization prevalent. Table 16.1 provides critical insights into the aforementioned international regulations of E-waste management.

The disparities amongst the developed and developing nation regarding the E-waste management practice are quite evident. Presently, the E-waste management practices are dependent on the regulations/policies/legal and institutional framework. The successful management of E-waste by the developed nations lays down the foundation for the developing nation to rectify the flaws with their system. Swiss nation can be considered as the pioneer of formulation of an E-waste management system which is comprehensive, whereas in terms of execution of E-waste policies, Europeans and Japanese are leading the world. Australia and Canada's E-waste management is heavily inspired by the Japanese E-waste management system. The credit for the successes of the E-waste management system in developed nations goes to three constitutional factors (UNEP, 2007):

1. National registry: It is an organization that keeps the record or registers regarding producers/recyclers/waste organizations, and a list of items that fall under the definition of E-waste has been defined as the national registry. It lays down the basic guidelines for the collection of E-waste by the responsible parties and ensures unbiased monitoring of the E-waste management system. It could be any government/nonprofit organization recognized/supported by the government for discharging the above-mentioned functions.

2. Collection systems: There are two subcategories of collection systems, viz. collective system (also referred to as monopoly) and clearinghouse system based on competition. Both systems aim to facilitate E-waste management services within the nation at a cheaper rate to both household and business consumers. They are usually nongovernmental, nonprofitable, and trade associations. However, in the case of the clearinghouse system, the service provider is multiple partners such as producers, waste business merchants, and recyclers. The government ensures the registration of such partners and provides legal support so as they abide by the norms/regulation regarding the allocation mechanisms, reporting and monitoring systems.

3. Logistics: The logistics of the E-waste management system comprises three channels; they are retailer's take-back, municipal collection sites, and producer take-back. Municipal collection sites primarily target households and provide free and unlimited access. The in-store retailer take-back policy means returning the E-waste for free to the store by the consumers. Lastly, the producer take-back policy directly involves large corporate associates with minimum public participation (Islam & Huda, 2018).

In developing nations, the E-waste management system is still in the second stage of its development and is evolving. A country is rated based on the various stages of its criteria and stages of its management system such presence of legal framework, national registry system, collection system, and infrastructure and recycling technologies available. There are huge disparities in terms of the presence and implementation of the legal framework. Whereas countries like India and China are highly rated for the impressive execution of their legal enforcement, Thailand has received an average rating due to poor implementation. In the case of Argentina, South Africa, and Indonesia they are planning to formulate a legal framework to deal with E-waste; however, there is a complete absence of any legal structure in Cambodia and the Philippines. Secondly, in terms of inventory system, the top position is shared by India and Taiwan, while Malaysia, South Africa, and Sri Lanka secure the bottom position. It has been observed that the Asian continent has active participation in terms of tracking waste entries and outflows compared to Africa. There is an absolute absence of collection centers in countries like Indonesia, the Philippines, and Argentina, whereas countries like Cambodia, China, Malaysia, Sri Lanka, Thailand, and India have local recyclers in absence of legal regulations or structure. In this case, Malaysia and South Africa can be rated as one of the top countries with not only an efficient collection system but some are operating at a pilot scale. The presence of advanced recycling technologies and infrastructure to support the E-waste management system is another parameter on basis of which countries are rated. Countries like Argentina, Sri Lanka, Indonesia, and the Philippines have an insufficient number of recycling facilities, while there is a complete absence of recycling facilities in Cambodia. Taiwan and South Korea are highly rated country, whereas India and China are undergoing certain constraints regarding full-scale operations of recycling facilities (Gollakota et al., 2020; Sthiannopkao & Wong, 2013).

Overall, it can be summarized that the major objective regarding the E-waste management of the developed nation is to shun the entry of toxins into the atmosphere. However, when it comes to developing countries the E-waste management is dependent on the well-being of their economies and ingrained sector, and its capabilities may be enhanced by emphasizing more on the recovery of valuables (Table 16.1).

16.8 Challenges and opportunities

To reduce the harmful effect of E-waste on the environment and human health, it is important to formulate an E-waste management plan which could tactfully and effectively handle E-waste. To enhance the effectiveness of any management plan, the plan must include meticulous collection and sorting of E-waste, its re-utilization whenever possible, and removal and decontamination of possible toxic compounds (Baldé et al., 2016; Namias, 2013; StEP, 2016). However, many times it is not

Table 16.1 Global legislations on E-waste management.

Name	Timeline	Objectives and directives
Waste Electrical and Electronic Equipment (WEEE) Directive	• The first directive was established in February 2003. • New WEEE directives were amended in the year 2012 and 2014. • In April 2017, implementing Regulation 2017/699 was adopted. • Implementing Regulation 2019/290 was adopted on February 19, 2019. • Implementing Decision 2019/2193 was adopted on December 17, 2019.	• Prevention of WEEE generation is the prime objective • To emphasize on efficient use of resources and resource recovery through reuse, recycling, and other forms of recovery • To improve the environmental performance of everyone involved in the life cycle of Electrical and Electronic Equipment (EEE) The Directive • Require the separate collection and proper treatment of WEEE and set targets for their collection as well as for their recovery and recycling • Help European countries fight illegal waste exports more effectively by making it harder for exporters to disguise illegal shipments of WEEE • Reduce the administrative burden by calling for the harmonization of national EEE registers and the reporting format
Restriction of Hazardous Substances (RoHS) Directive	• RoHS Directive was first introduced on January 27, 2003. • On July 21, 2011 new RoHS Directive came into force. • January 2, 2013 was set as a deadline for European Union (EU) countries to transpose provisions of the new RoHS Directive. • RoHS Directive was amended on November 15, 2017.	• Prevention of risk associated with environment and human health connected to E-waste management. • Restricts the utilization of certain hazardous substances such as heavy metals, flame retardants or plasticizers with safer alternatives in EEE • Promotes the recyclability of EEE and ensures to development field for manufacturers and importers of EEE in the European market

EU Directive on Energy-using Products (EuP)	• The first working plan of the Ecodesign Directive was adopted on October 21, 2008.	• To design EuP that will reduce energy consumption and other negative environmental impacts • It enforces other environmental considerations such as materials use, water use, polluting emissions, waste issues, and recyclability • It establishes a list of 10 product groups to be considered in priority for implementing measures in 2009–11
EU Directive on Registration, Evaluation and Authorisation of Chemicals (REACH)	• The European formally adopted REACH on December 18, 2006. • Came into force on June 1, 2007.	• Protects human health and the environment from the ill effects of chemical use • Enables free movement of substances on the EU market • Enhance innovation and the competitiveness of the EU chemicals industry • Reduce animal testing by promoting the use of alternative methods of assessing chemicals
Basel Convention(s)	• Opened for signature on March 21, 1989. • It came into force on May 5, 1992. • By October 2018, the total number of signatory states was 199.	• To restrict and regulate the transboundary movement of hazardous waste except for the countries meeting the norms of the environmentally sustainable management system • Reduction in production of hazardous waste as well as encouraging environmentally sound management system for hazardous wastes
Mobile Phone Partnership Initiative (MPPI)	• MPPI was launched in the year 2002 at the sixth meeting of the Conference of the Parties to the Basel Convention.	• To promote environmentally sound management of end-of-life mobile phones • To influence user behavior toward more environmentally friendly actions • To promote reuse, refurbishing, material recovery, recycling, and disposal options • To mobilize political and institutional support for sustainable management

(Continued)

Table 16.1 (Continued)

Name	Timeline	Objectives and directives
Partnership for Action on Computing Equipment (PACE)	• In June 2008 at the ninth meeting of the Conference of the Parties to the Basel Convention, PACE was launched.	• Promotes sustainable development through efforts to repair, refurbish and reuse computing equipment • To find incentives and methods to divert discarded computers from disposal and burning into environmentally sound commercial material recovery/recycling operations • To develop technical guidelines for suitable repair, refurbishing, and material recovery/recycling, transboundary movement of E-waste (testing, labeling and certification of repair, refurbishing and recycling facilities) • End shipment of used and waste computing equipment to countries where it is illegal to import this waste under their domestic laws
Solving the E-waste Problem (StEP) Initiative	• In 2004 the formulation of the StEP initiative emerged. • In March 2007 StEP Initiative was officially launched. • In 2009 first StEP E-waste Summer School took place. • In 2012 first E-waste Academy Management Edition in Ghana was launched. • StEP E-waste World Map was published in 2013. • New membership structure and new Memorandum of Understanding (MoU) were developed in 2018. • In 2019 StEP was registered as a standalone entity.	• It initiates and facilitates E-waste reduction approaches that are environmentally, economically, and socially sound and sustainable • Functions as a network of actors who share experiences and best practices • Carries out research and development projects • Disseminates experiences, best practices and recommendations
Regional 3R Forum in Asia	• It was launched in 2009 in Tokyo.	• Promoting 3Rs: reduce, reuse, recycle • Building a "sound-material-cycle society"

the case due to various complex and significant challenges that plague E-waste management. The absence of any consistent technical definition of E-waste is the first major hurdle that makes the precise estimation of waste volume an arduous task. Unregulated illegal trade of E-waste is another vital issue regarding the proper management of E-waste (Khan, 2016; Lepawsky, 2015). The dynamic attribute of E-wastes due to technological advancement makes the management process even more challenging (Khan, 2016; Magalini, 2016).

The persistent relentless manufacture of new electrical and electronic and growing E-waste volume is also in itself a considerable issue. It is estimated that about 4%−5% of E-waste is generated per annum which would amount to approximately 50 million metric ton (MMT) in the year 2018 (Baldé et al., 2016; Honda et al., 2016). According to Zeng et al. (2016), China alone will register an annual growth of 25.7% by the end of 2020 which is roughly equal to 15.5 MMT and will double by the year 2030. The planned shortening of the lifetime of electronic products due to fast-evolving technologies associated with low retail prices and strong marketing and advertising is the prime reason behind the rapid increase in E-waste (LeBel, 2016; Pickren, 2015; Zeng et al., 2017). Rapid advancement in the technologies involved with electronic equipment discourages its reuse which weakens the very principle of 3R toward management of waste. In addition to this, technological obsolescence leads to spare part scarcity, lack of public interest, incompatible tools, lengthy repairing process, the prevalence of auto-didacticism among repairers, high labor costs, and lack of initiative and effective logistics for testing and repairing (Sabbaghi et al., 2017). The faulty design of e-products makes the repair difficult and time-consuming which in turn discourages their repair and recycling (Karin, 2012; Pickren, 2015; Sabbaghi et al., 2017). According to Kumar et al. (2017), the rising throwaway culture over repair is due to changing esthetics, new technology appeal, and increasing technical failure. This is fueled by growing purchase power. It has been also identified that there is a positive correlation between the Gross Domestic Product (GDP) and E-waste generation in any country in the world, regardless of the number of inhabitants.

The lack of data regarding both legal and illegal transboundary E-waste is also one of the major reasons which restrict its proper estimation (Honda et al., 2016; Kumar et al., 2017). The absence of reliable data, consistently increasing volume of E-waste, geographies and the dynamics of E-waste trafficking are responsible for the complexity of the E-waste web. For example, various Asian countries are still engaged with E-waste trading among each other despite the international ban on E-waste imports (Lepawsky, 2015; Lines et al., 2016). In addition, due to the strict law to reuse E-waste the illegal trade continues through re-categorization of E-waste coupled with counterfeit custom declarations which enables the unregulated trade of hazardous E-waste as used e-products (Khan, 2016). The loopholes in the custom rules of various countries, lack of international information, human resources, training and logistics, and inability to differentiate between actual waste and still-functional goods complicate the issue even more. Some of the challenges associated with E-waste management are listed as follows (Yu et al., 2009):

1. Lack of proper and well-defined take back system
2. Lack of sound technological expertise in handling E-waste
3. Absence or limited availability of environmentally sound E-waste treatment technologies and infrastructure
4. Inefficient, lenient, and failed implementation of E-waste policies and legislation among the associate stakeholders, public and private spheres
5. Inadequate government support in the forms of incentives and subsidies
6. Lack of public awareness and initiatives
7. Lack of sustainable economic models

However, there are still some hopes due to growing consciousness among the public, NGOs, and producers regarding their environment and problems associated with it. Various educational institutes, NGOs, and businesses association are implementing campaigns and strategies to collect, reuse, and recycle E-waste. In this direction, several E-waste commercialization markets have come up recently which can be regarded as an opportunity for sustainable E-waste management. Apart from this, the existing recovery chain is a means for capital gains through material processing which is beneficial both environmentally and socially.

16.9 Future prospective and recommendations

In order to improve the sustainability of the existing management strategies revamping the legislation structure is a must. Some of the following recommendations listed address the measures to rectify such issues during the drafting of policies regarding recycling, disposal, and reuse.

- Providing financial assistance for safe and scientific collection, storage, and disposal of E-waste with mandatory establishment disposal and recycling systems.
- Use of advanced and environmentally and economically viable Encourage treatment methodologies to the full-scale extension.
- Integration of organized and unorganized recycling sector into the mainstream recycling of E-waste.
- Introduction of EPR where the producers, importers/retailers are bound to pay the cost of collection, recycling, and reuse of E-waste.
- Local and international governments must take up the initiative to support the transition to socioeconomically sustainable small-scale formal processing of E-waste

from informal recycling. Government should also organize timely training and workshop to encourage for proper handling of E-waste

- Strict monitoring of E-waste trading and processing practices in accordance with Basel Ban Amendment and maintenance of national registry inventory. The countries with inefficient facilities to deal with E-waste should restrict imports
- Establish a standard for the certification of the second-hand or reusable EEE to improve the trustworthiness among the consumers, and incentive-based reuse systems, awareness among the consumers, and to encourage the reuse rather than disposal.
- Comprehensive and detailed surveys regarding process emissions and health effects during E-waste processing should be conducted to formulate specific safety and standards and protocols.
- Encourage product life extension principle, that is, repairing the crashed EEE rather than repurchasing the new one. This is achieved by the improvement of repairing or service centers.
- A collaborative approach among developed and developing nations regarding E-waste management

16.10 Conclusion

This chapter intends to assess the regulations and policies regarding recycling prevalent around the world. It is probably one of the most important streams of rapidly growing waste that needs immediate stringent regulatory interventions. The rapidly evolving technologies, newer design, and growing demand for upgraded modes for transfer of information and communication are some of the prime causes of shorter end life which lead to increased rates of E-waste generation. To address this, countries (both developed and developing) have to come forward and adopt, formulate, and implement regulations and policies with immediate actions. This chapter also shed light on the disparities among different countries around the world when it comes to regulatory frameworks active in a particular region and their management issues. E-waste is a serious threat to both public health and the environment due to its unscientific dumping and open burning practices. The transformation of informal recyclers into micro-entrepreneurs will not only help us in meeting the goals of public safety but will also help in generating new employment opportunities, sustainability, and economic growth. The development of effective awareness campaigns along with a sustainable roadmap for a successful transition of the E-waste management system is the need of the hour.

References

Ahsan, S., Ali, M., & Islam, R. (2016). E-waste trading impact on public health and ecosystem services in of waste resources. *International Journal of Waste Resources*. Available from https://doi.org/10.4172/2252-5211.1000188.

Alghazo, J., & Ouda, O.K. (2016). Electronic waste management and security in GCC countries: A growing challenge. In *ICIEM International Conference*, Tunisia.

Argumedo-Delira, R., Gómez-Martínez, M. J., & Soto, B. J. (2019). Gold bioleaching from printed circuit boards of mobile phones by Aspergillus niger in a culture without agitation and with glucose as a carbon source. *Metals, 9*.

Arya, S., & Kumar, S. (2020). E-waste in India at a glance: Current trends, regulations, challenges and management strategies. *Journal of Cleaner Production*, 122707. Available from https://doi.org/10.1016/j.jclepro.2020.122707.

Awasthi, A. K., Wang, M., Awasthi, M. K., Wang, Z., & Li, J. (2018). Environmental pollution and human body burden from improper recycling of E-waste in China: A short-review. *Environmental Pollution*. Available from https://doi.org/10.1016/j.envpol.2018.08.037.

Awasthi, A. K., Zeng, X., & Li, J. (2016a). Environmental pollution of electronic waste recycling in India: A critical review. *Environmental Pollution, 211*, 259—270. Available from https://doi.org/10.1016/j.envpol.2015.11.027.

Awasthi, A. K., Zeng, X., & Li, J. (2016b). Relationship between E-waste recycling and human health risk in India: A critical review. *Environmental Science and Pollution Research*, 11509—11532. Available from https://doi.org/10.1007/s11356-016-6085-7.

Bakri, S., Nur, S., Surif, S., & Ramasamy, R. K. (2008). A case study of life cycle assessment (LCA) on ballast for fluorescent lamp in Malaysia. *2008 IEEE international symposium on electronics and the environment*. IEEE. Available from https://doi.org/10.1109/ISEE.2008.4562881.

Baldé, C. P., Wang, F., & Kuehr, R. (2016). Transboundary movements of used and waste electronic and electrical equipment. In Bonn, Germany: United Nations University, Vice Rectorate Europe—Sustainable Cycles Programme (SYCLE). Available from https://www.step-initiative.org/files/_documents/other_publications/UNU-Transboundary-Movement-of-Used-EEE.pdf.

Castillo, E., & Fernández-Canteli, A. (2009). *A unified statistical methodology for modeling fatigue damage*. (80). Springer.

Chao, G., Hui, W., Wei, L., Jiangang, F., & Xin, Y. (2011). Liberation characteristic and physical separation of printed circuit board (PCB). *Waste Management, 31*(9—10), 2161—2166. Available from https://doi.org/10.1016/j.wasman.2011.05.011.

Cooper, M. (2020). Updating the electronics cycle: Improving US e-waste management practices.

Cui, J., & Zhang, L. (2008). Metallurgical recovery of metals from electronic waste : A review. *Journal of Hazardous Materials, 158*, 228—256. Available from https://doi.org/10.1016/j.jhazmat.2008.02.001.

Dharini, K., Cynthia, J. B., Kamalambikai, B., Celestina., Sudar, J. P. A., & Muthu, D. (2017). Hazardous E-waste and its impact on soil structure. *IOP Conference Series: Earth and Environmental Science, 80*.

Ding, Z., Yi, G., Tam, V. W. Y., & Huang, T. (2016). A system dynamics-based environmental performance simulation of construction waste reduction management in China. *Waste Management, 51*, 130—141. Available from https://doi.org/10.1016/j.wasman.2016.03.001.

Duan, H., Eugster, M., Hischier, R., Streicher-porte, M., & Li, J. (2008). Life cycle assessment study of a Chinese desktop personal computer. *Science of the Total Environment, 407*(5), 1755—1764. Available from https://doi.org/10.1016/j.scitotenv.2008.10.063.

Esa, M. R., Halog, A., & Rigamonti, L. (2017). Developing strategies for managing construction and demolition wastes in Malaysia based on the concept of circular economy. *Journal of Material Cycles and Waste Management, 19*(3), 1144—1154. Available from https://doi.org/10.1007/s10163-016-0516-x.

European Commission. (2017). Environment waste. http://ec.europa.eu/environment/waste/weee/index_en.htm.

Everett, C. J., Frithsen, I., & Player, M. (2011). Relationship of polychlorinated biphenyls with type 2 diabetes and hypertension. *Journal of Environmental Monitoring*, 241—251. Available from https://doi.org/10.1039/c0em00400f.

Fatta, D., Papadopoulos, A., & Loizidou, M. (1999). A study on the landfill leachate and its impact on the groundwater quality of the greater area. *Environmental Geochemistry and Health, 21*, 175—190.

Fu, J., Zhou, Q., Liu, J., Liu, W., Wang, T., Zhang, Q., & Jiang, G. (2008). High levels of heavy metals in rice (*Oryza sativa* L.) from a typical E-waste recycling area in southeast China and its potential risk to human health. *Chemosphere*, *71*(7), 1269—1275. Available from https://doi.org/10.1016/j.chemosphere.2007.11.065.

Gangwar, C., Choudhari, R., Chauhan, A., Kumar, A., Singh, A., & Tripathi, A. (2019). Assessment of air pollution caused by illegal E-waste burning to evaluate the human health risk. *Environment International*, *125*, 191—199. Available from https://doi.org/10.1016/j.envint.2018.11.051.

Garfi, M., Tondelli, S., & Bonoli, A. (2009). Multi-criteria decision analysis for waste management in Saharawi refugee camps. *Waste Management*, *29*(10), 2729—2739. Available from https://doi.org/10.1016/j.wasman.2009.05.019.

Gharfalkar, M., Ali, Z., & Hillier, G. (2016). Clarifying the disagreements on various reuse options: Repair, recondition, refurbish and remanufacture. *Waste Management and Research*, *34*(10), 995—1005. Available from https://doi.org/10.1177/0734242X16628981.

Gollakota, A. R., Gautam, S., & Shu, C. M. (2020). Inconsistencies of E-waste management in developing nations — Facts and plausible solutions. *Journal of Environmental Management*, *261*, 110234.

Gramatyka, P., Nowosielski, R., & Sakiewicz, P. (2007). Recycling of waste electrical and electronic equipment. *Journal of Achievements of Materials and Manufacturing Engineering*, *20*, 535—538.

Grant, K., Goldizen, F. C., Sly, P. D., Brune, M., Neira, M., van den Berg, M., & Norman, R. E. (2013). Health consequences of exposure to E-waste : A systematic review. *The Lancet Global Health*, *1*(6), e350—e361. Available from https://doi.org/10.1016/S2214-109X(13)70101-3.

Hashmi, M. Z., & Varma, A. (2019). Electronic waste pollution. Switzerland: Springer. Available from https://doi.org/10.1007/978-3-030-26615-8_13.

Hischier, R., & Baudin, I. (2010). LCA study of a plasma television device. *The International Journal of Life Cycle Assessment*, 428—438. Available from https://doi.org/10.1007/s11367-010-0169-2.

Honda, S., Khetriwal, D. S., & Kuehr, R. (2016). *Regional E-waste monitor: East and Southeast Asia.* United Nations University and Japanese Ministry of the Environment.

Hsu, E., Barmak, K., West, A. C., & Park, A. H. A. (2019). Advancements in the treatment and processing of electronic waste with sustainability: A review of metal extraction and recovery technologies. *Green Chemistry*, *21*(5), 919—936. Available from https://doi.org/10.1039/c8gc03688h.

Huang, B., Wang, X., Kua, H., Geng, Y., & Bleischwitz, R. (2018). Construction and demolition waste management in China through the 3R principle. *Resources, Conservation & Recycling*, *129*, 36—44. Available from https://doi.org/10.1016/j.resconrec.2017.09.029.

Hula, A., Jalali, K., Hamza, K., & Skerlos, S. J. (2003). Multi-criteria decision-making for optimization of product disassembly under multiple situations. *Environmental Science and Technology*, *37*(23), 5303—5313.

Iannicelli-zubiani, E. M., Irene, M., Recanati, F., Dotelli, G., Puricelli, S., & Cristiani, C. (2017). Environmental impacts of a hydrometallurgical process for electronic waste treatment : A life cycle assessment case study. *Journal of Cleaner Production*, *140*, 1204—1216. Available from https://doi.org/10.1016/j.jclepro.2016.10.040.

Ilankoon, I. M. S. K., Ghorbani, Y., Chong, M. N., Herath, G., Moyo, T., & Petersen, J. (2018). E-waste in the international context — A review of trade flows, regulations, hazards, waste management strategies and technologies for value recovery. *Waste Management*, *82*, 258—275. Available from https://doi.org/10.1016/j.wasman.2018.10.018.

Islam, M. T., & Huda, N. (2018). Reverse logistics and closed-loop supply chain of Waste Electrical and Electronic Equipment (WEEE)/E-waste: A comprehensive literature review. *Resources, Conservation and Recycling*, *137*, 48—75. Available from https://doi.org/10.1016/j.resconrec.2018.05.026.

Jakimska, A., Konieczka, P., Skóra, K., & Namieśnik, J. (2011). Bioaccumulation of metals in tissues of marine animals, Part I : The role and impact of heavy metals on organisms. *Polish Journal of Environmental Studies*, *20*(5), 1117—1125.

Jayapradha, A. (2015). Review Article: Scenario of E-waste in India and application of new recycling approaches for E-waste management. *Journal of Chemical and Pharmaceutical Research*, *7*(3), 232—238.

Kannan, D., Govindan, K., & Shankar, M. (2016). Formalize recycling of electronic waste. *Nature*, *530*. Available from https://doi.org/10.1038/530281b.

Karin, L. (2012). *The global impact of E-waste: Addressing the challenge.* International Labour Office.

Khaliq, A., Rhamdhani, M. A., Brooks, G., & Masood, S. (2014). Metal extraction processes for electronic waste and existing industrial routes: A review and Australian perspective. *Resources*, 152–179. Available from https://doi.org/10.3390/resources3010152.

Khan, S. A. (2016). E-products, E-waste and the Basel convention: Regulatory challenges and impossibilities of international environmental law. *Review of European, Comparative and International Environmental Law*, *25*(2), 248–260. Available from https://doi.org/10.1111/reel.12163.

Kiddee, P., Naidu, R., & Wong, M. H. (2013). Electronic waste management approaches: An overview. *Waste Management*, *33*(5), 1237–1250. Available from https://doi.org/10.1016/j.wasman.2013.01.006.

Król, A., Nowakowski, P., & Mrówczyńska, B. (2016). How to improve WEEE management? Novel approach in mobile collection with application of artificial intelligence. *Waste Management*, *50*, 222–233. Available from https://doi.org/10.1016/j.wasman.2016.02.033.

Kumar, A., Holuszko, M., & Espinosa, D. C. R. (2017). E-waste: An overview on generation, collection, legislation and recycling practices. *Resources, Conservation and Recycling*, *122*, 32–42. Available from https://doi.org/10.1016/j.resconrec.2017.01.018.

LeBel, S. (2016). Fast machines, slow violence: ICTs, planned obsolescence, and E-waste. *Globalizations*, *13*(3), 300–309. Available from https://doi.org/10.1080/14747731.2015.1056492.

Lee, D., Offenhuber, D., Duarte, F., Biderman, A., & Ratti, C. (2017). Monitour: Tracking global routes of electronic waste. *Waste Management*. Available from https://doi.org/10.1016/j.wasman.2017.11.014.

Lepawsky, J. (2015). The changing geography of global trade in electronic discards: Time to rethink the E-waste problem. *Geographical Journal*, *181*(2), 147–159. Available from https://doi.org/10.1111/geoj.12077.

Li, J., Zeng, X., Chen, M., Ogunseitan, O. A., & Stevels, A. (2015). "Control-Alt-Delete": rebooting solutions for the E-waste problem. *Environmental Science & Technology*, *49*(12), 7095–7108.

Li, J., Zeng, X., & Stevels, A. (2015). Ecodesign in consumer electronics: Past, present, and future. *Critical Reviews in Environmental Science and Technology*, *45*(8), 840–860.

Lines, K., Graside, B., Sinha, S., & Fedorenko, I. (2016). *Clean and inclusive? Recycling E-waste in China and India*. International Institute for Environment and Development. http://pubs.iied.org/pdfs/16611IIED.pdf.

Llatas, C., & Osmani, M. (2016). Development and validation of a building design waste reduction model. *Waste Management*, *56*, 318–336. Available from https://doi.org/10.1016/j.wasman.2016.05.026.

Luo, C., Liu, C., Wang, Y., Liu, X., Li, F., Zhang, G., & Li, X. (2011). Heavy metal contamination in soils and vegetables near an E-waste processing site, south China. *Journal of Hazardous Materials*, *186*(1), 481–490. Available from https://doi.org/10.1016/j.jhazmat.2010.11.024.

Magalini, F. (2016). Global challenges for E-waste management: The societal implications. *Reviews on Environmental Health*, *31*(1), 137–140. Available from https://doi.org/10.1515/reveh-2015-0035.

Mddy, C., & Dowdall, T. (2014). *Green gadgets: Designing the future* (p. 52) Greenpeace International.

MeitY. (2019). *National Policy on Electronics*. India: Ministry of Electronics and Information Technology, Government of India.

Mielkel, H. W., & Reagan, P. L. (1998). Soil is an important pathway of human lead exposure. *Environmental Health Perspectives*, *106*, 217–229.

Namias, J. (2013). The future of electronic waste recycling in the United States: Obstacles and domestic solutions [Masters Dissertation, Columbia University].

Ngoc, N., Agusa, T., Ramu, K., Phuc, N., Tu, C., & Murata, S. (2009). Contamination by trace elements at E-waste recycling sites in Bangalore, India. *Chemosphere*, *76*(1), 9–15. Available from https://doi.org/10.1016/j.chemosphere.2009.02.056.

Nový, F., Činčala, M., Kopas, P., & Bokůvka, O. (2007). Mechanisms of high-strength structural materials fatigue failure in ultra-wide life region. *Materials Science and Engineering A*, *462*(1–2), 189–192. Available from https://doi.org/10.1016/j.msea.2006.03.147.

Ogunseitan, O. A., Schoenung, J. M., Saphores, J. D. M., & Shapiro, A. A. (2009). The electronics revolution: from e-wonderland to E-wasteland. *Science (New York, N.Y.)*, *326*(5953), 670–671.

Panwar, R. M., & Ahmed, S. (2018). Assessment of contamination of soil and groundwater due to E-waste handling. *Current Science*, *114*(1), 166−173. Available from https://doi.org/10.18520/cs/v114/i01/166-173.

Pickren, G. (2015). Making connections between global production networks for used goods and the realm of production: A case study on E-waste governance. *Global Networks*, *15*(4), 403−423. Available from https://doi.org/10.1111/glob.12071.

Qin, Q., Xu, X., Dai, Q., Ye, K., & Wang, C. (2019). Air pollution and body burden of persistent organic pollutants at an electronic waste recycling area of China. *Environmental Geochemistry and Health*, *41*(1). Available from https://doi.org/10.1007/s10653-018-0176-y.

Reck, B. K., & Graedel, T. E. (2012). Challenges in metal recycling. *Science*, *337*, 690−695.

Sabbaghi, M., Cade, W., Behdad, S., & Bisantz, A. M. (2017). The current status of the consumer electronics repair industry in the U.S.: A survey-based study. *Resources, Conservation and Recycling*, *116*, 137−151. Available from https://doi.org/10.1016/j.resconrec.2016.09.013.

Scruggs, C. E., Nimpuno, N., & Moore, R. B. B. (2016). Improving information flow on chemicals in electronic products and E-waste to minimize negative consequences for health and the environment. *Resources, Conservation and Recycling*, *113*(2016), 149−164. Available from https://doi.org/10.1016/j.resconrec.2016.06.009.

Shamim, A., Ali, M., & Islam, R. (2015). E-waste trading impact on public health and ecosystem services in developing countries. *International Journal of Waste Resources*, *5*(4). Available from https://doi.org/10.4172/2252-5211.1000188.

Shinkuma, T., Thi, N., & Huong, M. (2009). The flow of E-waste material in the Asian region and a reconsideration of international trade policies on E-waste. *Environmental Impact Assessment Review*, *29*, 25−31. Available from https://doi.org/10.1016/j.eiar.2008.04.004.

StEP. (2016). Step_WP_WEEE systems and legislation_final. 3576.

Stevels, A. (2007). *Adventures in ecodesign of electronic products*. Repository.Tudelft.Nl. Available from http://repository.tudelft.nl/assets/uuid:c7223473-bedb-4b01-a99e-b05865071acd/stevels.pdf.

Sthiannopkao, S., & Wong, M. H. (2013). Handling E-waste in developed and developing countries: Initiatives, practices, and consequences. *Science of the Total Environment*, *463*, 1147−1153.

Tuncuk, A., Stazi, V., Akcil, A., Yazici, E. Y., & Deveci, H. (2012). Aqueous metal recovery techniques from e-scrap : Hydrometallurgy in recycling. *Minerals Engineering*, *25*(1), 28−37. Available from https://doi.org/10.1016/j.mineng.2011.09.019.

UNEP. (2007). *E-waste volume II: E-waste management manual*. Osaka: Division of Technology, Industry and Economics, International Environmental Technology Centre; United Nations Environment Programme (UNEP).

Veit, H. M., & Bernardes, A. M. (2015). *Electronic waste: Generation and management. Electronic waste: Recycling techniques*. Springer. Available from https://doi.org/10.1007/978-3-319-15714-6.

Vetrivel, P., & Devi, P. K. (2012). A focus on E-waste: Effects on environment and human health. *International Journal of Novel Trends in Pharmaceutical Sciences*, *2*(1), 47−51.

Waite, R. (2013). Household waste recycling. Routledge.

Wang, H., Gu, Y., Li, L., Liu, T., Wu, Y., & Zuo, T. (2017). Operating models and development trends in the extended producer responsibility system for waste electrical and electronic equipment. *Resources, Conservation and Recycling*, *127*, 159−167. Available from https://doi.org/10.1016/j.resconrec.2017.09.002.

Wang, J., Ma, Y. J., Chen, S. J., Tian, M., Luo, X. J., & Mai, B. X. (2010). Brominated flame retardants in house dust from E-waste recycling and urban areas in South China: Implications on human exposure. *Environment International*, *36*(6), 535−541. Available from https://doi.org/10.1016/j.envint.2010.04.005.

Weerasundara, L., Mahatantila, K., & Vithanage, M. (2020). 5. E-waste as a challenge for public and ecosystem health. *Handbook of electronic waste management*. Butterworth-Heinemann. Available from https://doi.org/10.1016/B978-0-12-817030-4.00003-6.

Widmer, R., Oswald-krapf, H., Sinha-khetriwal, D., Schnellmann, M., & Bo, H. (2005). Global perspectives on E-waste. *Environmental Impact Assessment Review*, *25*, 436−458. Available from https://doi.org/10.1016/j.eiar.2005.04.001.

Wong, M. H., Wu, S. C., Deng, W. J., Yu, X. Z., Luo, Q., Leung, A. O. W., Wong, C. S. C., Luksemburg, W. J., & Wong, A. S. (2007). Export of toxic chemicals—A review of the case of uncontrolled electronic-waste recycling. *Environmental Pollution, 149*. Available from https://doi.org/10.1016/j.envpol.2007.01.044.

Yoshida, A., Tasaki, T., & Terazono, A. (2009). Material flow analysis of used personal computers in Japan. *Waste Management, 29*(5), 1602−1614. Available from https://doi.org/10.1016/j.wasman.2008.10.021.

Yu, J., Ju, M., & Williams, E. (2009). Waste electrical and electronic equipment recycling in China: Practices and strategies. In *2009 IEEE International Symposium on Sustainable Systems and Technology, ISSST '09* [in Cooperation with 2009 IEEE International Symposium on Technology and Society, ISTAS, 24, 2021]. <https://doi.org/10.1109/ISSST.2009.5156728>.

Zarei, M., Taghipour, H., & Hassanzadeh, Y. (2018). Survey of quantity and management condition of end-of-life tires in Iran: A case study in Tabriz. *Journal of Material Cycles and Waste Management, 20*(2), 1099−1105.

Zeng, X., Duan, H., Wang, F., & Li, J. (2017). Examining environmental management of E-waste: China's experience and lessons. *Renewable and Sustainable Energy Reviews, 72*, 1076−1082. Available from https://doi.org/10.1016/j.rser.2016.10.015.

Zeng, X., Gong, R., Chen, W. Q., & Li, J. (2016). Uncovering the recycling potential of "new" WEEE in China. *Environmental Science and Technology, 50*(3), 1347−1358. Available from https://doi.org/10.1021/acs.est.5b05446.

Zeng, X., & Li, J. (2016). Measuring the recyclability of E-waste : An innovative method and its implications. *Journal of Cleaner Production, 131*, 156−162. Available from https://doi.org/10.1016/j.jclepro.2016.05.055.

Zeng, X., Li, J., Stevels, A. L. N., & Liu, L. (2013). Perspective of electronic waste management in China based on a legislation comparison between China and the EU. *Journal of Cleaner Production, 51*, 80−87. Available from https://doi.org/10.1016/j.jclepro.2012.09.030.

Zhou, L., & Xu, Z. (2012). Response to waste electrical and electronic equipments in China: Legislation, recycling system, and advanced integrated process. *Environmental Science and Technology, 46*(9), 4713−4724. Available from https://doi.org/10.1021/es203771m.

Zhou, Y., & Qiu, K. (2010). A new technology for recycling materials from waste printed circuit boards. *Journal of hazardous Materials, 175*, 823−828. Available from https://doi.org/10.1016/j.jhazmat.2009.10.083.

CHAPTER 17

Development of strategic framework for effective E-waste management in developing countries

Somvir Arya, Ajay Gupta and Arvind Bhardwaj
Industrial and Production Engineering Department, Dr. B.R. Ambedkar National Institute of Technology, Jalandhar, Punjab, India

17.1 Introduction

Today, the usage of electrical and electronic equipment (EEEs) is growing rapidly and they are becoming an integral part of daily life, not just in the cities, but also in the remote parts of the world. Individuals, particularly the ones working in the cities, cannot imagine their life without day-to-day life products like computer systems, TVs, cell phones, etc. These equipment perform an important role in the economic and social improvement of any community as well as nation. A whole sector just like IT business has developed with the input and assistance of information and telecommunication equipment. Banking sector, nowadays, provides mobile phone and web banking as well. Clients of the industry are actually urged to turn to internet and mobile phone banking which is convenient, inexpensive, and mostly time-intensive. Nowadays, willingly or unwillingly, every individual has to rely on EEEs to a certain degree. Thus configuring the overall revolutionized instances of the world the mounts of E-waste can be estimated to be overwhelmed in the dumping areas and management systems.

E-waste is the term used to define the old and left unattended electronic gadgets. The discarding of E-waste is a complex process due to its formation. Waste generated from the electronic apparatus contains large amount of toxins as well as other hazardous substances which are not only harmful to the earth but also poses a threat to the human well-being. Waste from cathode ray tubes, photocopier cartridges, capacitors, selenium drums (copier), and electrolytes contains several pollutants as well as dangerous metals like lead, mercury, and hexavalent chromium (Rajya Sabha Secretariat Research Unit Report, 2011). In informal recycling sector of India, majority of E-waste management activities like collection, segregation, transportation, and recycling is being done manually with bare hands using screwdrivers and hammers. Also, the safety measures and human health concerns are not their priority as they are unaware of ill effects of E-waste. Nonworking products are then dismantled/disassembled to find out the reusable

Global E-waste Management Strategies and Future Implications
DOI: https://doi.org/10.1016/B978-0-323-99919-9.00006-4

components. In order to recover the valuable metals and nonmetals, parts having zero resale value are being burnt in the open environment. The state of Maharashtra produces maximum amount of waste materials from electrical and electronic apparatus in the nation. As per *United Nations Environmental Programme (2007, 2009)*, currently, the readily available information on increment in generation of E-waste is inadequate and more sophisticated estimation methods are required for analysis of known information for territorial comprehensive coverage. Management of solid waste becomes more complicated with the inclusion of e-squander in it, especially computer waste. Situation becomes more complex as the E-waste from the developed countries is being dispatched to poor and developing countries in the name of free trade (Saoji, 2012). More than 50,000 tons of discarded e-products have been exported to developing nations from developed countries for reuse, on charity basis. These imported products are finally discarded to informal recycling units immediately or after little reuse and only 3% of total e-products finds their final destination to formal recycling units. In 2005, **Greenpeace** study revealed the presence of high concentration of highly toxic materials and poisonous furans and dioxins in the E-waste recycling units. Workers in these facilities are having minimal or no protection during recycling processes. Due to lack of awareness, E-waste is being mostly disposed off along with household other wastes. Rag pickers collect such garbage and segregate electronic scrap from the whole waste. The rag pickers sell these to the native scrap dealers for earning their livelihoods as the E-waste contains reusable and valuable materials. The scrap dealers transfer the collected e-scrap to informal recycling sector. These informal recyclers use the traditional and dangerous techniques to recycle or to manage the E-waste (Gupta & Kumar, 2014).

In developing countries especially for China, Pakistan, India, Bhutan, Nepal, and Bangladesh the formal E-waste management is a difficult task (Arya and Kumar, 2020), as the roots of informal recycling sector have been grounded very deeply and also have a wide supply chain system to operate their units. The operating costs of these informal units are very low due to the absence of latest technology and machinery to manage e-scrap. For the same, there is a urgent need to divert and educating the consumers toward formal recycling. Financial incentives, rules and regulations play a key role toward this diversion. There is, also, a great need to develop a framework for effective collection and recycling of E-waste.

17.2 Glance of existing E-waste management options and its impact

Latest and feature-rich electronic gadgets have become a status symbol in our society. The rapid change in the technology in electronic industry along with the high demand of newly launched products results in the huge generation of E-waste. IT sector not only contributes to the financial growth of any country but also a key player in E-waste generation. But IT sector always neglects the proper disposal of E-waste produced therein. Vishwakarma et al. (2022) highlighted various challenges and initiatives during transportation and handling of E-waste faced by the IT industries. Authors concluded that the IT industries should involve the partners and consider their suggestions to develop a formal e-scrap recycling system. A framework, which helps in the conversion of waste into wealth, is being suggested by the authors for safe disposal of E-waste by considering various challenges faced by the IT sector. To deal with e-garbage in a way that is more environmentally friendly, an additional persistent and sustainable study must be carried out. Most of the E-waste in developing countries has been managed by informal sector. Rautela et al. (2021) discussed few challenges like lack of technical knowledge, poor framework, inadequate financial support from government, and lack of awareness of end users along with the solution to overcome these in managing discarded e-items. It starts from the inventory of rejected e-items which should be managed in a manner that the environment will be minimally affected. In developing countries, government bodies must enforce the proper application of rules and regulations laid by statutory bodies to encourage and maximize the formal E-waste recycling which in turn reduces the harmful effects of it. Most of the developing nations, including India, find it challenging to set up formal E-waste recycling facilities due to lack of administrative support and expensive recycling technologies and processes. Conversion of unplanned (informal) sector in to authorized channel is helpful to meet the goal of good health of well-being and surrounding, employment, sustainable solution of a serious issue, and economic growth. Also, awareness programs can play a vital role in enhancing the knowledge of end users toward the hazardous effects of E-waste. A framework is also essential to bridge the gap between approved and unapproved recycling sector.

Despite the existing effective E-waste management chain and framework the emergence of COVID-19 pandemic has shaken the entire E-waste value chain and existing strategic framework. Dutta et al. (2021) discussed the effects of COVID-19 pandemic on E-waste recycling sector. In lockdown people were working from home. Also, the lectures of schools/colleges and all the other educational institutions were delivered via online mode. It creates a sudden exponential demand of electronic products like computers, laptops, smartphones and tablet PCs, etc. in the market and this gives us a clear indication toward the exponential increase in e-garbage in the coming years. To avoid the transmission of virus, formal recycling units have to store the waste for some time and follow some other safety precautions like wearing of mask, gloves, use of sanitizer, social distancing, etc. during the handling of E-waste. Above all, unavailability and lack of proper collection, transportation, and disposal of E-waste, also worsen the situation. Formal recyclers tried to convert this complex problem into an opportunity by developing mobile applications for effective collection of e-scrap. These applications are also used to aware and educate the consumers toward the ill effects of E-waste and encourage them for proper disposal of e-scrap. There is an urgent need to increase the recycling capabilities along with the through collection of E-waste to cope up with its generation.

Also, the approach of adopting the 4R principles (Reduce, Reuse, Recycle, and Recover) is an effective way to manage the E-waste effectively. Anandh et al. (2021) stated that reuse of discarded electronic equipment is one of the best end-of-life (EoL) solution in connection with the environmental impact and socio-economic benefits. Based on the rising menace the authors highlighted and applied systematic literature review approach to find the existing knowledge base to show the crucial and emerging themes of the reuse assessment of E-waste. A total of 12,216 articles published from 2005 to 2019 in the Web of Science, ProQuest, and Google Scholar were studied, from which 331 articles were shortlisted for review purposes. The shortlisted research papers were further divided into two subperiods 2005−14 and 2015−19 to find the development of the research themes and the contribution of the recent research articles to the literature on E-waste reuse assessment. Authors used SciMat and VOSviewer software to conduct the bibliographic analysis. With the help of bibliographic analysis four major areas were identified firstly consumer behavior toward the use or handling, disposal, reuse, and repair/recycling of electronic scrap, secondly assessing the potential of E-waste for reuse, thirdly product recovery strategy and market analysis for discarded electronic waste remanufacturing, and in the last material flow analysis (MFA) of e-scrap in circular economy. In developing countries, latest research theme is the impact of government subsidy on discarded electronic equipment, product service system, and circular economy which require further attention. Despite of significant efforts the management system in the developing countries is still partially handicapped due to its continuous growth. Ahirwar and Tripathi (2021) stated that considerable growth in the generation of e-scrap has become a major environmental concern all over the world. E-waste recycling comprises systematic collection and treatment of E-waste to recover valuable materials. It offers a valuable tool to mitigate the escalating heap of E-waste and helpful in meeting the requirements of the spare parts in secondary market, and it also support the economy of the country. Authors further explained that E-waste, if not managed through formal channel, can be a source of hazardous substances like lead, mercury, etc. To gain the maximum benefits out of formal E-waste recycling involves improving the design of e-equipment, increase in formal recycling rate, and imposing the restrictions on the use of hazardous substances in e-gadgets. Authors, also, discussed various opportunities and plans to improve formal E-waste recycling. Further, recent global trends in e-scrap generation have been studied along with an overview of E-waste management channel and impact of hazardous substances from e-scrap on human health. Few strategies have been discussed for effective e-scrap recycling in environmentally safe manner. Such as by incorporating habits into a well-known model that predicts people's behavior (namely, the theory of planned behavior), researchers looked into the drivers of E-waste management behavior among young end users in an emerging economy environment. As frontrunners in the consumption, creation, and management processes, young consumers contribute considerably to the E-waste problem's expansion. The data collected from survey was analyzed using a multivariate statistical method which is known as Partial Least Squares Structural Equation Modeling. The results demonstrate that the integrated model can explain more than 47% of the variance in young consumers' desire to recycle E-waste and has excellent explanatory power and robustness. Recycling practices and attitudes have been shown to be major indicators of young individuals' desire to recycle E-waste. Surprisingly, variables from the theory of planned behavior including subjective standards and behavioral control had little effect on young consumers' intentions to recycle E-waste. Based on these findings, E-waste recycling initiatives should focus on changing people's attitudes and creating particular cues that activate behavior. In addition, the study offered a number of social and practical consequences that might help to promote E-waste recycling programs (Aboelmaged, 2021). A study conducted by Kalana (2010) in Shah Alam, Malaysia focused on two factors: *practices and awareness* of E-waste management system (EMS). The author identified the preferred methods of E-waste disposal adopted by the citizens of Shah Alam and found that, due to lack of availability of efficient take-back channels for end users, only 22% of e-scrap finds its way to formal recycling facilities. Currently, no formal structured mechanism is in place to collect the E-waste from households. It is clear from the study that majority of end users tend to store their discarded or damaged EEE for years before reselling or disposing off. Fathima et al. (2017) proposed a system to address the environmental problems associated with e-scrap in an effective and sustainable way to manage this stream of waste in a better way. Current E-waste management practices of India are considerably ineffective and ultimately affecting both the human beings and the environment. They suggested developing a web portal to assess the present condition of global E-waste management, considering the present regulations and guidelines. The developed web portal will act as a bridge between end users and recycling agencies to ensure proper disposal of E-waste. The end users register their discarded products on the website and the agencies can view the product details and quotation will be sent by the agencies to the end users. Information regarding the E-waste is collected from the companies/organization and further analysis is carried out which generates a proper report. The statistical view of the generated report will be shared among the agencies and the government organization. The analyzed report will be referred by the government and further steps for the electronic scrap recycling are undertaken. This is designed to serve as a knowledge based on E-waste recycling with the focus on the needs of the developing countries. Likewise, Ghosh et al. (2016) reviewed the policies for E-waste management of Brazil, Russia, India, China, and South Africa (BRICS) nations and noticed that BRICS nations are following the same type of framework to manage their E-waste. They analyzed and compared the existing waste electrical and electronic equipment (WEEE) management systems and Basel Convention. BRICS nations did not learn any lessons from well-developed systems, as implemented by the European Union (EU) and the United States. The study introduced a framework for discarded electronic items recycling system that is adopted in BRICS nations and disclosed that BRICS countries are facing similar types of challenges. They also identified important gaps associated with the recycling systems and dumping of WEEE (Kilica et al., 2015). Recovery of discarded electronic equipment is very important both from economic and environmental aspects of any country. Reverse supply chain plays a vital role in the movement of discarded products from end users to the manufacturers and recovery of valuable materials. They designed a reverse logistics system for discarded electronic equipment in Turkey. For each case, the optimal location of collection sites and recycling facilities is achieved, meeting the minimum recycling rates set out in the EU Directive for each product group. Jecton and Timothy (2013) studied the current e-scrap recycling practices of Kenya along with other countries also suggested a regulatory framework to manage the E-waste effectively in Kenyan context. They used qualitative and quantitative methods to collect the data using personal meetings/ interviews, review of literature and their direct observations. Electronic scrap is an emerging steam of solid waste in Kenya. Due to high volume generation, its ill effects on human health and surroundings and lack of regulatory policies make E-waste as a major concern. They identified the gaps in the field of level of awareness, e-scrap recycling technology, collection, segregation, disposal, financing and monitoring and collaboration of the various stakeholders. They emphasized that the E-waste recycling business is helpful to create job opportunities and sound environmental protection. They suggested best practices for socio-economic development and sustainability of environment. Rahman and Subramanian (2012) concerned that E-waste has become a crucial issue for the health of individuals and environment. They also suggested a solution by recovering the valuable materials from the discarded electronic items to minimize the quantity of final E-waste to be recycled. This paper identified prime elements affecting the discarded computer management program and established a relationship between these factors to develop a framework for effective computers management. Availability of resources, quantity and quality of E-waste, and coordination during recycling operation are the main factors identified by them. On the other side, legislation, attractive incentive schemes, and demand of the customers are the drivers that enforce the formal recycling of E-waste. They used Decision Making Trial and Evaluation Laboratory technique to establish cause-and-effect relationship between the various factors. Ramzy and Kavazanjian (2010) explored the quantities of e-scrap that keep growing-up around the world. It was observed from the data that more than 1.36 million metric ton (MT) of electronic squander was disposed off in the United States in landfills. They also addressed many issues in planning and managing E-waste systems in the United States and also reviewed the existing US recycling systems keeping the end users' responses. Different solutions have been proposed such as optimize use of refurbishing and reusing services, and also highlighted the significance of "end of life" of electronic products.

Brown-West (2010) explained the role of uncertainty in E-waste recovery system economics. Mass Flow and Economics model was developed to quantify impact of system variables on system economic performance. System managers need to control timings, quality and quantity of the returns. Model inputs, collection characteristics, product characteristics, secondary market characteristics, and EoL characteristics were clearly explained. As per Boma, age of product, distribution, depreciation rate, and reuse processing cost are used to determine reuse profit and material composition and commodity price, whereas product weight and recycling processing fee are used to determine recycling profit. Mutha and Pokharel (2009) emphasized on the importance of establishing reverse chain networks for various e-products manufacturers. Legislations are helpful to enforce the manufacturers to take back their old, EOL, or within warranty—period products to minimize the volume of waste and conserve the resources. Therefore the original manufacturers have to change the design of their products for optimum reuse, recycling, and also need to develop a network to collect the discarded electronic gadgets from the consumers to conserve the resources. Development of reverse supply network system depends not only on the quantity of returned items but also on the demand of the refurbished parts and products. They proposed a mathematical model for reverse supply network and assumed that initially the returned products are consolidated in the warehouse and then sent to the recycling units for further processing. Working parts/components are sent to the second-hand market to match the demand of the spare parts. Mitsutaka M. et al. (2009) conducted a study on the uncontrolled E-waste re-usage which can cause problems like environmental pollution, health issues, etc. and also a major issue in the area. The aim of this research is to provide effective framework for e-scrap management. The first step is the data which can be used for the assessment of flow of E-waste and is mainly for the Japanese cases. It includes quantity of E-waste flow, material content, life cycle assessment of electronic gadgets, and cost of reusing and to figure out the framework of electronic squander management. Lastly, they introduced the tele-inverse manufacturing model for future structure of E-waste reuse. Wang and Chou (2009) focused their study on PC waste processes in Taiwan. E-waste problems must be treated as emergency but there are many system dynamics that act like the obstacles in the E-waste management processes. They proposed a model for the effective E-waste management for Taiwan. Xuefeng et al. (2007) discussed the brand-new practices of E-waste recycling in China. The authors highlighted that how Chinese authorities can resolve the issue of e-scrap management from the view of complexity and recycling policy aspects. Authors have introduced the model of E-waste management, examined the components flow of WEEE along with managing technologies, and concentrated on pertinent e-scrap recycling policies. The study showed that flow of waste electronic and electric equipment was very low due to imbalance in the advancement and effective management model. Yoon (2006) carried out a study on electronic scrap recycling activities and associated problems with major focus on extended producer responsibility (EPR). The author tried to reflect an overview of the present recycling practices and management methods of e-scrap in Korea. The Yoon model highlighted the harmful effects of excessive and frequent use of e-gadgets which result in huge piling of E-waste in number and characteristics. Presently, reusing, refurbishing, and recycling are not well promoted in Korea, which results in substantial quantities of E-waste waiting for disposal. Additionally, dangerous materials like polybrominated biphenyl ethers, lead, along with dioxin emissions at e-squander recycling units, require close monitoring to minimize possible workplace threats to humans and atmosphere. Bullock J, C. (1995) performed search on the industries related to the electronics which were facing issues like legal responsibilities together with the take back system. Bullock came to know that, with the passage of time, computers do not deteriorate rather become outdated as per technical advancement at lower cost. Also, author described the hierarchy of components which are immediate reuse of PCs and working elements, also the recovery of microelectronics parts, and reclamation of the raw materials and disposal.

17.3 Concerns raised for strategic framework

On the basis of information extracted from various sources, literature review, and personal visits to formal and informal channels, a strategic framework has been proposed for effectively tackling the E-waste management issue.

17.3.1 Current framework

Before proposing the model for effective E-waste management in developing countries the existing management system is required to be understood. This is useful to find out the limitations and loop holes in the existing system. In the existing system of E-waste management in developing countries the formal E-waste recyclers are not playing a significant role in managing e-scrap. On the other side, scrap dealers are playing a vital role in the management of E-waste. Scrap dealers manage the E-waste in informal manner and cause high risk to human health and environment as well, so this type of channel should be avoided. In current framework, various stakeholders are identified and role of these stakeholders are explained below.

17.3.1.1 Electrical and electronics equipment manufacturers
They are responsible for the manufacturing of EEE products. EEE manufacturers sold their products to the end users through their well-established distribution network (importers, super stockiest and retail outlets).

17.3.1.2 Consumers
There are two types of end users: individuals and organizational. Consumers discard the electronic goods mainly due to two reasons:
- Either the product has become outdated due to frequent technological advancements.

- Or the product has completed its useful life.

The individual consumers are using one or more of the following choices for discarding their used electronics products:

- Sell it to service centers/repair shops
- Sell/gift/donate it to friends, relatives, or to some needy person
- Exchange with new products
- Directly sell it to scrap dealers

The organizational end users are using one or more of the following choices for discarding their e-items:

- Sell/gift/donate it to NGOs or other organizations
- Exchange with new products
- Sell it to scrap dealers or recyclers

17.3.1.3 Service centers

Service centers or repair shops extract the functional components/units from the discarded products and use them in repair or sell them in resale market. They sell nonrepairable units or parts to scrap dealers.

17.3.1.4 Scrap dealers

They collect the electronic scrap along with other domestic waste by providing door step service to the individuals and organizations. Scrap dealers are the backbone of informal E-waste recycling channel. The scrap dealers sell these collected electronic waste products in the unauthorized disposal market (dismantlers).

17.3.1.5 Dismantlers

They collect the E-waste from local scrap dealers. After collection, they segregate the working parts from the nonworking parts. Dismantlers sell the working parts in second-hand market and segregated parts to be disposed off are further processed for materials recovery. Material (metals and nonmetals) thus recovered is sold to the respective industries. The most harmful effect of E-waste processing using this process is that the residuals thus obtained which contain highly harmful and toxic materials are dumped to landfills.

Organizations like banks, colleges, etc. organize auctions for their electronic items after the completion of their useful life. The recyclers or scrap dealers buy e-scrap from these types of auctions. The focus of formal recyclers is on bulk quantity only; currently they are not collecting e-scrap from individual end users. The working of the current framework can be understood from Fig. 17.1.

Following limitations have been identified in the present E-waste management framework:

- Unavailability of formal channel for individual end users

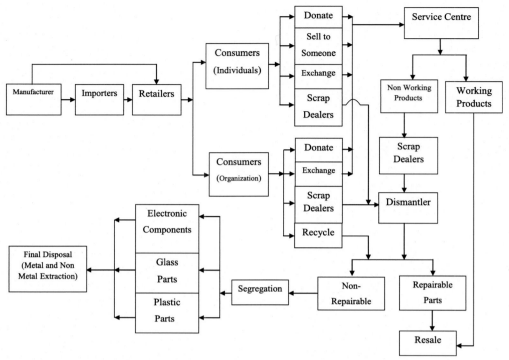

Figure 17.1 Current E-waste management framework in developing countries.

- Absence of buy-back centers
- Door-step pick up facility
- Absence of formal collection centers
- Absence of close monitoring system
- Absence of awareness program

17.3.2 Proposed framework for effective E-waste management system

Taking into consideration the present framework and its limitations, a few changes can be suggested for managing the E-waste management in a more efficient way as shown in Fig. 17.2. Consumer should have proper knowledge about the E-waste disposal through a proper structure. There should be a proper system for maintaining the record of bulk purchase. The same can be used for ensuring proper disposal at the end of useful life of e-products. Useful lives of various electronic products can be fixed in consultation with the manufacturers. In the effective E-waste management framework proposed below, government will have to play a vital role in managing e-scrap and

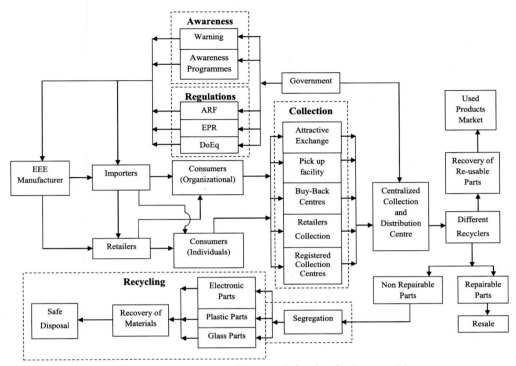

Figure 17.2 Proposed E-waste management framework for developing countries.

reducing its harmful effects. In this proposed framework government will have to ensure the tracking, collection, and disposal of electronic waste. In the present scenario different departments lack in the coordination and there is a huge communication gap between government bodies, manufacturers, and recyclers. It has been clear from the economic analysis conducted by us that end users stored a number of discarded electronic gadgets like mobile phones, TVs, etc. at their homes being unaware of negative effects of E-waste on human health and environment. One of the major problems with formal E-waste management is its collection and we have to find the alternate and effective ways to improve the collection of the discarded electronic products. Parallelly, it is required to work on the strategy to increase the awareness level of the end users as awareness supports the formal collection of E-waste.

17.3.2.1 Steps to improve formal collection
One of the major solutions to increase the formal E-waste collection is to offer attractive exchange schemes to the end users which motivate them to bring out the discarded and technological outdated e-products to the formal collection centers. Attractive exchange

offers will act as the panacea toward formal collection and one of the successful examples of the attractive exchange offer is the Jio feature phone exchange offer. The actual cost of the phone is Rs. 1500 and they offered this phone only for Rs. 500 with exchange of any other feature phone of any brand. Exchange offers like this can bridge the gap between formal recyclers and consumers.

In our suggested model, government needs to ensure that the manufacturers develop an effective system to provide door-step pick up service to consumers for the discarded electronic products. It may be through their retail outlets, buy-back centers, or registered collection centers. The collected E-waste will then be shifted to the centralized collection and distribution centers. Under the monitoring of government authorities, E-waste thus collected will be distributed to different authorized recyclers as per their capacity and category of waste.

To carry out this entire process, government may impose the advanced recycling fee (ARF) on the electronic products at manufacturer level. EPR fixes the responsibility of the manufacturers to manage the E-waste using the ARF collected. According to EPR, the manufacturer will be held responsible for the collection of discarded products from the end users and supply them to the recyclers.

17.3.2.2 Steps to increase the awareness level

Now, to increase the awareness level of the end users, the manufacturers will be required to mention warning about the adverse effects of E-waste on the packaging boxes of all the electronic items as is done on cigarette packaging, etc. The warning should be in English, Hindi, and/or other local languages as per region of circulation of products. Another option is to paste a warning sticker indicating the negative effects of E-waste on human health and on environment on any surface like back/front panel of mobile phones, LED TVs, etc. of the electronic gadgets. It acts like a reminder that though electronic gadgets are necessary for us, they also have adverse effects on our health and environment. The warning can be in graphical or written form. In case of mobile phones and LED TVs, when we switch ON them for the very first time, a no skippable short movie should be played highlighting the adverse effects of electronic gadgets after completion of its useful life. The short movie should also be repeated after particular period of time, it may be one month, two months, or six months. Along with these, a message should be sent to the buyer of electronic product(s) on their registered mobile number mentioning that this is the time to dispose off their electronic item(s). We may use preinstalled mobile application in the form of video games to teach the young blood of nation about the hazardous effects of electronic waste; also animated video CDs, comic books, etc. need to be provided along with the electronic gadgets. It is easier to mold the tiny tots in the era of technology through gaming application and animation videos which will pay off in the coming years. The outcome from comparison of polices can used by the government to strengthen their policies of E-waste management. Design of equipment (DoE) should be such that the electronic products are easily upgradable as per future trends. This concept will be helpful in increasing the shelf life of the products and in reducing the annual volume of electronic scrap generation. The comparison of Indian government's framework with the framework adopted by other countries helped in identifying the gray areas in Indian framework. The same was discussed with academicians and practitioners. After discussion, the following framework is proposed for effective electronic waste management in Indian context.

17.4 Conclusion

Strategic framework, for the effective electronic waste management system, is the result of detailed study of the EMS of different countries. Comparison of policies also helps to develop the effective strategic framework for Indian context. Government of India needs to develop a strong mechanism/system to collect the e-scrap through proper channels. Pick up facility (door step service), buy-back centers, collection at retailers' end, and registered collection centers need to be developed to strengthen collection of E-waste. As concluded earlier, the awareness level of the end users is not satisfactory and needs major improvements. For the same, awareness programs need to be organized; also the advertisements regarding negative impacts of e-trash should be run on TV, newspapers, and various social media platforms. DoE should be such that the electronic products can be easily upgraded to match the future technological advancements. EPR has to be implemented effectively along with an effective monitoring system.

References

Aboelmaged, M. (2021). E-waste recycling behaviour: An integration of recycling habits into the theory of planned behaviour. *Journal of Cleaner Production*, 278.

Ahirwar, R., & Tripathi, A. K. (2021). E-waste management: A review of recycling process, environmental and occupational health hazards, and potential solutions. *Environmental Nanotechnology, Monitoring & Management*, 15.

Anandh, G., Venkatesan, S. P., Goh, M., & Mathiyazhagan, K. (2021). Reuse assessment of WEEE: Systematic review of emerging themes and research directions. *Journal of Environmental Management*, 287.

Arya, S., & Kumar, S. (2020). E-waste in India at a glance: Current trends, regulations, challenges and waste management strategies. *Journal of Cleaner Production*, 271.

Brown-West, B. M. (2010). *A strategic analysis of the role of uncertainty in electronic waste recovery system economics: An investigation of the IT and appliance industries*. United States: Massachusetts Institute of Technology, Yale University, Thesis.

Bullock J.C. (1995). *Environmentally sound management of electronic scrap and the BaselConvention on control of trans-boundary movements of hazardous waste and their disposal*. In IEEE (pp. 192−197).

Dutta, D., Arya, S., Kumar, S., & Lichtfouse, E. (2021). Electronic waste pollution and the COVID-19 pandemic. *Environmental Chemistry Letters*, 20, 971−974.

Fathima, G., Apparna, L., Kusuma, V., & Nischitha, G. (2017). A framework for E-waste management. *International Journal of Latest Engineering and Management Research*, 2(3), 29−34.

Ghosh, S. K., Debnath, B., Baidya, R., & Debashree, D. (2016). Waste electrical and electronic equipments management and Basel Convention compliance in Brazil, Russia, India, China and South Africa (BRICS) nations. *Waste Management & Research, 34*(8), 693–707.

Gupta, V., & Kumar, A. (2014). E-waste status and management in India. *Journal of Information Engineering and Applications, 4*(9), 41–48.

Jecton, A. T., & Timothy, M. W. (2013). Towards an E-waste management framework in Kenya. *Info, 15*(5), 99–113.

Kalana, A. (2010). Electrical and electronic waste management practice by households in Shah Alam, Selangor, Malaysia. *International Journal of Environmental Science, 1*(2), 132–144.

Kilica, H. S., Cebecib, U., & Ayhanaa, M. B. (2015). Reverse logistics system design for the waste of electrical and electronic equipment (WEEE) in Turkey. *Resources, Conservation and Recycling, 95,* 120–132.

Mitsutaka M., Nozomu M., Keijiro M., & Kondoh, S. (2008), Proposal and feasibility assessment of "tele-inverse manufacturing," In *IEEE International Symposium on Electronics and the Environment* (pp. 1–6), 2008.

Mutha, A., & Pokharel, S. (2009). Strategic network design for reverse logistics and remanufacturing using new and old product modules. *Computers & Industrial Engineering, 56,* 334–346.

Rahman, S., & Subramanian, N. (2012). Factors for implementing end-of-life computer recycling operations in reverse supply chains. *International Journal of Production Economics, 140,* 239–248.

Rajya Sabha Secretariat Research Unit Report (2011). E-waste in India.

Ramzy, F. K., & Kavazanjian Jr., E. (2010). Preliminary feasibility study on the use of monodisposal landfills for E-waste as temporary storage for future mining. In *IEEE international symposium on sustainable systems and technology (ISSST 2010)* (Vol. 1, pp. 1–5). IEEE.

Rautela, R., Arya, S., Vishwakarma, S., Lee, J., Kim, K.-H., & Kumar, S. (2021). E-waste management and its effects on the environment and human health. *Science of The Total Environment, 773,* 1–16.

Saoji. (2012). E-waste management: An emerging environmental and health issue in India. *National Journal of Medical Research, 2*(1), 107–110.

United Nations Environmental Programme (2007). E-waste-Volume I: Inventory assessment manual.

United Nations Environmental Programme, (UNEP). (2009). Roles and responsibilities of stakeholders. Philippines: Cebu.

Vishwakarma, S., Kumar, V., Arya, S., Tembhare, M., Rahul., Dutta, D., & Kumar, S. (2022). E-waste in Information and Communication Technology Sector: Existing scenario, management schemes and initiatives. *Environmental Technology and Innovation, 27.*

Wang, C.-S. & Chou, T.-S. (2009). Personal computer waste management process in Taiwan via System Dynamics Perspective. In *2009 International conference on new trends in information and service science* (pp. 1227–1230). IEEE.

Xuefeng, W., Xiaohua, Z. & Hualong H. (2007) *The New Process in Integrated E-wasteManagement in China,* IEEE (pp. 1–6).

Yoon, H., & Jang, Y.-C. (2006). The practice and challenges of electronic waste recycling in korea with emphasis on extended producer responsibility (EPR). In *IEEE* (pp 326–330).

CHAPTER 18

Resilient E-waste management system in emergencies like COVID-19 pandemic

Unnikrishna Menon, Anjaly P Thomas and Brajesh Kumar Dubey
Environmental Engineering and Management, Department of Civil Engineering, Indian Institute of Technology Kharagpur, Kharagpur, West Bengal, India

18.1 Introduction

Electronic waste (E-waste) is becoming the fastest-growing municipal waste stream and accounts for the majority of waste disposed of by the world. The E-waste generation rate depends upon time, space, population, lifestyle preferences, and socioeconomic level (Rautela et al., 2021). According to ASSOCHAM-cKinetics study (2016), India was expected to generate 5.2 Mt of E-waste by 2020(ASSOCHAM, 2016). Forti et al. (2020) reported that India generated 3.2 Mt of E-waste as of 2019 (Fig. 18.1). The global E-waste generation is expected to be 74.3 Mt by the end of 2030, without considering the effect of COVID-19 pandemic as shown in Fig. 18.2. However, a reduction of 4.9 Mt of E-waste in the future could be estimated due to lower consumption of Electrical and Electronic Equipment (EEE) in the first and second quarter of the year 2020. Low and middle-income nations contributed to 30% reduction, while high-income countries reduced the E-waste generation by 5% during the first three quarters of 2020 (Baldé & Kuehr, 2020). But some reports also show that there has been a surge in E-waste generation during this period since the immediate need for social distancing and lockdown led to a greater dependency of human life on EEE while staying at home in the absence of commutation (Kumar et al., 2020). Usage of energy-consuming appliances (Chen et al., 2020), and online shopping in the restricted mobility services were increased during the lockdowns. This may be a major cause of concern since the purchase of electronic gadgets during the pandemic can lead to higher E-waste generation in the future. Due to the closure of E-waste recycling centers during the pandemic period, people started to discard old gadgets and switch to the new ones for office works, which also led to the increase in waste generation during COVID-19. Waste management practices also took a hit due to constraint mobility as well as national lockdowns, specifically in Indian contexts (Cuff, 2020b).

Global E-waste Management Strategies and Future Implications
DOI: https://doi.org/10.1016/B978-0-323-99919-9.00010-6

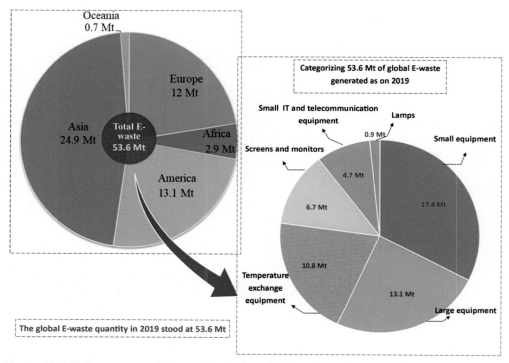

Figure 18.1 Major sources of Waste Electrical and Electronic Equipment and its categorization from different countries as on 2019.

Even when the E-waste forecasting and estimation are done, it needs to be managed scientifically. When this continues to be the standard procedure for E-waste management, an unexpected disaster like COVID-19 comes into the picture, disturbing the entire waste management ecosystem. In this chapter, we are considering COVID-19 as the case study to understand its impact on each element of E-waste management and the measures that need to be adopted during such emergencies. Exposure to E-waste as well as EEE can also lead to COVID-19 infection as long as the virus can survive on the surface of the equipment. In order to prevent the spread of infection via E-waste, it is preferable to store such wastes at the source itself during the pandemic. Also, the periphery of gadgets is usually made of plastics. The virus can survive on the plastic surfaces of the electronic equipment for 3 days (Kampf et al., 2020; Sharma et al., 2020; Taylor et al., 2020). So, an asymptomatic COVID-19-infected person using equipment or an electronic gadget may not be aware that he/she may be infected. The same gadget may be used by another person and unknowingly the virus could be transmitted among the family, at workplaces, shops, etc., ultimately resulting in a virus breakout in the society which also

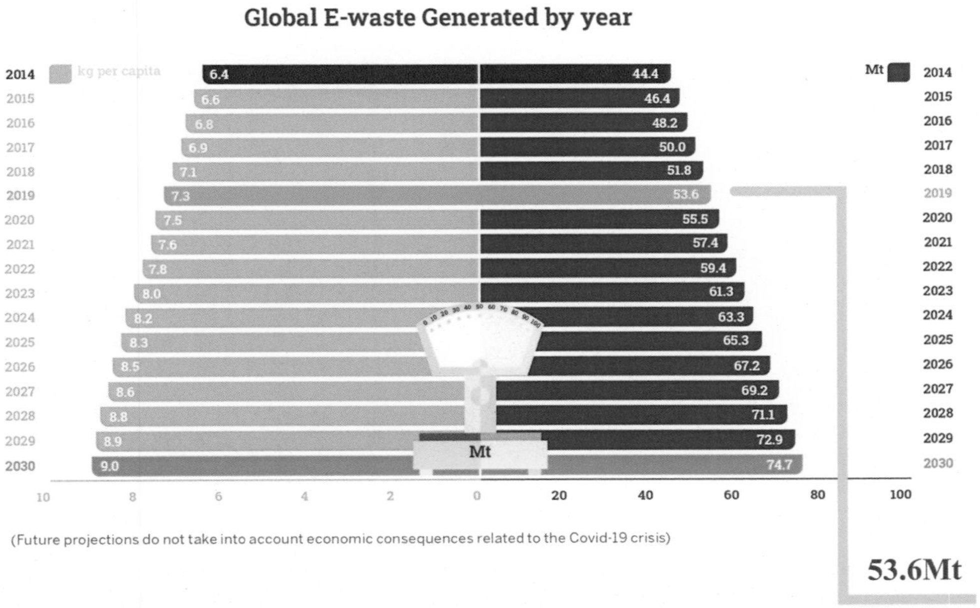

Global E-waste Generated by year

	kg per capita		Mt	
2014	6.4		44.4	2014
2015	6.6		46.4	2015
2016	6.8		48.2	2016
2017	6.9		50.0	2017
2018	7.1		51.8	2018
2019	7.3		53.6	2019
2020	7.5		55.5	2020
2021	7.6		57.4	2021
2022	7.8		59.4	2022
2023	8.0		61.3	2023
2024	8.2		63.3	2024
2025	8.3		65.3	2025
2026	8.5		67.2	2026
2027	8.6		69.2	2027
2028	8.8		71.1	2028
2029	8.9		72.9	2029
2030	9.0		74.7	2030

(Future projections do not take into account economic consequences related to the Covid-19 crisis)

53.6Mt

Figure 18.2 Yearly E-waste generation and trend projection (Forti et al., 2020).

makes it difficult for the authorities to identify the primary source of the virus, as represented in Fig. 18.3.

18.2 Impacts of COVID-19 in the electronic industry and E-waste

The pandemic outbreak in early 2020 had a substantial impact on the electrical and electronics industry, which showed considerable variances from previous trends due to changes in demand and consumer purchasing patterns during that period. In the first three quarters of 2020, the EEE consumption decreased by 6.4% or 4.9 Mt. Slowdown of logistics and lack of workforce led to supply chain disruption, resulting in shipment delays and collapsing consumption (Nayak et al., 2022). The nationwide lockdowns forced both the employers and employees to stay at their respective houses until the lockdowns were lifted. However, the pandemic was spreading across the globe, which created uncertainty among the workers, irrespective of the field they work in. Many corporates came out with the "Work from Home" concept which enabled the workers to carry out their jobs at their houses. This has led to the purchase of gadgets like smartphones, laptops, game consoles, Wi-Fi routers, dongles, electric ovens, IT equipment, and refrigerators and freezers for food storage (Grantham, 2020; Rachel, 2021), mostly toward the last quarter of 2020. The increase

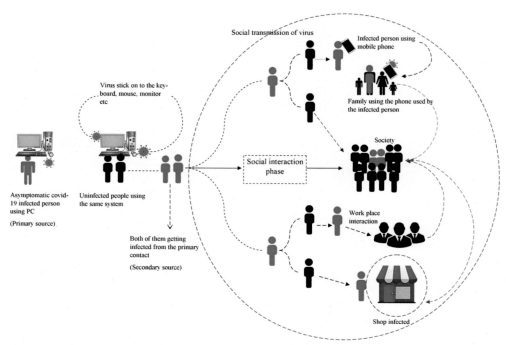

Figure 18.3 Covid-19 exposure pathways via electrical/electronic equipment.

in production was 0.3 Mt which will drive future E-waste generation for such products (Baldé & Kuehr, 2020). A similar trend was observed for the electric vehicle (EV) market in the first half of 2020 where the sales decreased by 14%. However, there has been a dramatic increase in the sales of automobiles (also EVs), also pertaining to COVID-19, as everyone started to depend on private vehicles for emergencies (Gruenwald, 2020). The EV sales in the first six months of 2021 rose to 197% in China, 166% in the United States, and 157% in Europe, while other markets contributed 95% toward EV sales (Irle, 2021). This is not due to the pandemic alone but also contributed to the EV revolution taking place across the world. Even though this seems to be good for the public, this can also be a concern for the E-waste recycling industries as the automobile sector can contribute toward the disposal of more Li-ion batteries in addition to the electronic gadget—related wastes which can result in Lithium fires in the future (Christensen et al., 2021; Saloojee & Lloyd, 2015).

The E-waste management has been a concern even before most of the countries announced lockdown on March 2020 due to COVID-19 (Rachel, 2021). During this period, the pandemic has aggravated the collection, recycling, and disposal issue further due to the closure of formal Waste Electrical and Electronic Equipment (WEEE) recycling plants. The quantitative impact on E-waste collection is not well researched since the monthly or quarterly data are not available. However, at the global level, it is understood that the closing of collection points, lack of workforce has decreased the collection of E-waste (Baldé & Kuehr, 2020). In the United Kingdom, E-waste collection was halved in the second quarter of 2020 (Baldé & Kuehr, 2020), Household Waste Recycling Centers were shut down, which resulted in WEEE falling short by 80% (Cuff, 2020a). The interference of COVID-19 has created an impact on the workers in both formal and informal sectors across the world. Waste pickers, being considered to play an important role in maintaining circular economy in developing nations, were no longer able to collect the wastes as the governments announced lockdowns in different nations with an idea to keep workers away from the secondary transmission of the virus and instructed people to maintain social distancing. They used to collect reusable and recyclable E-wastes that are inevitable for the integration onto economic production before their income plummeted to 50%–75% from the pre-COVID-19 levels (Asumadu et al., 2020; Bharat et al., 2021).

E-waste that will be generated in the future due to a disaster like COVID-19 can have disproportionately large health, economic, and environmental impacts if not handled in the right way. The most difficult aspect of postdisaster waste management is separating and estimating the volume of various waste streams so that they can be treated independently. In the case of E-waste, there are several chances for recycling and recovery that are frequently overlooked due to a lack of quantitative analyses of postdisaster recycling feasibility and sufficient preparation that must be completed prior to the disaster to make recycling a more viable option. When calculating the amount of E-waste generated after a disaster, it is important to include electronics that may deteriorate as a result of the disaster, as well as old and abandoned electronics that are stored in households. As a result, the volume of E-waste that ends up in waste streams will be higher than that would be expected under normal circumstances and therefore the challenges associated with postdisaster E-waste management will continue to grow (Leader et al., 2018). So a well-planned disaster recovery policy is required to address these concerns. Such policies necessitate effective communication between local governments and commercial recyclers, as well as efficient data collection on the amounts of E-waste generated during disasters to forecast future projections which can be accomplished through formal and informal sector cooperation.

18.3 Role of formal and informal sectors and effective E-waste management strategies

Formal and informal are considered as two sectors in any waste management system. When the formal sector uses appropriate equipment and specially constructed facilities for the extraction of valuables (from E-wastes) under a safe environment, informal sectors involve people managing wastes with little or no protective equipment in an unregulated, unregistered, illegal, structure lacking system. In formal recycling, facilities ensure that the end–of–life (EoL) products are treated in a way that does not cause harm to both humans and the environment. Therefore building and running these facilities can be expensive that it may be not affordable for all the nations. Even then, there may be a probability that the workers can be still exposed to low–dose exposure during recycling. Informal E-waste dismantling and recycling are done by an individual, family, or communities at their scrap yards, thereby getting exposed to various components in E-waste. Also, the number of informal workers in the E-waste sector is still unknown (Bel et al., 2019). Tracking them can be costly as well as difficult (Perkins et al., 2014). Populations of different age groups are expected to get exposed to E-waste under unprotected environments in the informal sector. Among them, children are more likely to get exposed to many components in E-waste due to child labor, which is popular in the informal sector (Perkins et al., 2014). Besides this, the risk of COVID-19 susceptibility increases with age, where people with a history of comorbidities may have severe effects (Scott & Beach, 2020). It shall be aggravated if the person is exposed to an unsafe environment in E-waste management sector.

During the period of any disaster, it is important to have an emergency response phase where it requires immediate response to deal with the situation. For example, temporary licensing of facilities can be done which is capable of providing waste management services. Also, when it comes to the informal sector, waste pickers should be instructed not to open the waste bags, especially during the pandemic. The workers involved in collection, transportation, and recycling shall be provided with the Personal Protection Equipment (PPE) kit by the local authorities. This includes N95 masks, gloves, shoes/boots, and disposable work clothing. The supervisors and handlers should be instructed on the nature of waste they are going to deal with and brief the required procedures to be followed to carry out their tasks, whether it is collection, transportation, dismantling, etc. During the pandemic, it is also important to avoid the spread of the COVID-19 virus and therefore social distancing is to be maintained. The collection workers should be instructed to maintain a distance of at least 1 m during their house-to-house collection activities. The PPE used one day should be disposed of on the same day and a new PPE should be worn on the next day during the E-waste collection activity. They should also be provided with sanitizers and soaps to wash their hands every time they pick up the E-wastes from the quarantine locations or houses. The E-waste should be segregated from all other wastes into a separate bin which is devoid of all garbage, radioactive, explosive, biomedical/infectious waste, ammunition, etc. The consumers shall be provided a bin or box or a demarcated area by the organization/firm dealing with the collection to deposit the E-wastes. If the E-waste collection is to be done from a quarantine location or containment zone with suspected or confirmed presence of COVID-19-positive patient, it is recommended to keep the recyclable E-wastes at the source for a certain period until the patients are cured from COVID-19 (United Nations Environment Programme (UNEP), 2020). The E-waste Management Rules 2016 allows the storage of E-waste for a period of maximum 180 days and this can be well satisfied as the consumer takes only a few weeks to recover from COVID-19 which will be certainly below 180 days. This shall be the most preferable solution during the pandemic. However, if the municipal authorities in some countries are collecting E-wastes during this scenario, not sure of whether the E-waste is contaminated by the virus, the bags and E-waste boxes shall be disinfected using 1% sodium hypochlorate solution or 70% alcoholic disinfectant regularly during collection (UNEP, 2020). The collection and transportation should be done in such a way that no wastes are allowed to scatter or escape from the vehicle, and it shall be deposited at the designated site that does the dismantling or recycling activity. The vehicle should be disinfected at the end of every trip.

Besides the rules and precautions mentioned, awareness raising and capacity building shall be considered as an emergency response. It also includes contingency plan development as well as preparedness and "build back better" strategies, which should be strengthened among all stakeholders like public and private sectors, informal sectors, and communities. In this context, the following actions can be taken as part of the training and capacity building:

- **Bringing formal and informal sectors together to form a single unit** that can be beneficial for both sectors, the environment, and society. In many developing countries, waste pickers help in maintaining high recycling rates as they collect solid wastes including E-wastes and sell them for their living. Since the informal sector is living in an unprotected environment, they are more prone to COVID-19 infection and transmission. Formalizing them may reduce the spread to some extent and also cuts down the operational costs in the formal sector.
- **Introduce education and training programs** for workers in the field of E-waste on the possible viral transmission and that may occur during the collection, transportation, and storage along with the potentially hazardous environment the workers may be exposed to while dismantling and recycling activities.
- **At the workplace**, awareness shall be raised for workers on the importance of using masks, sanitizers, and maintaining social distance, thereby creating a healthy working environment. For example,
  - Employees showing symptoms of COVID-19 like fever, tiredness, dry cough, etc., should be restrained from the collection, transportation, or recycling activity. They should be advised to take care of themselves by undergoing quarantine.
  - Healthy employees should notify their supervisors/managers if any of their family members are sick and employees should inform other employees if an employee is suspected or confirmed to have COVID-19 symptoms.

- Strategies to lessen human interaction among the workers and ensure that they keep distance (at least 1 m) between each other and consider revising work shifts. If this is not possible, provide ventilation at the facilities, encourage washing hands, and recommend wearing masks.
- Ensuring that the workers are equipped with a PPE kit at all the stages of work.
- Provide user-friendly communication materials such as public service announcements on radio, social media, television, or websites with frequent displays to increase the awareness among the public on safe handling of wastes.

The effectiveness of the above strategies should be assessed at frequent intervals. The governments also insist on the installation of thermal scanners at the entry of the workplace, considering it as a measure to prevent and control the virus spread. However, relying on thermal scanners alone would not prevent its spread as it need not necessarily detect COVID-19 in individuals in the early stages of infection. It is also essential to clean and disinfect the surfaces that the workers may come in contact with, like switches, handles of equipment, doors, etc. (Ijjasz-Vasquez & Kaza, 2020). Moreover, WHO recommends workers to self-monitor their health regularly, if possible through questionnaires and if the workers are safe and healthy, they shall be allowed to work, also suggesting them to follow the regulation to manage the E-waste in a sustainable manner that creates less environmental footprint and protects human health.

18.3.1 Adopting circular economy to manage E-waste

A circular economy approach would enable the electronics for a reuse opportunity by sending the E-waste components from EoL products back to production. This needs to be done in developing countries through E-waste legislation like Extended Producer Responsibility (EPR) that may include buyback policies, Deposit Refund System, exchange schemes, etc., and therefore it requires proper channelization through E-waste collection centers placed either by the municipality or by the IT firms or respective corporates that produce electronic gadgets. EPR makes a producer, manufacturer, or exporter responsible for the safe, sound handling, recycling, and disposal of the E-waste form of products they produce. Producer Responsibility Organizations hold responsible for managing and strengthening EPR. This is the principal idea on which E-waste Management 2016 strongly rests on. It is also recommended to have a recycling unit within each EEE manufacturing cluster which shall be facilitated by the state governments. Many states, especially in India, transport the collected E-waste to other states which are very far, thereby increasing transportation costs, which is not favorable for the producer. The different stages of E-waste management require proper data. Data quality forms an inevitable part of planning and policy development. However, obtaining accurate data during any natural disaster or even at normal times is very difficult, since most of the waste materials remain in the hands of informal sector, which is not accounted or regulated. However, cooperating with academic institutions may aid in gathering good quality data in the waste management sector (UNEP, 2020).

There are legislations to deal with E-wastes which have been placed in 67 countries, covering two-thirds of the world population. EPR is the main practice in these countries, where a small charge on new electronic devices covers its EoL collection and recycling costs. Canada and the United States now prefer to adopt EPR and product stewardship for E-waste recycling. However, in some other regions like South East Asia, Africa, Latin America, etc., E-waste is not regulated or enforced (Bel et al., 2019). The transboundary movement of hazardous E-waste from developed countries to Asian countries like India, Bangladesh, Pakistan, Malaysia, Thailand, Sri Lanka, The Philippines, Indonesia, Vietnam, Cambodia, etc., should be reduced, which the Basel Convention sticks on to (Arya & Kumar, 2020b). But whether this is happening in these countries is still a question. However, China has already taken the initiative not to accept E-waste from developed countries. In developed countries like Germany, curbside collection is showing success in the field of E-waste recycling and management. Municipalities in some European countries collect E-waste by themselves, while some others contract with other agents/parties to collect it on their behalf.

In Switzerland consumers are required to pay an Advanced Recycling Fund, while the people purchase EEEs for the operations like collection and transport recycling/disposal (Mmereki et al., 2012). Authorized dealers shall collect these wastes and transport them to the sorting and dismantling facility. The Swiss Association for Information, Communication and Organisational Technology and Stiftung Entsorgung Schweiz hold the responsibility to operate and manage E-waste on behalf of their member producers in Switzerland (Sinha-Khetriwal et al., 2005). The material and financial flows in Swiss E-waste are depicted in Fig. 18.4. In Australia there exists National Television and Computer Recycling Scheme 2011 which is supported by the National Waste Policy and the Product Stewardship Act 2011. It provides framework for managing televisions, computers and their outer casings. However, this scheme also targets DVD drives, hard drives, printers, keyboard, mouse etc. In Fig. 18.5, liable parties (A) includes producers, importers, distributors, and users of EEE above thresholds of 15,000 units of computer peripherals or 5000 units of TVs, computers, or printers. Coregulatory arrangements (CRAs) (B) organize collection as well as recycling, also responsible to meet the scheme's outcomes. These CRAs give contracts to the E-waste recyclers (C) who perform first-stage recycling which includes dismantling and sorting. The Australian government (D) estimates the arising E-waste data based on the imports, ensures the compliance of the liable parties and that the scheme outcomes are met by the CRAs. The wastes collected outside this scheme are dealt by the state and territory governments (E). They shall work with CRA, if required (Pablo et al., 2018). Generally,

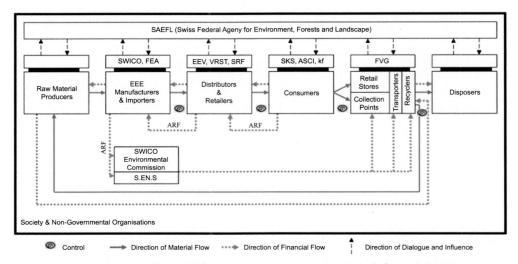

Figure 18.4 Material and financial flows in Swiss E-waste. *Adapted from Sinha-Khetriwal, D., Kraeuchi P., & Schwaninger, M. (2005). A Comparison of electronic waste recycling in Switzerland and in India. Environmental Impact Assessment Review, 25 (5 Spec. Iss.), 492—504. https://doi.org/10.1016/j.eiar.2005.04.006, with permission from Elsevier, licensee number 5157570645922.*

municipalities shall also consider promoting small and medium-sized enterprises and business models that can recover and thrive in E-waste management after a pandemic phase. This requires educating, training, and skill development initiated by the government, thereby providing green jobs for the people.

Japan has door-to-door E-waste collection since it is mandatory to collect E-waste to achieve material recovery targets set under the legislation. Retailers collect the EoL products used by the consumers and transfer them to the recycling facilities of the manufacturer. The government has also appointed Association for Electric Home Appliances to collect the E-waste that are not collected by the municipality or retailers (Chaudhary & Vrat, 2017).

In India obsolete electronic products are sold to waste collectors by consumers for a particular price. These waste collectors in turn sell the E-wastes to the traders who accumulate and sort different kinds of waste. They then sell it to the recyclers who recover metals. There exist a chain of collectors, traders, and recyclers where jobs are created at every point of the chain whose main focus is on the financial benefit that they get by selling the EoL products rather than an environmental or social concern (Sinha-Khetriwal et al., 2005). This needs to be changed by adhering to E-waste Management Rules 2016 that holds the producers, consumers, and dealers responsible for the recycling and disposal of E-waste. The recoverable materials shall be sent to the Material Recovery Facility (MRF) for sorting and recycling (Fig. 18.6).

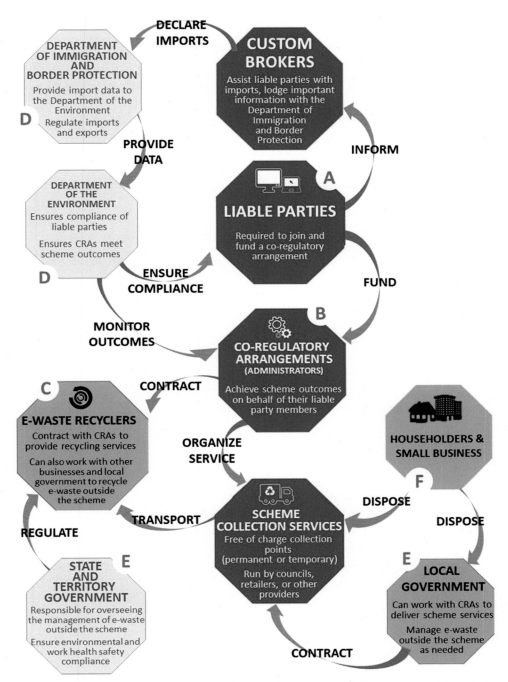

Figure 18.5 Roles and responsibilities under Australia E-waste recycling system. *Adapted from, Pablo, D., Bernardes, A.M., & Huda, N. (2018). Waste electrical and electronic equipment (WEEE) management: An analysis on the Australian E-waste recycling scheme.* Journal of Cleaner Production, 197, *750–764. https://doi.org/10.1016/j.jclepro.2018.06.161, with permission from Elsevier, licensee number 5157730190218.*

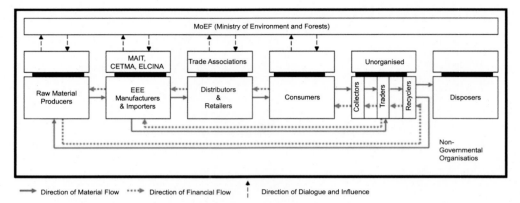

Figure 18.6 Material and financial flows in Indian E-waste. *Adapted from Sinha-Khetriwal, D., Kraeuchi P., & Schwaninger, M. (2005). A Comparison of Electronic Waste Recycling in Switzerland and in India.* Environmental Impact Assessment Review, *25 (5 Spec. Iss.), 492–504. https://doi.org/ 10.1016/j.eiar.2005.04.006, with permission from Elsevier, license number 5166600407353.*

18.3.2 Recycling

Collection, sorting, and recovery determine the efficiency of recycling in which collection and transportation are the most expensive stage toward reuse and recycling of electronic devices. The collected E-wastes are transported to the MRF for testing and sorting. This is considered to be the most critical step in E-waste recycling. The equipment and parts that can be reused are sorted and the remaining materials will become either scrap or recycled (Kang & Schoenung, 2005). The simplified flow of E-waste recycling is shown in Fig. 18.7.

Electronic equipment usually consists of metals, glass, plastics, etc., which are to be dismantled before recycling and it is recommended to carry out the sorting and dismantling activity by the authorized recyclers that specialize in reconditioning. The dismantling can be manual, semimanual, or mechanical. Manual dismantling involves the use of hammers, screwdrivers, and labeled containers, whereas mechanical dismantling requires magnets, conveyor belts, mechanical shredders, etc. The dismantled components other than plastics (especially that forms the outer casing of products will be considered after disinfection) need to be repaired and refurbished which is performed in a factory setting with operational specifications. Repairing is an act of correcting damage or fault, bringing the used product back to working condition. Refurbishment provides a clean and repaired product in the functional state with minimal or no visual flaws by the use of solvents and surface treatments. The products recovered during refurbishment allow the re-utilization of the component that was E-waste before, extending the life of a used product by donating or selling it in the

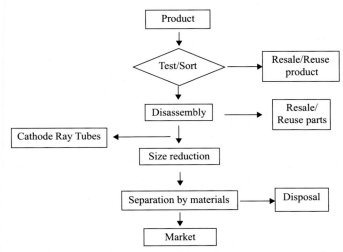

Figure 18.7 Flow diagram of E-waste recycling. *Adapted from Kang, H. Y., & Schoenung, J. M. (2005). Electronic waste recycling: A review of U.S. infrastructure and technology options. Resources, Conservation and Recycling, 45 (4): 368–400. https://doi.org/10.1016/j.resconrec.2005.06.001, with permission from Elsevier, licensee number 5166360164819.*

market which may be used in new products (Ijomah & Danis, 2019). However, during COVID-19, the plastic components dismantled from E-waste shall be disinfected at the biomedical waste treatment facility before transporting it to the recycling facility as shown in Fig. 18.8 since its surfaces may contain virus and this is applicable only if authorities decide to collect and sort/recycle the E-wastes of different sources where the surfaces of E-waste may contain virus, which itself is uncertain. This approach is not necessary if the E-wastes are collected and kept at a temporary storage facility for a few weeks.

The plastic materials collected at the treatment facilities will undergo different types of recycling processes based on its properties and the options available. For example, plastic shredding is done in mechanical recycling to create new plastic products. Chemical recycling is used when the waste plastics are used as raw materials for petrochemical processes. Generally, thermoplastics are more recyclable than thermosetting plastics since thermosetting plastics cannot be re-melted and form new products. Therefore shredding is preferred while recycling thermosetting plastics. This type of plastic is used in electrical motor components, switchboards, etc., whereas thermoplastics are used in automobile parts, pipework and fittings, microwave ovens, electronic gadgets, and devices like computers, etc. The major concern in plastics recycling is the identification and separation of plastics from EoL electronics. Acrylonitrile butadiene styrene, high impact polystyrene, polyphenylene oxide, polyethylene, polyvinyl, etc., are the resins mainly used in electronic products which serve respective purposes. It is

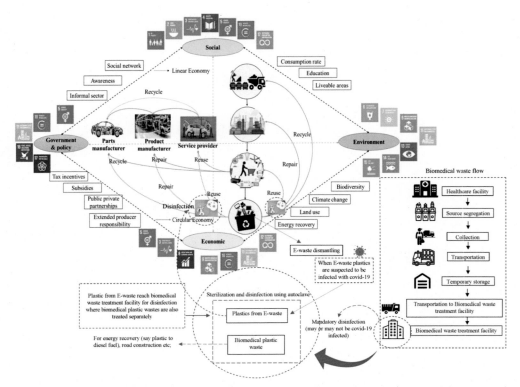

Figure 18.8 A suggestible approach toward e-waste management during COVID-19 (Sharma et al., 2021). *Modified from Sharma, H.B., Vanapalli, K.R., Samal, B., Sankar Cheela, V.R., Dubey, B.K., & Bhattacharya, J. (2021). Circular economy approach in solid waste management system to achieve UN-SDGs: Solutions for post-COVID recovery.* Science of The Total Environment 800, *149605. https://doi. org/10.1016/j.scitotenv.2021.149605, with permission from Elsevier, licensee number 5135250054988.*

necessary for manufacturers to reduce the number of plastic resins in electronics for better recycling (Kang & Schoenung, 2005).

Glasses are also found in electronic products like cathode-ray tubes (CRTs). After EoL, the glasses from televisions and computer monitors can serve as raw materials for ceramic glazes. Generally, there are two methods for glass recycling: glass-to-glass and glass-to-lead recycling. In glass-to-glass recycling, the whole glass is ground without separating the panel and funnel glass. This cullet can be sent to CRT manufacturers to make new CRT. The recycled cullet can be a substitute for virgin materials so that it improves the efficiency of the furnace by consuming less energy to manufacture CRT glass (Kang & Schoenung, 2005). However, the presence of thin mercury lamps, liquid crystals in LCD television screens makes it difficult to recycle glass (Dutta & Goel, 2017). In glass-to-lead recycling, separation and recovery of metallic lead (Pb) and copper (Cu) are done through smelting, before which the plastics and metallic parts need to be separated.

18.3.3 Heavy metal recovery techniques

Additional focus should be given to E-wastes [e.g., batteries, printed circuit board (PCBs), CRTs, etc.] that require special treatments to recover more groups of homogeneous materials like metals in recovery and recycling (Lucier & Gareau, 2012). This can be achieved by chemical, thermal, or metallurgical operations, popularly called end-processing. A high recovery rate with low environmental impact can be obtained using various metallurgical techniques, even though it requires significant investment for these technologies. Some of the metallurgical techniques are as follows:

1. Hydrometallurgy
2. Pyrometallurgy
3. Biometallurgy
4. Electrometallurgy

18.3.3.1 Hydrometallurgy

Hydrometallurgy can be effectively used for the recovery of metals from E-waste, especially spent Li-ion batteries, and it involves the use of chemical reagents like aqueous and organic solvents, mineral acids, some bases, and mixed acids to dissolve the metals from E-waste (Dutta & Goel, 2017). This needs to be purified using methods such as ion exchange, cementation, activated carbon adsorption, etc. Acid leaching has attained a popularity for its efficiency in heavy metal leaching. Cyanide leaching is used for the recovery of gold (Au) but one should be extremely cautious while dealing with it because of its toxicity. However, due to less risk of emission of dust, this method has an upper hand over the pyrometallurgy. Organic acids like citric acid, oxalic acid, malic acid, ascorbic acid, etc., are also gaining popularity for their leaching properties during heavy metal recovery (Djoudi et al., 2021). Ag, Cu, Pb, Cu, etc. are some other metals that can be recovered using hydrometallurgy. Since this method consumes a large amount of acids and solvents, there is also a possibility of water pollution while discharging waste liquid remains onto the rivers (Fig. 18.9).

18.3.3.2 Pyrometallurgy

Pyrometallurgy is a thermal treatment used to recover heavy metals from E-waste, which results in physical and chemical transformation of metals like gold, silver, etc. (Ramanayaka et al., 2019). However, it requires high temperature, where the E-waste is usually heated above 1500°C in a closed reactor to ensure zero emissions. The different stages involved in this process are incineration, smelting, drossing, sintering, melting, and gas-phase reactions at high temperatures. More than 70% of E-waste is

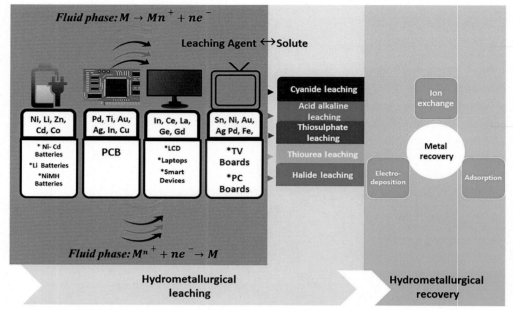

Figure 18.9 Hydrometallurgical recovery of metals from E-waste (Ashiq et al., 2019). *Adapted and modified from Ashiq, A., Kulkarni, J., & Vithanage, M. (2019). Hydrometallurgical recovery of metals from E-waste. In Electronic waste management and treatment technology. Elsevier Inc. https://doi.org/10.1016/B978-0-12-816190-6.00010-8, with permission from Elsevier, licensee number 5157571046815.*

treated in smelters at high temperatures to liberate valuable heavy metals like Cu, Au, Ag, etc.

Do you know? Metals like Al and Fe cannot be recovered in pyrometallurgy since these metals get converted to their oxide forms.

18.3.3.3 Biometallurgy

Biometallurgy employs diverse microorganisms in an aqueous environment to recover metals like Cd, Ni, Cu, U, etc. It is receiving attention due to the less capital requirement for the treatment, environmental friendliness, and low energy requirement. Bioleaching and biosorption are the two steps involved in this process. However, factors like bacterial attachment, microbial consortia, and microbial tolerance influence metal extraction. The interaction between the charged surface groups of microorganisms including algae, bacteria, yeasts, and fungi, and ions in solution takes place in the bioabsorption (Ramanayaka et al., 2019). This metallurgical operation is currently suitable for small scale and laboratory levels (Fig. 18.10).

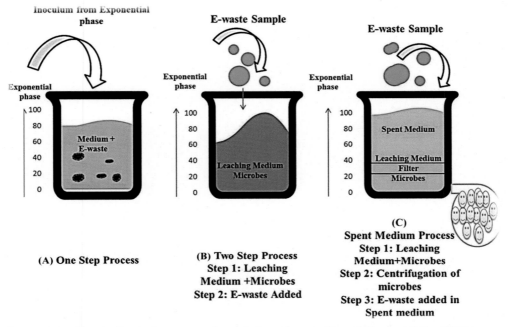

Figure 18.10 Biometallurgical recovery of metals from E-waste (Arya & Kumar, 2020a, 2020b).

18.3.3.4 Electrometallurgy

Electrometallurgy is a process of recovering metals like Cu, Zn, Pb, Ni, Ti etc. by electro-deposition or electroplating in the form of cathodes. Different processes involved in electrometallurgy are electrowinning, electro-refining, and electro-forming (Dutta & Goel, 2017). In electrowinning, current is passed through the leaching solution containing the metal of interest, thereby depositing metal at the cathode. Electro-refining is required to get the electrowon metal to its purest form. When current is passed through an electrolytic cell containing anode made of impure metal and acid electrolyte, the anode dissolves into the electrolyte, later from which the pure metal gets deposited at the cathode. In electro-forming, thin metal parts are molded on preformed cathodes and manufactured through electroplating.

The recovery methods mentioned above are performed conventionally. However, those methods are highly infrastructure demanding, expensive, and energy-consuming (Arya & Kumar, 2020a). A comparative assessment of hydrometallurgy, pyrometallurgy, and biometallurgy along with the criteria index is given in Fig. 18.11.

18.3.4 Difficulties and risks involved in recycling and recovery

Even though the recycling and recovery processes sound good, many E-waste workers are not aware of the potential health hazards that can be caused due to E-wastes. The

Criteria Index					
			Metallurgical Processes		
High	Medium	Low	Pyrometallurgy	Hydrometallurgy	Bio-metallurgy
Background study					
Investment required					
Manpower and Skill					
Energy Consumption					
Time					
Average Resource Recovery					
Loss of Resources during the process					
Corrosive					
Hazardous gaseous emissions					
Wastewater generation					
Dust generation					
Compatibility with environment					

Figure 18.11 Comparative assessment of treatment options for resource recovery from E-waste (Arya & Kumar, 2020a).

workers may be exposed to varieties of direct and indirect hazardous substances during the E-waste recycling process in industries. Direct exposure refers to the ingestion of contaminated air, water, and food, and inhalation of coarse and fine-grained particles. Furans and dioxins may be released due to incomplete combustion when E-wastes are burned at lower temperatures (Noel-Brune et al., 2013; Perkins et al., 2014). Heavy metals like mercury, cadmium, and lead are released during incineration. Strong acids are used for leaching; reagents like cyanides are used during the heavy metal recovery process. Lead saturated fumes are released during the de-soldering of circuit boards. Inhalation of these chemical fumes results in adverse effects to the workers. People are prone to indirect exposure through contaminated water, soil, and air, when E-waste recycling is not managed properly. In many developing countries, the door-to-door collection is done by a group of self-employed individuals in the informal sector. Some groups of this sector pay marginal fees to E-waste owners and then sell those wastes to the E-waste retailers or dealers, while the others take these E-wastes, burn components like polyvinyl chloride in the open air to obtain copper from cables; acid/caustic leaching of PCBs exemplifies the recovery of metals in the informal sector (Kumar et al., 2017; Mmereki et al., 2012). Toxic elements can enter into the blood streams of informal workers while open burning of E-waste (Bel et al., 2019). If the waste collected does not possess any value, then it is either incinerated or dumped in a landfill, causing harm not just to the environment but also humans (Kumar et al., 2017). Around 70% of heavy metals ending up in the US landfills come from E-waste (Widmer et al., 2005). The release of hazardous chemicals from E-waste leads to bioaccumulation, thereby allowing toxins to enter the food web, which can be either through soil, air, and water. This can also cause mental impairment, damage to liver and kidney, and lung problems when carcinogens are released.

18.4 Promising efforts for E-waste diversion

Many E-waste management activities across the world divert E-waste from being dumped in a landfill which are inspiring and worth learning. When it eliminates the environmental problems on one side, the creative use of recycled E-waste shall also be appreciated. Two such examples are shown in the following subsections:

18.4.1 2020 Tokyo Olympic medals from E-waste

It was an encouraging move from the Japan municipal authorities to initiate the E-waste collection drive in 2017 which was used to make medals for Tokyo Olympics 2020. According to Leader et al. (2017), it was estimated that Japan would require 2.4—3.0 million cell phones, around 1.2—1.5 million laptops, or 0.33—0.41 million desktop computers are to be used to produce approximately 5000 medals (including gold, silver, and bronze). However, according to the latest reports 78,985 tons of E-waste were received including laptops, digital cameras, and 6.21 million cell phones to produce 5000 medals by extracting 32 kg gold, 3500 kg silver, and 2200 kg bronze. The collection drive was started in April 2017 and completed by March 2019 (Northey, 2020; Krishnan, 2021; Times of India, 2019). The collected E-wastes were then dismantled, and smelted via pyrometallurgical process to recover the required elements (Tokyo, 2020 Medal Project: Towards an Innovative Future for All, 2020). However, the bronze medals were not actually made from copper—tin alloy, but from red brass, which is an alloy of copper and zinc and therefore a significant amount of copper and zinc had to be extracted as well (Northey, 2020). If one medal weighs 500 g, it would contain 1.2% gold and 98.8% silver for gold medals, 100% silver for silver medals, and 95% copper and 5% zinc for bronze medals (Leader et al., 2017). According to the life cycle analysis done by Hischier et al. (2005), it was found that recycling E-waste creates a significant environmental footprint. However, it was much less than the primary mining of raw materials and its waste incineration.

18.4.2 Cathode-ray tube (CRT) as a construction material

While focusing on E-waste, it should not be just the metallic components that need to be focused. There are components of plastic present inside or outside a gadget, say mobile phone casing, CRT monitor casing, etc. One of the most significant fractions of E-waste is plastic which accounts for 20% and is capable of reuse in various fields. Open-loop recycling of CRT monitors has been studied in which the plastic from the monitors are separated, crushed, and utilized as coarse aggregate in concrete mix (Hamsavathi et al., 2020). On inclusion of up to 15% E-waste resulted in outstanding compressive and flexural strength. As a result, concrete with this optimal proportion of E-waste displayed equivalent strength when compared to conventionally used concrete beams. This research justifies how E-waste can be considered as a potential coarse aggregate replacement material in concrete with promising structural qualities. Furthermore, continued usage of rocks for construction would deplete the natural resource and create an ecological imbalance. Thus replacing coarse aggregate in concrete with E-waste has the dual benefit of minimizing mineral exhaustion in land resources as well as lowering E-waste pollution by a significant percentage.

18.5 Conclusion

E-waste was considered to be a major threat even before the COVID-19 pandemic. Since the pandemic altered people's lifestyle across the world, it demanded a rise in the electronic gadget and EV consumption in the last quarter of 2020. This surge in consumption will result in significant volumes of E-waste ending up in municipal landfill in the near future unless an effective waste management system is set up to deal with it in order to protect human health and the environment from contamination. Even though developed countries contribute more toward WEEE, currently Asia contributes the highest amount of E-waste across the world in terms of the total E-waste generated. Developing countries like India, Japan, and Bangladesh should also control the flow of E-wastes. India should stick on to E-waste Management Rules 2016, in which producers, consumers, and importers are held responsible for the recycling and disposal of EoL products. Consumers need to be more aware of the importance of segregating E-wastes, clean and dry in a separate container, and this can make the job easy for the recyclers. At the same time, schemes like EPR and buy-back policy should be made mandatory by the government and so that the producers can get back the used products through recyclers who repair them and send them to the production facilities where new electronic gadgets or appliances are manufactured, thereby diverting E-waste from landfills. The governments should encourage the participation of enterprises that can perform well in plastic and glass recycling for long term. During a disaster like COVID-19, if E-waste at a source is contaminated by the virus, it is recommended to keep it at the source itself as the virus can survive on its surface for 4–5 days. However, if E-waste collection is done, one must be more cautious to deal with E-waste as it may or may not be infected with the virus and hence disinfection has to be done using any cleansing agents before collection, even when the worker is using other PPEs. This shall be transported to a temporary storage facility for a few weeks after which recycling activities can be resumed. As an additional precaution, the plastics dismantled from E-waste shall be first disinfected separately at the biomedical waste treatment facility before sending it for recycling, which is optional. The informal sector and formal sector should work together for efficient E-waste management. Though the initial investment may be at the higher side, optimal recovery of heavy metals like Li, Co, Mn, etc. from E-wastes can be used for many industrial applications and this recovery strategy may be necessary for the future as there is already a spike in the demand for EEEs and EVs to reduce the dependency on fossil fuel vehicles. The consumption of refurbished electronic gadgets and equipment, recycled plastics, glass, and other recoverable materials helps to reduce environmental footprint and maintain a circular economy. Finally, incineration or disposal in landfills shall be considered only if the materials remaining of E-waste are nonrecoverable.

References

Arya, S., & Kumar, S. (2020a). Bioleaching: Urban mining option to curb the menace of E-waste challenge. *Bioengineered, 11*(1), 640–660. Available from https://doi.org/10.1080/21655979.2020.1775988.

Arya, S., & Kumar, S. (2020b). E-waste in India at a glance: Current trends, regulations, challenges and management strategies. *Journal of Cleaner Production, 271*, 122707. Available from https://doi.org/10.1016/j.jclepro.2020.122707.

Ashiq, A., Kulkarni, J., & Vithanage, M. (2019). *Hydrometallurgical recovery of metals from E-waste. Electronic waste management and treatment technology.* Elsevier Inc. Available from https://doi.org/10.1016/B978-0-12-816190-6.00010-8.

ASSOCHAM. (2016). *India's E-waste growing at 30% per annum: ASSOCHAM-CKinetics Study.*

Asumadu, S., Phebe, S., & Owusu, A. (2020). Impact of COVID-19 pandemic on waste management. *Environment, Development and Sustainability.* Available from https://doi.org/10.1007/s10668-020-00956-y.

Baldé, C. P., & Kuehr, R. (2020). *Impact of Covid19 pandemic on E-waste: The first three quarters of 2020.* United Nations University (UNU)/United Nations Institute for Training and Research (UNITAR) — co-hosting the SCYCLE Programme, Bonn (Germany).

Bel, G., van Brunschot, C., Easen, N., Gray, V., Kuehr, R., Milios, A., Mylvakanam, I., & Pennington, J. (2019). *A new circular vision for electronics: Time for a global reboot.* World Economic Forum. Available from http://www.weforum.org.

Bharat, G. K., Adam, H. N., & Noklebye, E. (2021). *A sustainable development agenda: Plastic and biomedical waste post COVID-19.* The Energy and Resources Institute.

Chaudhary, K., & Vrat, P. (2017). Case study analysis of E-waste management systems in Germany, Switzerland, Japan and India: A RADAR chart approach. *Benchmarking: An International Journal*, 25.

Chen, C.-F., Zarazua de Rubens, G., Xu, X., & Li, J. (2020). Coronavirus comes home? Energy use, home energy management, and the social-psychological factors of COVID-19. *Energy Research & Social Science*, *68*, 101688. Available from https://doi.org/10.1016/j.erss.2020.101688, 32839705.

Christensen, P. A., Anderson, P. A., Harper, G. D. J., Lambert, S. M., Mrozik, W., Rajaeifar, M. A., Wise, M. S., & Heidrich, O. (2021). Risk management over the life cycle of lithium-ion batteries in electric vehicles. *Renewable and Sustainable Energy Reviews*, *148*, 111240. Available from https://doi.org/10.1016/j.rser.2021.111240.

Cuff, M.. (2020b). *Surge in lockdown E-waste heading for landfill, experts fear, as Recycling figures halve*. INews.

Cuff, M.. (2020a). *Coronavirus in the UK: Fly-tipping on the rise as waste collections are cut*. INews.

Djoudi, N., Le Page Mostefa, M., & Muhr, H. (2021). Hydrometallurgical process to recover cobalt from spent Li-ion batteries. *Resources*, *10*(6). Available from https://doi.org/10.3390/resources10060058.

Dutta, D., & Goel, S. (2017). *Advances in solid and hazardous waste management* (pp. 1–371). Springer, https://doi.org/10.1007/978-3-319-57076-1.

Forti, V., Baldé, C.P., Kuehr, R., & Bel, G.. (2020). The global E-waste monitor 2020. <http://ewaste-monitor.info/>.

Grantham, L.. (2020). *The impact of Covid-19 on the UK WEEE sector*. Resource.

Gruenwald, H. (2020). Covid-19 and electric cars. *Energy Research & Social Science*. Available from https://doi.org/10.13140/RG.2.2.29553.71527.

Hamsavathi, K., Soorya Prakash, K., & Kavimani, V. (2020). Green high strength concrete containing recycled cathode ray tube panel plastics (E-waste) as coarse aggregate in concrete beams for structural applications. *Journal of Building Engineering*, *30*, 101192. Available from https://doi.org/10.1016/j.jobe.2020.101192.

Hischier, R., Wäger, P., & Gauglhofer, J. (2005). Does WEEE recycling make sense from an environmental perspective? The environmental impacts of the Swiss Take-Back and Recycling Systems for Waste Electrical and Electronic Equipment (WEEE). *Environmental Impact Assessment Review*, *25*(5 Spec. Iss.), 525–539. Available from https://doi.org/10.1016/j.eiar.2005.04.003.

Ijjasz-Vasquez, E., & Kaza, S. (2020). *Better trash collection for a stronger recovery: Solid waste management as a pillar of urban change*. TheCityFix.

Ijomah, W. L., & Danis, M. (2019). *Refurbishment and reuse of waste electrical and electronic equipment. Waste electrical and electronic equipment (WEEE) handbook*. Elsevier Ltd. Available from https://doi.org/10.1016/B978-0-08-102158-3.00009-4.

Irle, R. (2021). *Global EV sales for 2021 H1*. EV-Volumes. Available from https://www.ev-volumes.com/.

Kampf, G., Todt, D., Pfaender, S., & Steinmann, E. (2020). Persistence of coronaviruses on inanimate surfaces and their inactivation with biocidal agents. *Journal of Hospital Infection*, *104*(3), 246–251. Available from https://doi.org/10.1016/j.jhin.2020.01.022.

Kang, H. Y., & Schoenung, J. M. (2005). Electronic waste recycling: A review of U.S. infrastructure and technology options. *Resources, Conservation and Recycling*, *45*(4), 368–400. Available from https://doi.org/10.1016/j.resconrec.2005.06.001.

Krishnan, N.. (2021). *Tokyo Olympics 2020*. Moneycontrol.

Kumar, A., Holuszko, M., & Espinosa, D. C. R. (2017). E-waste: An overview on generation, collection, legislation and recycling practices. *Resources, Conservation and Recycling*, *122*, 32–42. Available from https://doi.org/10.1016/j.resconrec.2017.01.018.

Kumar, A., Luthra, S., Mangla, S. K., & Kazançoğlu, Y. (2020). COVID-19 impact on sustainable production and operations management. *Sustainable Operations and Computers*, *1*, 1–7. Available from https://doi.org/10.1016/j.susoc.2020.06.001.

Leader, A., Gaustad, G., Tomaszewski, B., & Babbitt, C. W. (2018). The consequences of electronic waste post-disaster: A case study of flooding in Bonn, Germany. *Sustainability (Switzerland)*, *10*(11), 1–14. Available from https://doi.org/10.3390/su10114193.

Leader, A. M., Wang, X., & Gaustad, G. (2017). Creating the 2020 Tokyo Olympic medals from electronic scrap: Sustainability analysis. *The Journal of The Minerals, Metals & Materials Society (TMS)*, *69*(9), 1539–1545. Available from https://doi.org/10.1007/s11837-017-2441-4.

Lucier, C. A., & Gareau, B. J. (2012). *Electronic waste recycling and disposal: An overview. Assessment and management of radioactive and electronic wastes*. IntechOpen. Available from https://doi.org/10.5772/intechopen.85983.

Mmereki, D., Li, B., Baldwin, A., & Hong, L. (2012). *The generation, composition, collection, treatment and disposal system, and impact of E-waste. In E-waste in transition — From pollution to resource*. IntechOpen. Available from https://doi.org/10.5772/61332.

Nayak, J., Mishra, M., Naik, B., Swapnarekha, H., Cengiz, K., & Shanmuganathan, V. (2022). An impact of COVID-19 on six different industries: Automobile, energy and power, agriculture, education, travel and tourism and consumer electronics. *Expert Systems, 39*, 1−32.

Noel-Brune, M., Goldizen, F. C., Neira, M., van den Berg, M., Lewis, N., King, M., Suk, W. A., Carpenter, D. O., Arnold, R. G., & Sly, P. D. (2013). Health effects of exposure to E-waste. *The Lancet Global Health, 1*(2), 2012. Available from https://doi.org/10.1016/S2214-109X(13)70020-2.

Northey, S. A. (2020). Going for bronze: Will mineral resource depletion make it harder to get an Olympic medal in the future? *Energy and Resources, 41*(5).

Pablo, D., Bernardes, A. M., & Huda, N. (2018). Waste electrical and electronic equipment (WEEE) management: An analysis on the Australian E-waste recycling scheme. *Journal of Cleaner Production, 197*, 750−764. Available from https://doi.org/10.1016/j.jclepro.2018.06.161.

Perkins, D. N., Brune Drisse, M. N., Nxele, T., & Sly, P. D. (2014). E-waste: A global hazard. *Annals of Global Health, 80*(4), 286−295. Available from https://doi.org/10.1016/j.aogh.2014.10.001.

Rachel. (2021). *How has Covid-19 had an impact on electronic waste?* DCW.

Ramanayaka, S., Keerthanan, S., & Vithanage, M. (2019). *Urban mining of E-waste: Treasure hunting for precious nanometals. Handbook of electronic waste management: International best practices and case studies*. Butterworth-Heinemann, https://doi.org/10.1016/B978-0-12-817030-4.00023-1.

Rautela, R., Arya, S., Vishwakarma, S., Lee, J., Kim, K. H., & Kumar, S. (2021). E-waste management and its effects on the environment and human health. *Science of The Total Environment, 773*, 145623. Available from https://doi.org/10.1016/j.scitotenv.2021.145623.

Saloojee, F., & Lloyd, J. (2015). *Lithium battery recycling process. Desktop Study* (p. 27) CM Solutions − Metallurgical Consultancy & Laboratories.

Scott, D., & E D T Miami Beach. (2020). *The Covid-19 risks for different age groups, explained*.

Sharma, H. B., Vanapalli, K. R., Samal, B., Sankar Cheela, V. R., Dubey, B. K., & Bhattacharya, J. (2021). Circular economy approach in solid waste management system to achieve UN-SDGs: Solutions for post-COVID recovery. *Science of The Total Environment, 800*, 149605. Available from https://doi.org/10.1016/j.scitotenv.2021.149605.

Sharma, H. B., Vanapalli, K. R., Shankar Cheela, V. R., Ranjan, V. P., Jaglan, A. K., Dubey, B., Goel, S., & Bhattacharya, J. (2020). Challenges, opportunities, and innovations for effective solid waste management during and post COVID-19 pandemic. *Resources, Conservation and Recycling, 162*, 105052. Available from https://doi.org/10.1016/j.resconrec.2020.105052.

Sinha-Khetriwal, D., Kraeuchi, P., & Schwaninger, M. (2005). A comparison of electronic waste recycling in Switzerland and in India. *Environmental Impact Assessment Review, 25*(5 Spec. Iss.), 492−504. Available from https://doi.org/10.1016/j.eiar.2005.04.006.

Taylor, D., Lindsay, A. C., & Halcox, J. P. (2020). Aerosol and surface stability of SARS-CoV-2 as compared with SARS-CoV-1. *The New England Journal of Medicine, 382*(16), 0−2.

Times of India. (2019). *How 62.1 lakh mobile phones were used to make 2020 Olympics medals*. Times of India.

Tokyo 2020 medal project: Towards an innovative future for all. Olympics.com.

United Nations Environment Programme (UNEP). (2020). *Waste management during the COVID-19 pandemic: From response to recovery* (p. 21). https://doi.org/10.1093/PM/PNAA200.

Widmer, R., Oswald-Krapf, H., Sinha-Khetriwal, D., Schnellmann, M., & Böni, H. (2005). Global perspectives on E-waste. *Environmental Impact Assessment Review, 25*(5 Spec. Iss.), 436−458. Available from https://doi.org/10.1016/j.eiar.2005.04.001.

Index

Note: Page numbers followed by "*f*" and "*t*" refer to figures and tables, respectively.